Protein Microarray Technology

Edited by
Dev Kambhampati

Related Titles:

Reiner Westermeier, Tom Naven

Proteomics in Practice
A Laboratory Manual of Proteome Analysis
pp. 342, 2002

ISBN: 3-527-30354-5

Douglas T. Gjerde, Christoph P. Hanna, David Hornby

DNA Chromatography
pp. 243, 2002

ISBN: 3-527-30244-1

Jennifer van Eyk, Michel J. Dunn (eds.)

Proteomic and Genomic Analysis of Cardiovascular Disease
pp. 527, 2003

ISBN: 3-527-30596-3

Günter Kahl

The Dictionary of Gene Technology
Genomics, Transcriptomics, Proteomics
pp. 958, 2001

ISBN: 3-527-30100-3

Protein Microarray Technology

Edited by
Dev Kambhampati

WILEY-VCH

WILEY-VCH Verlag GmbH & Co. KGaA

Editor
Dr. Dev Kambhampati
1777 Shoreline Drive, Apt 332A
Alameda, CA 94501
USA

Library of Congress Card No.: applied for

British Library Cataloguing-in-Publication Data: A catalogue record for this book is available from the British Library.

Bibliographic information published by Die Deutsche Bibliothek
Die Deutsche Bibliothek lists this publication in the Deutsche Nationalbibliografie; detailed bibliographic data is available in the Internet at <http://dnb.ddb.de>

Printed in the Federal Republic of Germany
Printed on acid-free paper

Composition TypoDesign Hecker GmbH, Leimen
Printing Strauss Offsetdruck GmbH, Mörlenbach
Bookbinding Litges & Dopf Buchbinderei GmbH, Heppenheim

ISBN 3-527-30597-1

Preface

We currently live in an age of fast-paced discovery that is geared towards deciphering the intricacies of the human body. With the completion of the Human Genome Project and the advent of new technologies for monitoring biomolecular interactions, a revolution is currently underway in the field of Functional Genomics and Proteomics. Screening the expression behavior of genes, proteins (and their post-translationally modified products) and small molecule drug targets are currently being conducted at a frenetic pace to discover novel therapeutics and biomarkers for a wide range of diseases. The outcome of such screening activities can influence our fundamental understanding of life's cellular processes at the molecular level.

To fulfill some of the above challenging objectives, new technologies such as microarrays have been developed to increase the overall throughput of biomolecular screening (by using increasingly smaller volumes) at reduced cost. While microarrays have been extensively used in gene expression and genotyping studies during the past decade, numerous challenges still persist when it comes to understanding the biological significance of the screened data and various practical issues pertaining to quality control and reproducibility. In addition, DNA arrays have not been able to effectively provide a direct correlation between mRNA and protein expression levels within the cell. Since proteins are the work-horses of the cells, understanding their rate of formation, biomolecular interactions and modifications, and ultimately elimination from the cell, are all vital for monitoring the development of diseases and which cellular targets to focus on while designing drugs.

Monitoring protein interactions is highly complex, and an approach which involves direct extrapolation of DNA microarray strategies for proteomic studies is bound to have limited success. Proteins, unlike nucleic acids which are relatively homogeneous in terms of structural and electrostatic properties, are a few orders of magnitude more abundant than genes and can have extremely diverse chemical, structural and biological properties. In addition, the relative abundance of proteins within the body also varies greatly (there can be anywhere in the region of 6 to 10 orders of magnitude between the most and least abundant proteins) and can affect the design of the assay and the detection conditions. Preventing protein denaturation and maintaining structural conformations will be key issues while conducting

Protein Microarray Technology. Edited by Dev Kambhampati

ISBN: 3-527-30597-1

protein interaction studies on solid substrates in addition to circumventing the already existing pitfalls of conventional DNA microarrays.

This book will address novel strategies for constructing highly functional and biocompatible protein microarrays. Leading world experts have been assembled to address the above issues and offer a roadmap for solving the complex challenges pertaining to monitoring protein–protein interactions on a wide range of microarray platforms. A comprehensive overview of microarray surface chemistry, detection technologies, fabrication options for array development, and data analysis of a plethora of protein interactions will also be provided.

An outline of the various topics which will be covered in detail within this book is shown below.

- **Types of Interactions**
  - Protein–Protein
  - Peptide–Protein
  - Antibody–Antigen
  - Protein–Peptide
  - Protein–Carbohydrate
  - Protein–DNA
  - Aptamer–Protein

- Surface Chemistry (on various substrates)
  - Gold
  - Glass
  - Plastics/Polymer Membranes
  - Semiconductor

- Detection Technologies
  - Atomic Force Microscopy
  - Fluorescence Spectroscopy/Scanners
  - Surface Plasmon Resonance (SPR)
  - SPR enhanced Fluorescence Spectroscopy (SPFS)
  - Surface Plasmon Microscopy
  - Mass Spectrometry

- Spotting Technologies
  - Microarrayer (contact/non-contact)
  - Surface Patterning (micro-contact printing)

- Bioinformatics/Data Analysis
  - Microarray pitfalls
  - Data analysis
  - Commercial software

This pioneering book is the first of its kind in the new field of Protein Microarrays. By implementing the above technologies and strategies, rapid advances are expected in the field of protein-based therapeutics, discovery of innovative drugs based on antibodies and small molecules, and in the field of clinical diagnostics for screening diseases. Some of the above-mentioned technologies can also be used in single-molecule screening of biomolecules. The protein immobilization strategies that are described in this book can also be directly implemented in developing microfluidic devices (for point-of-care applications) and biosensors for monitoring pathogens that are present in food, water and the environment. As a result, the impact of this technology can also have a potential effect on the areas of Food Diagnostics, Environmental Monitoring and National Security.

This book is expected to serve as a useful reference for researchers and students embarking on the exciting field of Proteomics, Drug Discovery and Clinical Diagnostics.

Dev Kambhampati
Berkeley, October 2003

Contents

Protein Microarray Technology. Edited by Dev Kambhampati

ISBN: 3-527-30597-1

List of Authors

MARKO K. BALLER
Veeco Instruments, Inc.
112 Robin Hill Road
Santa Barbara, CA 93117
USA

LISA R. BOOTH
Prolinx Inc.
22322 20th Avenue SE
Bothell, WA 98021
USA

JULIAN F. BURKE
Genetix Ltd.
Queensway
New Milton, Hamshire BH25 5NN
UK

ROBERT M. CORN
Department of Chemistry
University of Wisconsin
1101 University Avenue
Madison, WI 53706
USA

GARY FRANKLIN
Biacore AB
Rapsgatan 7
754 50, Uppsala
Sweden

JÜRGEN FRITZ
International University Bremen
School of Engineering and Science
P.O.Box 750 561
28725 Bremen
Germany

KARIN A. HUGHES, PH. D.
Senior Product Manager Calbiochem
Biochemicals and Immunochemicals
EMD Biosciences
10394 Pacific Center Court
San Diego, CA 92121
USA

ROBERT J. KAISER
Prolinx Inc.
22322 20th Avenue SE
Bothell, WA 98021
USA

DEV KAMBHAMPATI
1777 Shorline Drive, Apt 332A
Alameda, CA 94501
USA

WOLFGANG KNOLL
Max-Planck-Institute for Polymer Research
Ackermannweg 10
55128 Mainz
Germany

HYE JIN LEE
Department of Chemistry
University of Wisconsin
1101 University Avenue
Madison, WI 53706
USA

STEFAN LÖFÅS
Biacore AB
Rapsgatan 7
75450 Uppsala
Sweden

Lee O. Lomas
Ciphergen Biosystems Inc
Fremont CA, 94555
USA

Kevin P. Lund
Prolinx Inc.
22322 20th Avenue SE
Bothell, WA 98021
USA

Christopher J. Mann
Genetix Ltd.
Queensway
New Milton, Hamshire BH25 5NN
UK

Alan McWhirter
Biacore AB
Rapsgatan 7
754 50, Uppsala
Sweden.

Björn Persson
Biacore AB
Rapsgatan 7
75450 Uppsala
Sweden

Anton Petrov
BioDiscovery Inc.
100 N. Sepulveda Blvd.
Suite 1230
El Segundo, CA 90245
USA

Michael Schaeferling
Institut für Analytische Chemie
Chemo- und Biosensorik
Universität Regensburg
93040 Regensburg
Germany

Soheil Shams
BioDiscovery Inc.
100 N. Sepulveda Blvd.
Suite 1230
El Segundo, CA 90245
USA

Douglas A. Spicer
Prolinx Inc.
22322 20th Avenue SE
Bothell, WA 98021
USA

Sarah K. Stephens
Genetix Ltd.
Queensway
New Milton, Hamshire BH25 5NN
UK

Greta J. Wegner
Department of Chemistry
University of Wisconsin
1101 University Avenue
Madison, WI 53706
USA

Fang Yu
Max-Planck-Institut
für Polymerforschung
Materialwissenschaften
Ackermannweg 10
55128 Mainz
Germany

Chapter 2

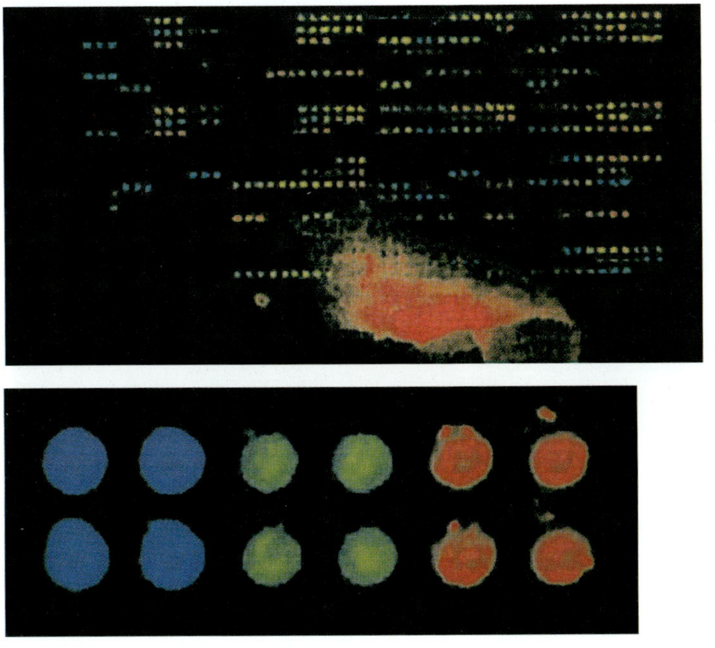

Figure 2.1 (a) Smearing effects on antibody-immobilized poly-lysine arrays [8]. (b) Snapshot of microarray spots with poor morphology and surface chemistry (within the individual spots). Differences in spot homogeneity can result in varying signal intensities within the same spot [9].

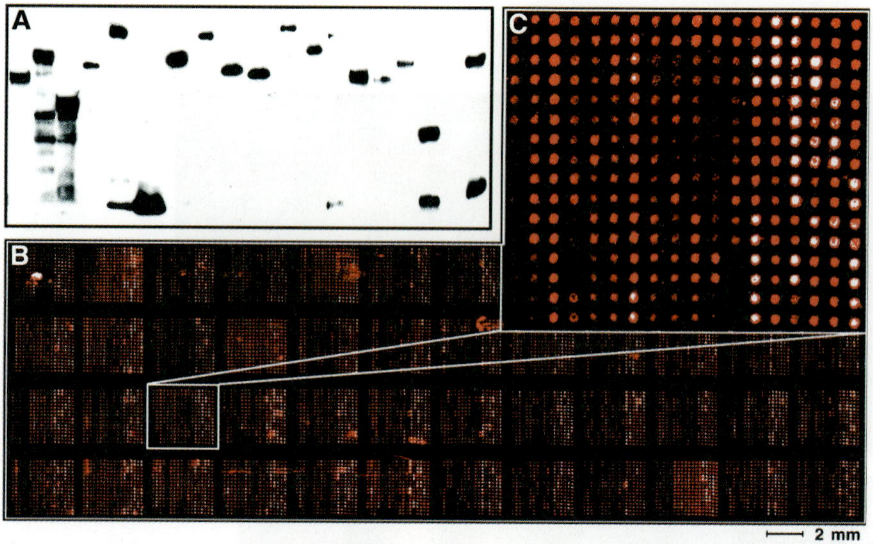

Figure 2.2 GST yeast protein analysis. (A) 60 samples were examined by immunoblot analysis using anti-GST; 19 representative samples are shown. (B) 6566 protein samples representing 5800 unique proteins were spotted in duplicate onto a single nickel-coated microscope slide which was then probed with anti-GST. (C) Enlarged image of one of the 48 blocks.

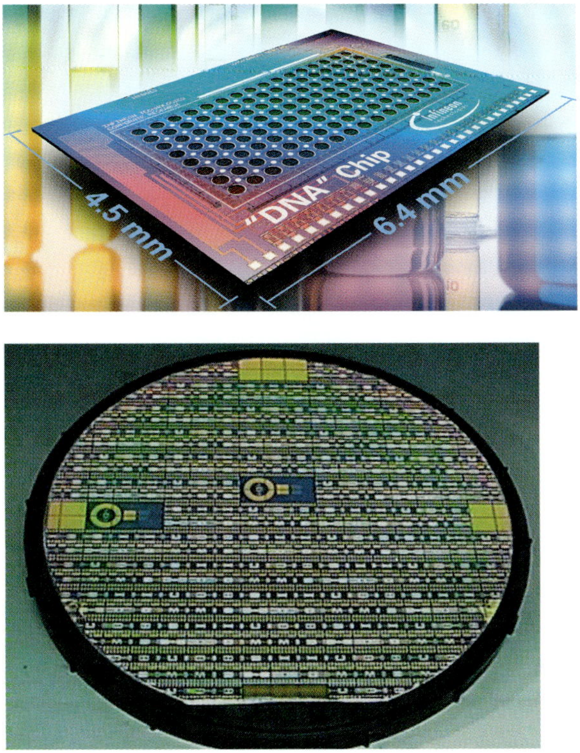

Figure 2.6 CMOS-based biochip platform with integrated electronic microstructure (gold on silicon) suitable for an electrochemical or optical chip read-out by Infineon (top) and Bionchip (bottom). Images by coutesy of Infineon and Bionchip BV.

Chapter 3

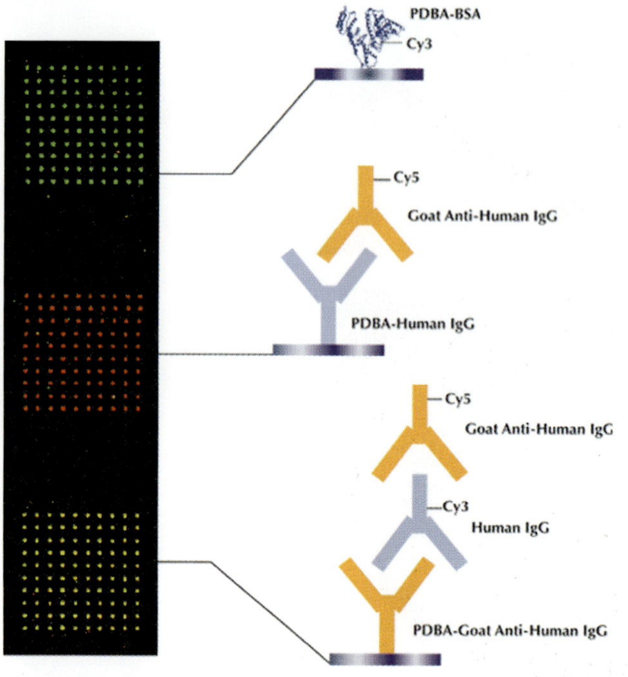

Figure 3.3 Protein microarray immunoassay on a 3-D SHA-coated slide. Fluorescence false-color composite image of 3–10 x 10 sub-arrays corresponding to PDBA-BSA-Cy3, PDBA-human IgG antigen detected with Cy5-labeled goat anti-human IgG and PDBA-goat anti-human IgG detected with Cy3-labeled human IgG antigen and Cy5-labeled goat anti-human IgG. Green indicates signal detected in the Cy3 channel, red in the Cy5 channel and yellow in both the Cy3 and Cy5 channels.

Chapter 4

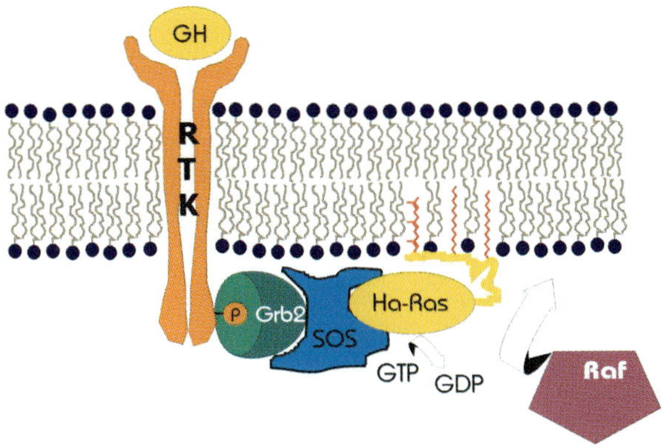

Figure 4.11 Signal transduction occurs in the plasma membrane. Ras is anchored in the plasma membrane via one farnesyl and one or two palmitoyl modifications at its C-terminus, where activation by GEF-proteins (here Sos) and recruitment of downstream effectors such as Raf-kinase occurs.

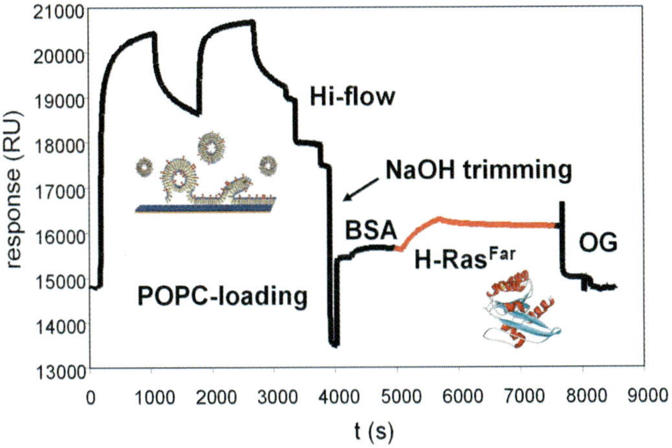

Figure 4.12 Insertion of lipid-modified Ras protein into an artificial membrane on Sensor Chip HPA. Sensor Chip HPA was treated with POPC (palmitoyl oleoyl phosphatidyl choline)-containing lipid vesicles and conditioned by treatment with NaOH and bovine serum albumin at high flow rates. Farnesylated Ras is inserted into the artificial membrane, while all non-covalently attached molecules are easily displaced from the lipid surface by the detergent octylglycoside.

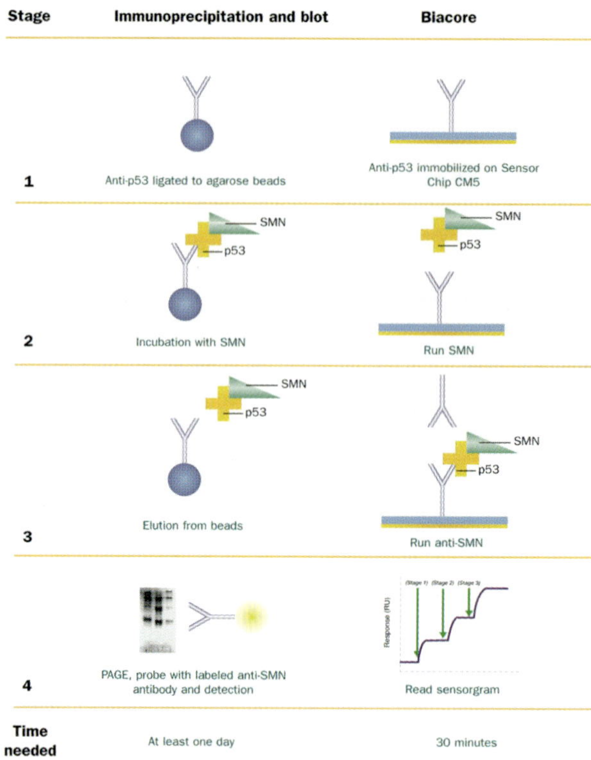

Figure 4.14 Comparison between immunoprecipitation/blotting and Biacore in the detection of molecular complexes. Both assays reveal that SMN protein derived from whole cell extracts is associated with p53 in the cell. Results achieved by immunoprecipitation are available within 1 day whereas the Biacore assay takes minutes.

Chapter 5

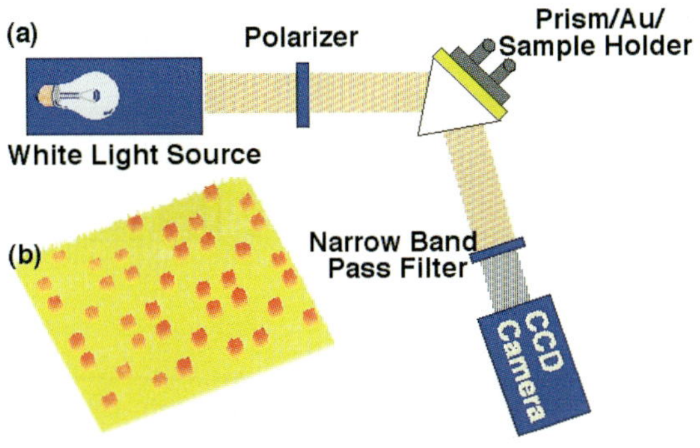

Figure 5.2 Schematic diagram of the surface plasmon imager apparatus. P-polarized white light impinges on a prism sample apparatus at a fixed optimal angle, near the surface plasmon angle. The light is then passed through a narrow band pass filter and images are collected by a CCD camera. The analyte is delivered to the array fabricated on a gold thin film using a simple 500 μL flow cell. An SPR difference image of the sequence-specific hybridization of 18-mer DNA to a two-component DNA array is shown in the bottom left corner of the figure. A maximum of 160 array elements can be studied on one SPR chip with a total surface area of 0.8 cm^2 using square elements with 500 μm widths.

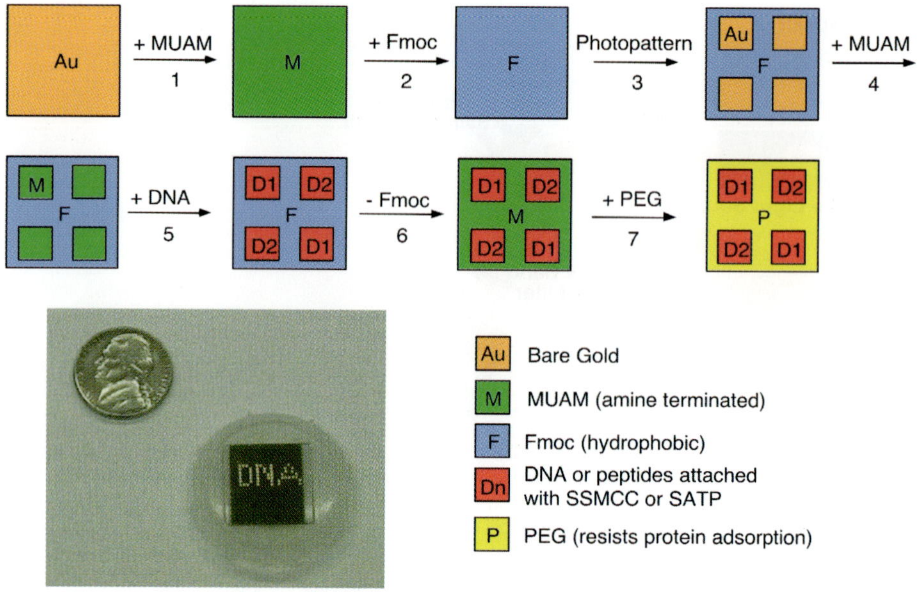

Figure 5.4 Multi-step array fabrication process using protection/deprotection chemistry and photopatterning techniques to prepare DNA and peptide arrays. The bottom left corner of the figure shows individually addressable hydrophilic drops of DNA pinned to the surface by a hydrophobic background. First, a bare gold surface is modified with a self-assembled monolayer of 11-mercapto-undecylamine (MUAM). This amine-terminated surface is reacted with the hydrophobic protecting group Fmoc. UV-light is used to break the gold–thiol bond and create bare gold pads on the surface. These pads are subsequently filled with MUAM. A bifunctional linker is used to attach DNA or peptide probes to create an array. Finally, the Fmoc is removed with base and replaced with a polyethylene derivative to prevent the non-specific adsorption of biopolymers to the background. Reprinted with permission from *Journal of the American Chemical Society*, **121,** 8044–8051. Copyright 1999 American Chemical Society.

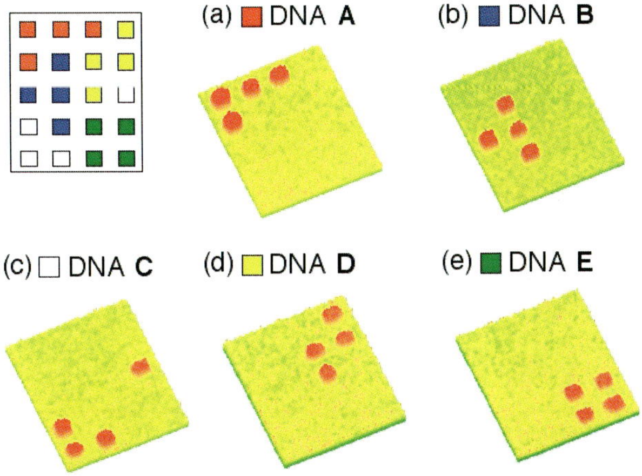

Figure 5.5 SPR difference images showing the hybridization of perfect-match DNA complements to a DNA array containing five different probes. The sequences were immobilized on 500 by 500 μm array elements in the pattern shown in the figure. (a) First, the array was exposed to a 100 nM solution of the DNA complement to probe A for 15 min. Hybridization adsorption onto the array is indicated by a change in the percentage reflectivity of incident light. (b) After rinsing with 8 M urea to regenerate the surface, the experiment was repeated with the DNA complement of probe B. Successive rounds of denaturation and hybridization of the remaining DNA complement probes resulted in images c–e. The same region of the array is shown for all images.

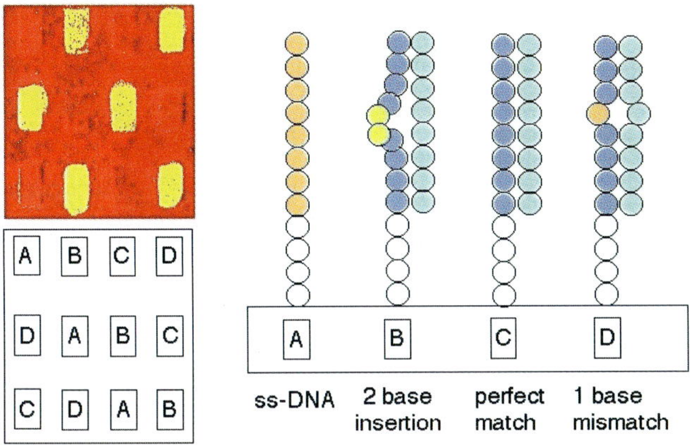

Figure 5.6 SPR imaging measurements of E. *coli* mismatch binding protein MutS adsorption onto a DNA array. This array was created with DNA probes A through D immobilized on 500 by 500 μm array elements in the pattern shown on the left of the figure. The array was then exposed to a solution containing the sequence Z. Z does not bind at all to probe A leaving it single strandedbut binds to probe B to create a duplex containing a two-base insertion, to probe C in a perfectly complementary manner, and to probe D to form a duplex containing a single-base mismatch. An SPR difference image of the binding of MutS to the array is shown on the right of the figure. The image shown is the difference between two images collected before and after exposure of the surface to MutS. Only 12 array elements from the total patterned surface area of 0.8 cm2 are presented in these images. Reprinted with permission from *Annual Review of Physical Chemistry* **51,** pp 41–63. Copyright 2000 by Annual Reviews www.annualreview.org.

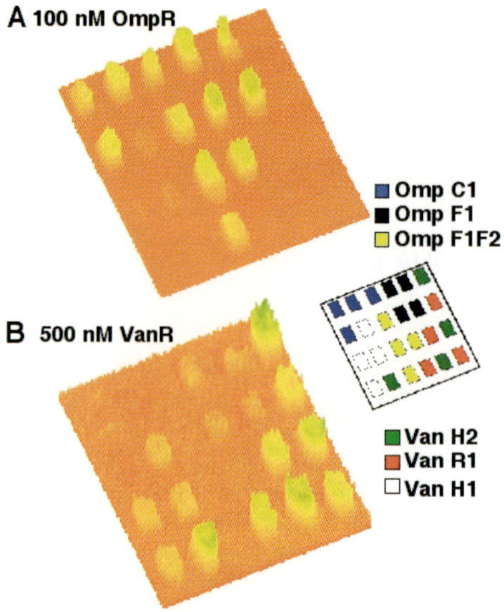

Figure 5.7 Surface plasmon imaging difference images of response regulator adsorption to double-stranded DNA immobilized on a photopatterned array composed of 500 by 500 μm array elements. (A) Specific adsorption of a 100 nM solution of OmpR to the DNA sequences OmpF1, F1F2, and C2. (B) VanR (500 nM) adsorbs to the DNA sequences VanH1, H2, and R1. OmpR is known to bind to the DNA sequences OmpF1, F1F2, and C2 and VanR is known to bind VanH1, H2, R1. There is little non-specific adsorption of the response regulators to the other sequences on the chip. DNA probe sequences for VanR and OmpR response regulators can be found in the literature. Reprinted with permission from *Langmuir*, Vol. 19, 1486–1492. Copyright 2003 American Chemical Society.

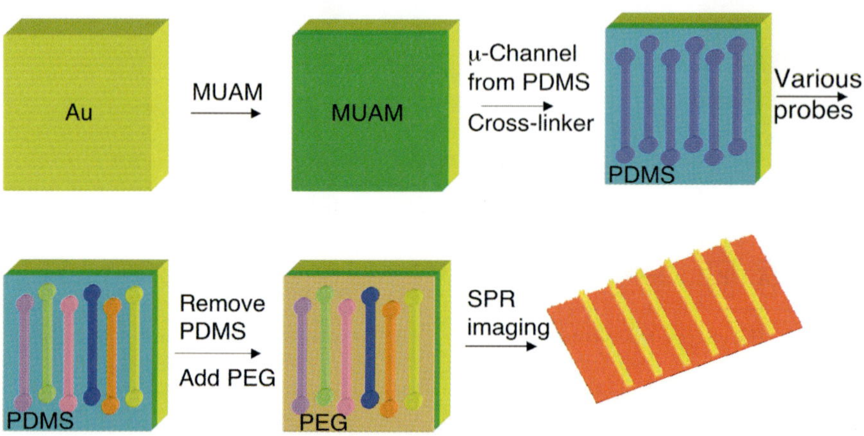

Figure 5.9 Schematic representation of the microfabrication strategy used to create DNA, peptide and protein linear arrays. A self-assembled monolayer of MUAM was formed on a clean gold surface. Polydimethyl siloxane (PDMS) microchannels were used to deliver bifunctional linkers and probe biomolecules to the surface. The microchannels were removed and the background was protected with a polyethylene glycol derivative. Adapted with permission from *Analytical Chemistry*, **73** 5525–5531. Copyright 2001 American Chemical Society.

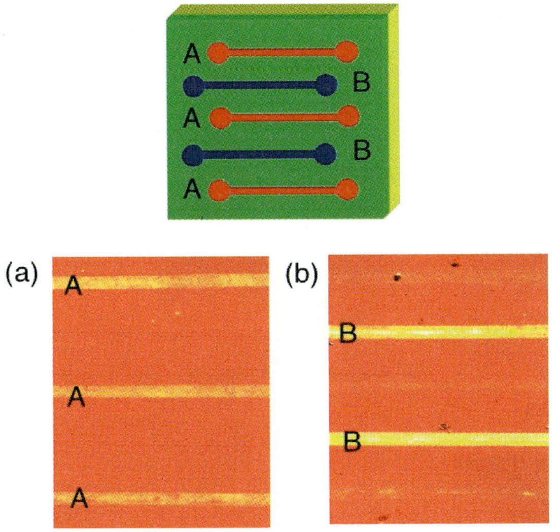

Figure 5.10 SPR difference images showing the sequence-specific binding interactions of antibodies to a peptide linear array containing two probes A and B corresponding to the peptide sequences Myc and HA. The peptides were immobilized in the pattern shown in the figure. (a) First, the array was exposed to a 25 nM solution of anti-Myc. Specific adsorption of anti-Myc to probe A is observed. (b) After rinsing with pH 11.5 buffer to regenerate the surface, the experiment was repeated 100 nM anti-HA. Anti-HA primarily bound to probe B, with little non-specific interaction with probe A.

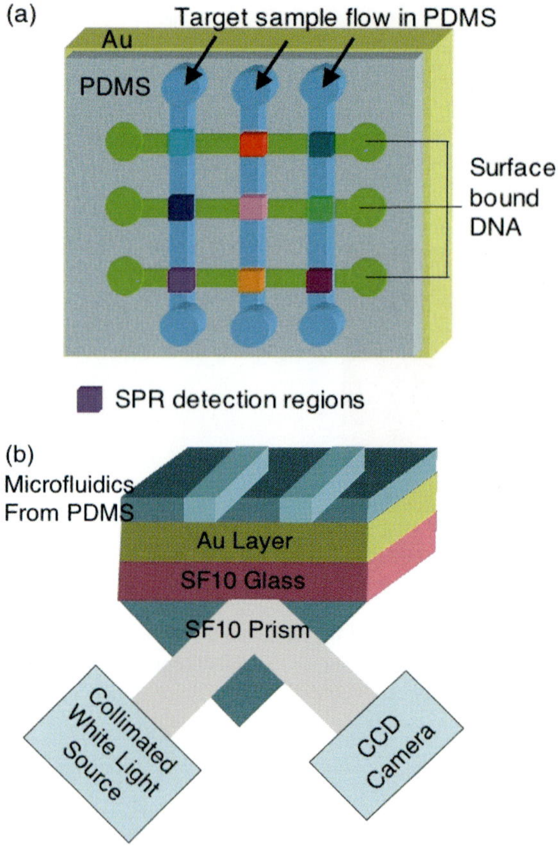

Figure 5.11 (a) Schematic diagram of parallel polydimethyl siloxane (PDMS) microchannels used to deliver small volumes of analyte (1 μl) to peptide and DNA linear arrays. (b) Schematic representation of the SPR imaging experimental set-up incorporating PDMS microfluidics to deliver a small volume of target sample. This configuration allows different analytes to be delivered through each channel.

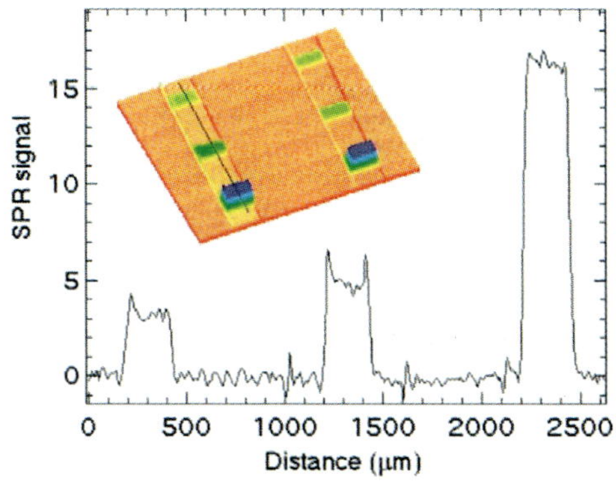

Figure 5.12 An SPR difference image showing adsorption of GUS gene ssRNA onto DNA probes. Each channel was 300 μm wide, 35 μm deep, and 1.2 μm long with 900 μm spacing between channels. Reprinted with permission from *Analytical Chemistry*, **73** 5525–5531. Copyright 2001 American Chemical Society.

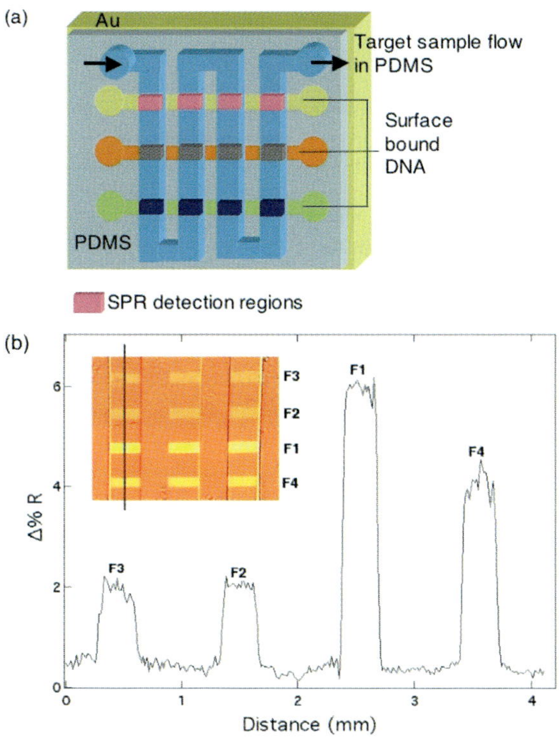

Figure 5.13 (a) A worm microchannel is used to deliver 5 μl aliquots of antibody solution to the peptide linear array. This microfluidic configuration is suitable for the delivery of a small volume of a single analyte to the array with a detection area equivalent to a configuration using a large volume flow cell. The worm channel was 500 μm wide, 35 μm deep, and 12 mm long with 500 μm spacing between channels.
(b) SPR difference image showing the adsorption of 100 nM anti-FLAG to a peptide array composed of four epitopes, differing by a single amino acid, based on the FLAG peptide sequence. The line profile reveals that the greatest adsorption occurs at elements containing the original sequence, F1. Diminished adsorption is observed at sequences, F3 and F4, containing alanine substitutions for essential residues of the binding motif. Peptide probe sequences can be found in the literature. Reprinted with permission from *Analytical Chemistry*, **74,** 5161–5168. Copyright 2002 American Chemical Society.

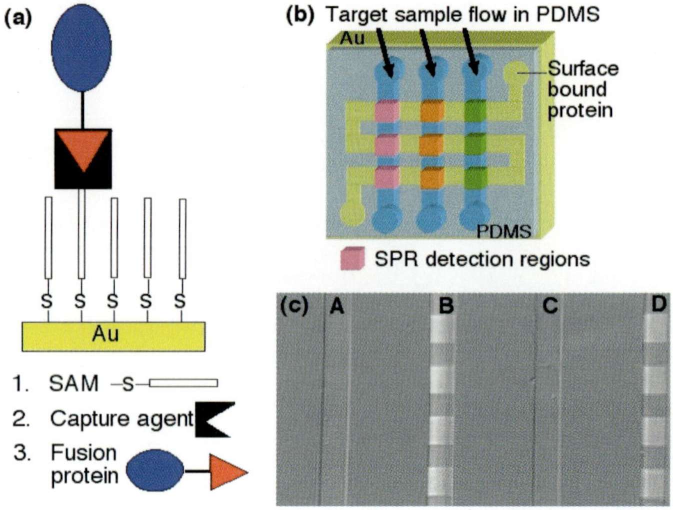

Figure 5.15 (a) Schematic representation of a general approach for the creation of an oriented fusion protein array on gold surfaces. (b) The NTA capture agent was immobilized through a worm microchannel. Next, a parallel set of PDMS channels was used to simultaneously deliver multiple histidine-tagged analytes to the NTA elements. (c) An example of an SPR difference image showing the immobilization of histidine-tagged ubiquitin **B** and polyhistidine D to NTA-modified monolayers in the presence of nickel ions. In the absence of nickel ions there is little non-specific adsorption of his-tagged ubiquitin **A** or polyhistidine **C** to the NTA surface, suggesting that a specific chelation interaction is occurring.

Chapter 8

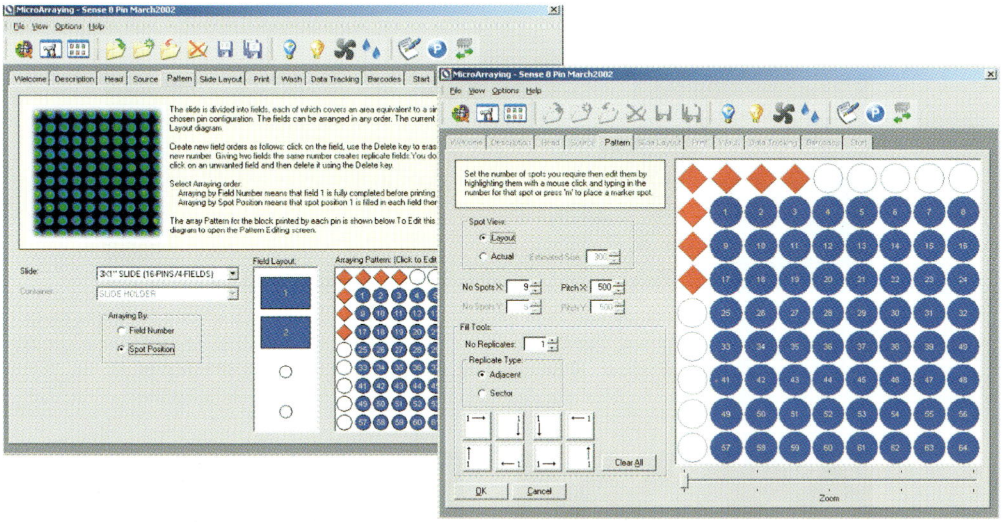

Figure 8.3 Illustrated step-by-step arraying instructions of QArray software.

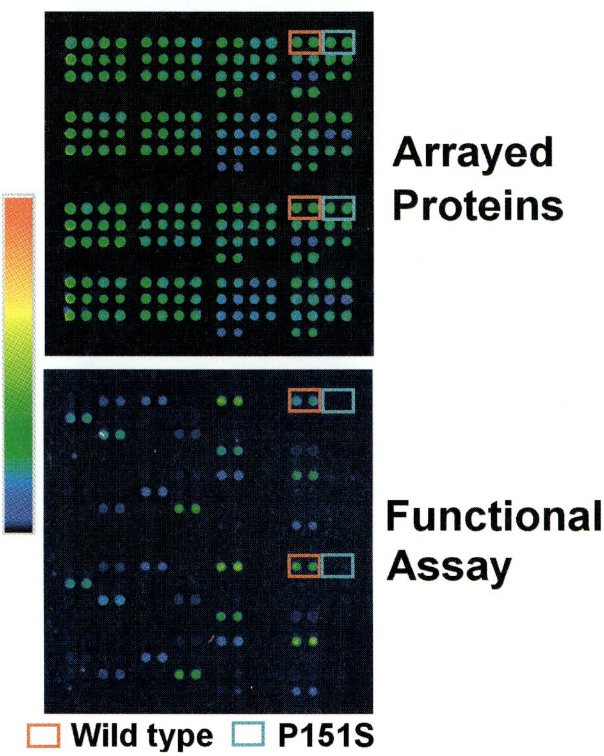

Figure 8.5 Procognia's p53 functional protein array manufactured at Genetix. Wild type and mutant p53 proteins were arrayed using 300-μm pins in adjacent duplicate spots and in two separate fields (= quadruplicate). The upper panel shows the immobilized array probed with a cy5-labeled antibody specific to a tag common to all proteins. The lower panel shows the functional assay, in this case the ability of the different mutants to bind cy3-labeled GADD45 dsDNA probe. Both assays were conducted using the same array. Images have been pseudo-colored to better visualize the intensity levels. Wild type and P151S mutant p53 are highlighted to show interaction differences. Images courtesy of Procognia Ltd., Maidenhead, UK.

Chapter 10

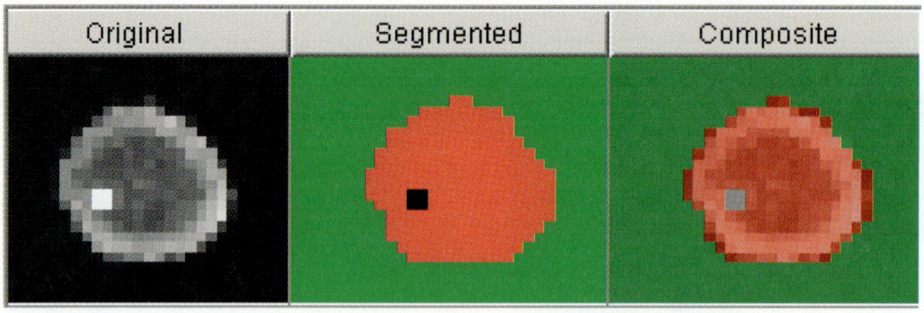

Figure 10.4 Results of the segmentation process using a "hybrid" method. Note that the pixels associated with the small amount of contaminant have been successfully removed from the signal area which has been matched to the non-circular spot shape to include all signal pixels.

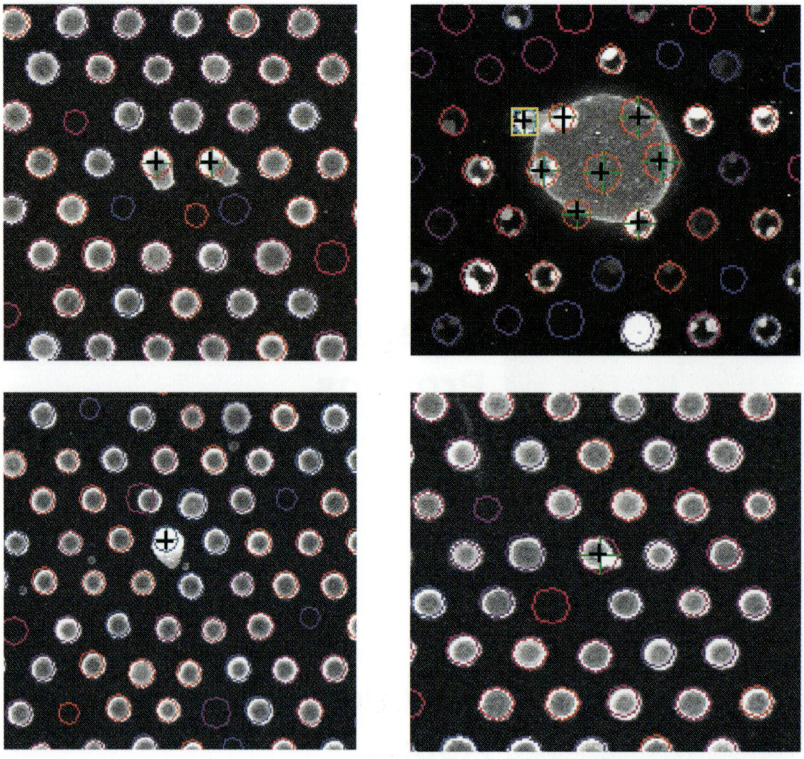

Figure 10.5 Various defects flagged by ImaGene 5.5.

1
Protein Microarrays: From Fundamental Screening to Clinical Diagnostics

Dev Kambhampati

Abstract

This chapter will focus on the driving forces that are causing a revolution in the field of protein biomolecular interactions and the corresponding microarray screening tools that are used for extracting their expression behavior. Numerous current applications of protein microarrays will be analyzed and their current problems and challenges will be discussed. An overview of those key technologies (described in detail in this book) which have the potential to overcome the above limitations and drive the industry towards diagnostic and other high-value drug discovery applications will also be provided.

1.1
Potential Need for Protein Microarrays

Proteins are the work-horses of the cell and have important functions in both healthy and diseased states. In order to understand how proteins interact and regulate various cellular processes, it is important to understand their expression behavior under a wide range of experimental conditions. Unlike the genome which contains a fixed number of genes, the levels of protein within the cells are highly dynamic. Proteins are constantly processed within the body in response to external stimuli and undergo a wide range of post-translational modifications. As a result, it is hard to accurately determine the exact number or quantities of proteins which are present within humans or animals (or other biological systems). In addition, protein families are extremely diverse and have considerable differences in their physical sizes, chemical and structural properties, affinity constants and relative abundance within the cells. As a result, accurately characterizing such interactions is extremely challenging.

One attractive method for monitoring biomolecular interactions in a high throughput (HTP) fashion is the use of microarrays. Microarrays have been successfully used in the area of gene expression [1] and genotyping research[2] using extremely limited volumes of biological samples. While the HTP goals have been achieved in the case of DNA arrays, quality control (QC) issues still remain on such

Protein Microarray Technology. Edited by Dev Kambhampati

ISBN: 3-527-30597-1

established microarray platforms. Microarrays can be used for screening proteins although there will still be challenges relating to QC and protein denaturation (refer to following subsections).

In addition to the above-mentioned goal of understanding the basic functions of proteins, the following are the three main catalyzing events that have resulted in a rapid interest and need for protein screening tools, especially protein microarrays.

1.1.1 Protein Therapeutics

The use of antibody-based therapeutics in drug discovery is expected to explode during this decade. Numerous products have reached clinical trials and it is predicted that there could be several tens of monoclonal antibodies on the market within the next several years [3]. Currently, antibodies account for 20 % of marketed biological products [4]. Thus, the potential for screening these high affinity molecules using microarrays and integrating them into drug discovery pipelines is huge. Another driving force has been the sequencing of the Human Genome, which has increased dramatically the number of molecules that can potentially be targeted as drug-like candidates. Screening of these molecules is still at a very early stage and will require extensive characterization before their potential can be harnessed for drug applications.

1.1.2 Clinical Diagnostics

Infectious diseases such as herpes, *lower case,* hepatitis C (HCV) and HIV are increasing rapidly around the world, with several million new cases (in each disease category) annually. This is reaching epidemic proportions in both developed and developing nations, and is affecting all sections of society. In the case of HCV [5], it is estimated that there are over 170 million people affected chronically from this disease globally (according to World Health Organization estimates) and the number of new cases is increasing at a rate of 3–4 million annually. Most diagnostic assays are based on ELISA or PCR formats. While PCR-based testing is highly sensitive (superior to ELISA), the cost factor prohibits it from widespread use across all nations. ELISA methods, with sensitivities ranging between 70 and 95 %, continue to be one of the standard technologies used in this sector. Microarray-based platforms which involve effective screening of antigen-antibody interactions using smaller amounts of samples, can play an important role in screening of infectious diseases. Again, a protein microarray containing antigens for a variety of diseases will also be helpful in reducing the costs. As the number of infected cases increases, the demand for such screening platforms is bound to grow (provided costs are kept to a minimum). In addition to infectious diseases, other areas such as screening of markers for autoimmune (diabetes type I, etc.) and cardiac diseases are also propelling a strong interest in the area of protein microarrays.

1.1.3
National Security

A growing demand for methods to detect the presence of pathogens in the environment (including urban settings), food and in battlefield conditions is currently driving a need for new technologies which can detect bacteria and other toxins rapidly [6]. The general principles of assembling proteins on solid substrates (although not necessarily in conventional microarray format) coupled with microfluidics (and similar fluid-handling systems) are expected to play an important role in the detection process. Protein mimics such as aptamers or synthetic molecules [7] are also being used for detection purposes in situations that are too harsh for antibodies or proteins to survive using similar microarray surface immobilization chemistries. With western nations committed to security, additional funding in this sector will encourage basic research and the development of novel sensing tools in the near future. All of these activities will positively impact the development of protein microarray technologies as such approaches are complementary.

1.2
Current Applications of Protein Microarrays

Most of the current applications of protein arrays are focused on disease proteomics [8] using conventional glass-based arrays. Other versions [9] of protein arrays are available although glass-based arrays still dominate this market due to cost benefits and considerable experience gained by the research community from using DNA arrays. Proteins have been detected in practically all kinds of human fluids, including serum [10], synovial fluids [11] and sweat [12]. Oncoproteomic studies have also been reported in the area of gynecology for studies on the placenta [13]. The use of this type of neonatal analysis is expected to grow although ethical concerns still remain. Most research has been focussed on various cancers, which is not surprising since the majority of pharmaceutical and government programs devote a large proportion of their funding to this disease.

Peptide [14] and oligosaccharide [15] microarrays have also been developed for screening antigen-antibody and carbohydrate-protein interactions. These studies, especially those focussed on carbohydrate moieties are still at an early stage although the results are promising. Analysis of protease substrates [16], cytokine activity [17] and detection of bacteria and toxins [18] have also been successfully carried out using various combinations of protein and peptide microarrays. Some preliminary studies on cell membrane [19] and GPCR-related activities [20] have been reported recently. The efficacy of such studies is still in question as it is very difficult to assemble an adequate biological membrane layer (which accurately mimics the natural system) on solid substrates, let alone monitor delicate receptor interactions. The stability of such membranes is still in question since such surfaces are easily affected by small changes in pH, temperature and humidity, among other factors. Some preliminary cell analysis can also be conducted on peptide-based platforms. Such integrated systems can be used in biomedical/bioelectronic applications.

Detection of biomarkers [21] using protein arrays is currently one of the most active areas of research directed towards clinical diagnostic applications. While numerous studies are currently underway in academic and industrial laboratories, none of these assays are ready for the diagnostics market as of today. But, this situation is expected to change in the next 3–5 years, when advances in technologies (most of which are mentioned in this book) are expected to result in the development of clinical assays.

1.3
Problems and Challenges

The ultimate goal of any screening experiment is to accurately determine the kinetics (affinity constants, k_{on}, k_{off}, etc.) and thermodynamic properties of how biomolecules interact, and thereafter integrate such data into downstream applications such as drug discovery or biomarker analysis. Rational design and discovery of pharmaceutical drugs and tools for diagnostics requires sound knowledge of biomolecular interactions.

For a given affinity between two interacting molecules (e.g. proteins), a rate constant can always be generated on any screening platform. The real question (particularly for drug discovery purposes) is – How relevant is the data and does it closely mimic the processes occurring in biological systems in their pristine environments? In diagnostic applications, it is important that the intended assay kit, which relies on the interaction between the surface-immobilized probes and biomarker targets present in human or animal specimens, accurately indentify the diseased state without any false results (positive or negative).

Modern drug discovery efforts require the screening of millions of compounds in a high throughput format before novel therapeutic entities can be identified. The same is true in the case of diagnostic assays (which are performed in several hundreds of thousands in certain disease categories). If there are fundamental flaws in the screening platform (e.g. microarray based), it defeats the overall benefits obtained from a high throughput format and the objectives of the proposed experimental effort.

Potential pitfalls (Figure 1.1) in microarray-related screening include (although not limited to):

- Sample preparation and handling (probe and target)
- Microarray platform
- Detection technology
- Data analysis

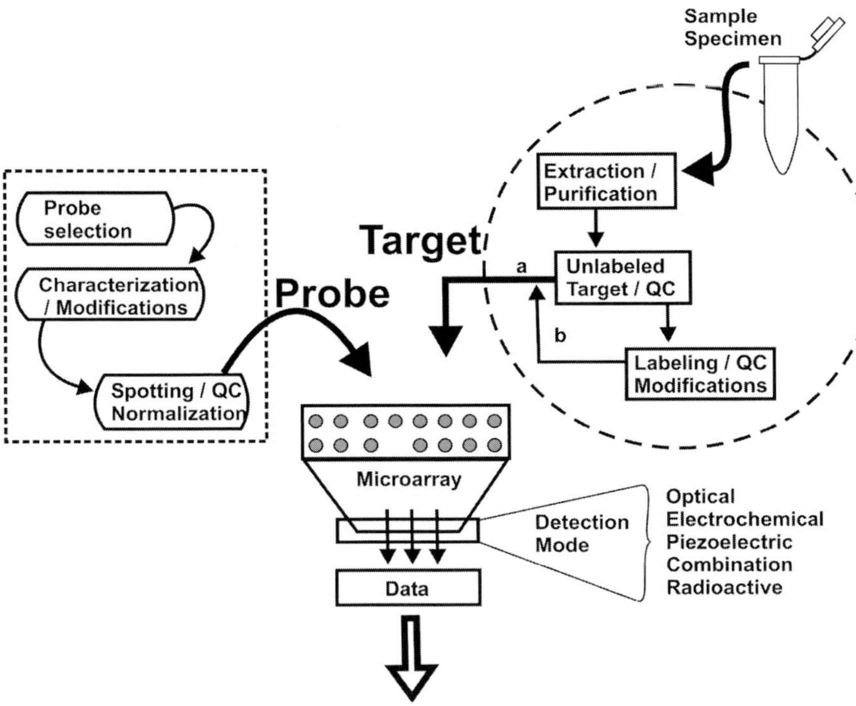

Figure 1.1 Conventional microarray processing flowchart. All the processes occurring during the probe and sample preparation steps are shown within the square and the circle. Significant - challenges are present at all stages of the microarray experiment starting from sample preparation, interactions on microarrays to detection and data analysis.

1.3.1
Sample Preparation and Handling (Probe and Target)

Even before conducting a single microarray experiment, numerous challenges are present at the sample collection and processing stages. The entire purpose of the experiment can be defeated if the selected probes and targets are not properly characterized. In addition, appropriate quality control techniques should be selected, especially during certain modifications (labeling or conjugation of surface-anchoring groups) of the probes and targets. Since the data analysis is based on the detected signal (which is correlated to expression levels, etc.), it is important to accurately normalize each and every probe and target sample so that the end result is representative of the true biomolecular expression level. From Figure 1.1, it is clearly evident that most of the initial sample preparation steps involve numerous unit operations which are laborious. While some detection techniques can work with unlabeled targets (Figure 1.1, a), most microarray formats require signal amplification. As a result, additional labeling of the targets is required (Figure 1.1, b). In certain sandwich assay studies (applicable to diagnostic studies where direct labeling of the biomarker target present in patient samples is not advisable), the use of secondary-labeled targets is necessary. At each and every stage, basic unit operation principles of extraction and purification apply and this tends to slow down the overall process. In multiplex experiments involving numerous targets, such processes will be quite time consuming and the need for incorporating other automation systems will become fairly critical. Spotting the probes and targets is also complicated as some probes are easily denatured during the process (drying of spots is also a major challenge although certain types of humidity chambers can be used to minimize such problems). The most critical aspect however, will be the experimental design and how well versed the end-user is with the above technical challenges.

1.3.2
Microarray Platform

The cost of the microarray platform is relatively insignificant when compared to the cost of the expensive reagents (probes, buffers, etc.) and valuable targets that are used for conducting biomolecular interaction studies. Surprisingly, the research community has still failed to fully realize the significance of microarray surface chemistry (which is essentially the foundation of a microarray) and its efficient characterization prior to the use of such surfaces in conducting protein (or any biomolecule) interaction studies. Most of the current problems facing microarrays such as spot heterogeneity, denaturation of biomolecules, non-specific interactions, microarray smearing and cross-talk can all be effectively addressed if the microarray surface is appropriately prepared.

1.3.3
Detection Technologies

A variety of detection technologies are available based on optics, electrochemistry, radioactivity and piezoelectric principles. In protein microarrays, it is predicted that optical (fluorescence-based detection systems [22], SPR [23], waveguide [24], etc.) and piezoelectric techniques [25] will dominate the detection techniques. In order to improve the sensitivity of these techniques, improvements in all the other aspects of biomolecular interactions will have to be considered (e.g. affinity of interactions, use of labels with excellent optical propeties and microarray surfaces with low backgrounds, resulting in high signal-to-noise ratios). A combination of the above techniques for simultaneously characterizing the optical, viscoelastic and electronic properties of biomolecular interactions can provide additional information regarding the orientation of proteins, packing densities and also kinetic information, which is normally not possible from a single detection technology.

1.3.4
Data Analysis

It is physically impossible for an individual to accurately monitor the interactions occurring on several hundreds or thousands of individual spots. In addition, while conducting statistical analysis and group comparisons, several additional degrees of data sets are involved. Although advanced data analysis software [26] is available, it is still difficult to accurately comprehend the biological significance of the data obtained. Simple events such as the presence of particulate dust or improper orientation of microarrays can throw the data analysis software program into disarray. Again, data analysis is at the mercy of all the steps that occur prior to data collection (these were mentioned in Sections 1.3.1–1.3.3). In addition, regression analysis software, particularly for kinetic studies, needs to be improved as the selection of data points can greatly influence the end k_{on}/k_{off} rates. For end-point experiments, where real-time kinetics are not important, additional challenges are currently being addressed (see Chapter 10).

1.4
Potential Solutions: Enabling Technologies and Advancements

A number of interesting technologies are currently available which can be used to effectively overcome the pitfalls of microarrays (both protein and DNA). Self-assembly techniques are excellent for improving the biocompatibility of protein microarrays on gold and oxidic surfaces. Such technologies will be reviewed in detail by Schaeferling's group (University of Regensburg, Germany; see Chapter 2). They also provide a comprehensive overview of all available surface chemistry options on glass, semiconductors and plastic substrates for efficiently coupling proteins and other biomolecules. Small molecule chemical affinity systems developed by Prolinx can also be used for developing protein arrays in both two- and three-dimensional formats.

Excellent reports by Biacore and the research groups headed by Professors Corn and Knoll will highlight some of the interesting optical technologies such as Surface Plasmon Resonance (SPR), Surface Plasmon Microscopy (SPM) [27] and Surface Plasmon enhanced Fluorescence Spectroscopy (SPFS) [28] that are currently available for monitoring real-time biomolecular interactions on gold-based substrates. These technologies offer both labeled and label-free detection modes, and as a result, are useful when dealing with delicate proteins and systems where adding a label might significantly alter the properties of any biomolecular interaction. The SPFS technique is also currently being investigated for the development of a real-time diagnostic detection instrument for the analysis of blood analytes. As a result, this technology is already pushing biomolecular screening into the realm of clinical diagnostics. Fluorescence-based detection systems still continue to dominate the microarray field and in future, it is expected that LED-based scanners will significantly lower the cost of detection. The use of novel nanoparticle-based systems such as quantum dots [29] and the principles of resonance light scattering [30], can also be used to enhance the signal, although the cost of such materials will still be an issue.

One of the most exciting detection technologies that is driving the proteomics revolution is mass spectrometry. This label-free technology when coupled with SPR and gel-based systems is very powerful. Ciphergen is a market leader in this sector and has undertaken major programs in both disease proteomics and biomarker discovery. Chapter 7 by Lomas will highlight the various options of screening proteins on array formats using the SELDI [31] mass spectrometry technique. This technology is expected to play an important role in all aspects of protein screening since there are numerous surface chemistry options available on their protein arrays.

Chapter 8 by Mann et al. from Genetix highlights the significant challenges of arraying biomolecules during the fabrication of protein microarrays. The research community is tackling such complex proteins even with "relatively simple" antibody-based systems. When complex molecules such as membrane proteins are arrayed, the problems will take on an entirely new dimension, where smart surface technologies will have to be integrated with biocompatible and "gentle" liquid-handling systems.

A technology that is currently causing much interest in the area of biosciences and nanotechnology is the use of cantilever-based systems for monitoring label-free biomolecular interactions. Functionalization of cantilevers with biomolecules can be used to probe multiple interactions in parallel. Such technologies can also be incorporated into microfluidic assemblies for highly sensitive detection of pathogens. Fritz and colleagues (Chapter 9) will provide an overview of AFM-based sensors for protein screening. This technology has great potential and can also be used for screening and detecting single molecules. As advances in nanotechnology continue, the use of nanoparticles and similar systems will play a major role in the design of molecular scale sensors for certain applications.

Finally, data analysis procedures and the common pitfalls are described in detail in Chapter 10 by Petrov and Shams from BioDiscovery. Since the outcome of data analysis from a microarray experiment will determine the end application, the importance of data quality and analysis in screening experiments cannot be overemphasized.

The implementation of the technologies that have been mentioned in this book will lead to better screening of proteins and other biomolecules. This is expected to impact the field of drug discovery in the short term and once the quality control and cost issues are optimized, it is also expected to play an important role in the area of diagnostics in the near future.

References

1. Brown, P.O, and Botstein, D., *Nature Genetics* (Suppl) 1999, **21,** 33–37.
2. Grant, D.M., and Phillips, M.S., In *Pharmacogenomics* Vol. 113, Kalow, Meyer & Tyndale (Eds). Marcel Dekker Inc. 2001, 183–190.
3. *Business Week* 2000 and *PhRMA,* 2002.
4. *PhRMA* 2002 (http://www.phrma.org)
5. World Health Organization, *Hepatitis C Factsheet* (http://www.who.int/inf-fs/en/fact164.html)
6. Delehanty, J. B., and Ligler, F. S., *Anal. Chem.* 2002, **74,** 5681–5687.
7. *Materials Research to Meet 21st Century Defense Needs* – Interim Report, National Research Council, USA. (http://www.nap.edu/html/materials_research/interim_report.pdf)
8. Hanash, S., *Nature* 2003, **422,** 226–232.
9. Schaeferling, M., Schiller, S., Paul, H., Kruschina, M., Pavlickova, P., Meerkamp, M., Giammasi, C., and Kambhampati, D., *Electrophoresis* 2002, **18,** 3097–3105.
10. Pavlickova, P., Jensen, N. M., Paul, H., Schaeferling, M., Giammasi, C., Kruschina, M, Du, W. D., Ibba, M., Ortigao, F., and Kambhampati, D., *J. Proteome Res.* 2002, **1,** 227–231.
11. Uchida, T., Fukawa, A., Uchida, M., Fujita, K., and Saito, K., *J. Proteome Res.* 2002, **1,** 495–499.
12. Flad, T., Bogumi, R., Tolson, J., Schittek, B., Garbe, C., Deeg, M., Mueller, C. A., and Kalbacher, H., *J. Immunol. Methods* 2002, **270,** 53–62.
13. Batorfi, J., Ye, B., Mok, S. C., Cseh, I., Berkowitz, R. S., and Fulop, V., *Gynecol. Oncol.* 2003, **88,** 424–428.
14. Reimer, U., Reineke, U., and Schneider-Mergener, J., *Curr. Opin. Biotechnol.* 2002, **13,** 315–320.
15. Fukui, S., Feizi, T., Galustian, C., Lawson, A. M., and Chai, W., *Nature Biotechnol.* 2002, **10,** 1011–1017.
16. Salisbury, C. M., Maly, D. J., and Ellman, J.A., *J. Am. Chem. Soc.* 2002, **124,** 14868–14870.
17. Whiteside, T. L., *Biotechniques* 2002, **10** (Suppl. 4–8), 12–15.
18. Howell, S. W., Inerowicz, H. D., Regnier, F. E., and Reifenberger, R., *Langmuir* 2003, **19,** 436–439.
19. Groves, J. T., *Curr. Opin. Drug Discov. Devel.*
20. Fang, Y., Frutos, A. G., and Lahiri, J., *J. Am. Chem. Soc.* 2002, **124,** 2394–2395.
21. Krieg, R. C., Paweletz, C. P., Liotta, L. A., and Petricoin, E. F., *Technol. Cancer Res. Treat.* 2002, **1,** 263–272.
22. Martinez, M. J., Aragon, A. D., Rodriguez, A. L., Weber, J. M., Timlin, J. A., Sinclair, M. B., Haaland, D. M., and Werner-Washburne, M., *Nucleic Acids Res.* 2003, 31, e18.
23. Stenlund, P., Babcock, G. J., Sodroski, J., and Myszka, D. G., *Anal. Biochem.* 2003, **316,** 243–250.
24. Pawlak, M., Schick, E., Bopp, M. A., Schneider, M. J., Oroszlan, P., and Ehrat, M., *Proteomics* 2002, **2,** 383–393.
25. Guanghua, W., Datar, R. H., Hansen, K. M., Thundat, T., Cote, R. J., and Majumdar, A., *Nature Biotechnol.* 2001, **19,** 856–860.
26. www.biodiscovery.com, and http://rana.lbl.gov (excellent source for software and relevant publications).
27. Smith, E. A., Erickson, M. G., Ulijasz, A. T., Weisblum, B., and Corn, R. M., *Langmuir* 2003, **19,** 1486–1492.
28. Liebermann, T., and Knoll, W., *Colloids Surfaces A.* 2000, **171,** 115–130.
29. Watson, A., Wu, X., and Bruchez, M., *Biotechniques* 2003, **34,** 296–303.
30. Yguerabide, J., and Yguerabide, E. E., *J.Cell Biochem.* 2001, **37** (Suppl.), 71–81.
31. Issaq, H. J., Conrads, T. P., Prieto, D. A., Tirumalai, R., and Veenstra, T. D., *Anal. Chem.* 2003, **75,** 148A–155A.

2 Protein Microarray Surface Chemistry and Coupling Schemes

Michael Schäferling and Dev Kambhampati

Abstract

Efficient immobilization of biomolecules is a key factor in determining the overall success of a microarray or, for that matter, any highthroughput (HTP) screening tool. If the immobilized probes are not correctly oriented on the microarray surface or are denatured, it can dramatically affect the downstream biomolecular interaction events. Here, various schemes for immobilizing biomolecules on glass, gold, polymer and semiconductor surfaces are described. In addition to tailoring the surface chemistry on the above substrates, various schemes for manipulating the probes with tags will also be addressed to achieve efficient and biocompatible screening platforms.

2.1 Introduction

This review will focus on various strategies for constructing protein microarrays on a wide range of substrates. The way in which proteins are immobilized on solid substrates will determine the functional properties of a protein microarray. While protein interactions are currently conducted on a wide variety of formats, it is still a major challenge to obtain high quality biomolecular interaction data that can be used to predict how proteins behave in their biological environments. Two approaches for developing protein microarrays are currently underway. In the first approach the properties of the microarray surface are tailored to functionalize unmodified (pristine) biomolecules and in the second biomolecules are modified with tags to effectively anchor them to existing microarray or sensor substrates. Irrespective of which approach is chosen for a particular application, fundamental challenges still remain when it comes to interpreting the data. Extrapolation of protein immobilization strategies from one system to another for different classes of proteins and applications is difficult since proteins have a wide range of structural, chemical, and biological properties. Even within a single protein array application system, normalization studies (corresponding to probe surface concentration and activity) of a wide range of probes with different functional properties have to be considered. Sample preparation and quality control issues plague almost all

Protein Microarray Technology. Edited by Dev Kambhampati

ISBN: 3-527-30597-1

microarray formats that are currently available. These processes are extremely time consuming and counteract the overall claims and advantages of such microarrays in functional high throughput screening.

A "one-approach-fit-all" scheme for screening all the various types of proteins is improbable. The current (and the most practical) approach is to generate surfaces that can be used for broad subsets of protein families without compromising their functional properties. In this chapter, protein chips constructed by using different immobilization chemistries on glass, gold, polymer and other special formats (semiconductor surfaces and lab-on-a-chip) will be discussed in detail. Various strategies for increasing the degree of probe immobilization and preventing protein denaturation by the use of effective tags will also be described.

2.1.1
Background and Current State of Biomolecule Libraries

Most large-scale biotechnology, pharmaceutical and research laboratories currently have existing libraries of both pristine proteins and those conjugated with various tags (biotin, His-Tag, Strep-Tag, Avi-Tag, etc.). Since these molecules have been extensively characterized, such laboratories will be reluctant to alter them to suit the surface immobilization requirements of microarray platforms. As a result, surface immobilization strategies on microarrays have to be designed on the basis of the existing state of biomolecule libraries where there is limited flexibility for operation. Most of the microarray schemes that will be described here will focus on the above-mentioned existing biomolecule libraries. Almost all kinds of proteins have reactive functional groups in their side chains (Table 2.1) which can be used as a starting point for anchoring these molecules on microarrays. In the case of biomolecules conjugated with tags, the immobilization approach proceeds with a higher surface coupling efficiency (when compared to pristine probes). The drawback in this case is that the inherent biomolecular interaction properties of the conjugated probes might be altered, thereby affecting their kinetic and thermodynamic properties. Irrespective of the nature of the probes, the overall goal is to target reactive sites which do not denature the molecules or affect the complementarity-determining regions (CDRs) of the probes during surface immobilization. Numerous surface chemistry approaches are currently available to address such issues.

2.2
Microarrays Based on Glass Substrates

Most of the current analysis of protein interactions on microarrays are based on surfaces and formats that have been used for DNA arrays. Glass slides coated by epoxy- or aldehyde-terminated alkylsilanes or by polylysine can also be used to immobilize proteins, antibodies and antigens. Such activated glass (microscope slide format) slides can be spotted with biomolecules by the end-user in any format and density. The detection mode for such systems is based mainly on optical techniques such as fluorescence. Table 2.2 lists some examples of commercially available glass slides with reactive surfaces.

2.2.1
Surface Modification

The reactive coatings of the glass surface are mainly based on self-assembly techniques. Self- assembled monolayers (SAMs) of alkylchlorosilanes or alkylalkoxysilanes require hydroxylated surfaces as substrates for their formation. The driving force for this self-assembly is the in-situ formation of polysiloxane, which is connected to surface silanol groups (SiOH) via Si-O-Si bonds (Scheme 2.1). These thin organic films enhance the biocompatibility of the surface and protect the proteins from denaturation and structural changes during the immobilization step.

A great variety of silanization conditions pertaining to temperature and reaction time, which are also strongly influenced by the stability of the terminal functional

Table 2.1 Various methods of coupling proteins on a microarray platform.

Functional side group on peptide	*Available surface for derivatization*	*Type of binding*
Natural		
–COOH (carboxylic acid)	Amino	Electrostatic
Asp		Covalent amide (after carboxy activation)
–NH2 (amino)	Carboxylic acid, active ester, epoxy, aldehyde	Electrostatic
Lys, Gln, Arg		Covalent amide
–SH (thiol)	Maleimide	Covalent thio ether
Cys		
–OH	Epoxy	Covalent ether
Ser, Thr		
Synthetic		
His-Tag	Ni-NTA complex	Coordination complex
Strep-Tag	Strep-Tactin	Supramolecular complex
Biotin	Streptavidin	Supramolecular complex

Table 2.2 Examples of commonly available glass-based microarray platforms.

Coating	*Supplier*	*Advantages*	*Disadvantages*
Poly-L-lysine	Sigma, Fisher	Cheap, easy to handle	Only electrostatic absorption, low stability
Aminopropylsilane	Corning, Erie, Xomics		
Aldehyde-, NHS-ester- or epoxy-silane	GeneScan, Xomics Genetix NoAb BioDiscoveries Telechem, Quantifoil	Highly reactive, great versatility	Low spot homogeneity, susceptible to contamination

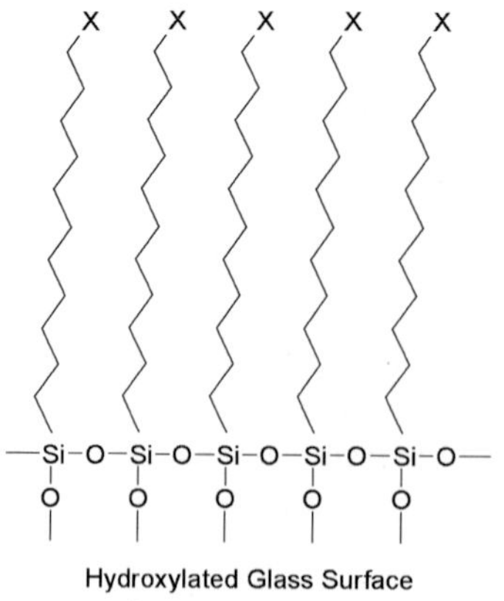

Scheme 2.1 2D schematic description of a polysiloxane monolayer on a glass surface (X = terminal functional

groups have been described in the literature [1]. Monolayers of aminopropylsilanes, for example, can be formed under mild conditions (room temperature for 30 min) [2]. Scheme 2.2 shows some commonly-used silane-reagents used for the derivatization of glass surfaces [3]. Substrates on which these organosilicon monolayers have been successfully prepared include silicon oxide, aluminium oxide, quartz, glass, mica, zinc selenide and germanium oxide [4,5]. Silicon substrates can also be covered with silane monolayers [6], resulting in a new approach for tuning the electronic properties of semiconductor chips [7] and for challenging new interfaces in the field of bioelectronics.

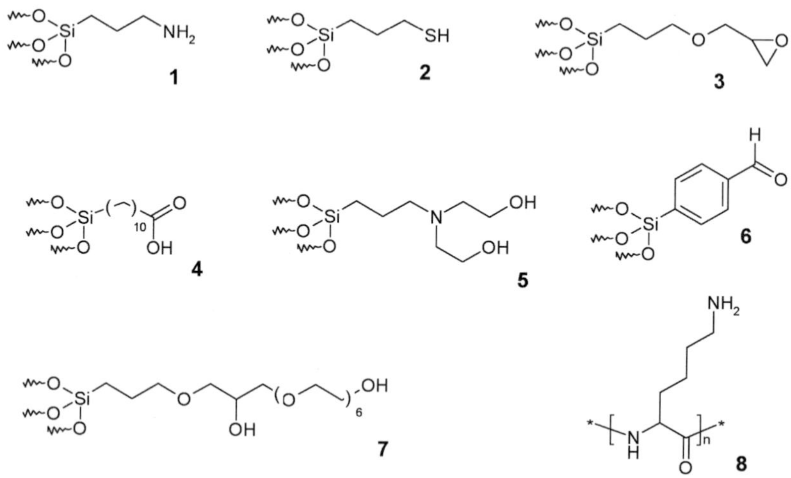

Scheme 2.2 Reagents for derivatization of glass surfaces. **1** APTES = aminopropyltriethoxysilane; **2** MPTS = 3-mercaptopropyltrimethoxysilane; **3** GPTS = glycidoxypropyltrimethoxysilane; **4** TETU = triethoxysilane undecanoic acid; **5** HE-APTS = bis(hydroxyethyl)aminopropyltriethoxysilane); **6** 4-trimethoxysilylbenzaldehyde; **7** GPTS/HEG = glycidoxypropyltrimethoxysilane-hexaethylene glycol; **8** poly(lysine).

2.2.2
Current State of Glass-Based Protein Microarrays

As in the case of DNA-arrays, complex issues pertaining to spot morphology, homogeneity of signal across the spot, microarray smearing and non-specific adsorption are not adequately addressed in the case of protein microarrays. In addition to the above challenges, it is difficult to compare the experimental results obtained in two different laboratories using the same set of arrays. Reproducibility of data is a major concern and if this current trend continues, it will become extremely difficult to accurately validate the biomolecular interactions that form the basis of many drug discovery and research studies. Preliminary studies [8] pertaining to the use of polylysine arrays in monitoring the interactions of 115 antigen/antibody pairs using complex protein mixture solutions was recently reported. While the success rate of the interaction data was between 20 and 50 %, concerns pertaining to microarray smearing and background still remain (Figure 2.1a). Microarray smearing is a common artifact found in most glass-based arrays and it is one of the methods by which cross-talk between the adjacent probes occurs, leading to false results. In addition to microarray smearing, the electrostatic charges present on the glass can also lead to the denaturation of proteins, although the exact level of electrostatic charge is difficult to estimate.

In another study, large-scale protein–protein interaction studies were also successfully conducted on aldehyde-coated glass slides [9]. While the slides were successful in discriminating between the specific and non-specific interactions, issues pertaining to "quenching" or neutralizing the remaining reactive groups on the slides still remain. BSA was used to effectively block the free surface of the chip to minimize non-specific interactions. This could pose a problem on such slides if small molecules are used which can potentially be masked by the bulky BSA molecules. While spot morphology is not a major concern in normal laboratory-scale research studies, it becomes critical when investigating several thousands of protein samples spotted on a single microarray slide. Fluorescence intensities emanating from the spots (Figure 2.1b) and their adjoining areas (if the spots are not homogeneous) can affect the data acquisition process. In addition, if the surface chemistry within the spots is not homogeneous, varying intensities of fluorescence signal can be observed within the same spot. Unless the data analysis software is highly sophisticated, such spots can lead to poor averaging effects during the analysis step which will in turn result in misleading protein expression levels for the protein of interest.

In another study, Joos et al. [10] reported the direct parallel detection of autoantibodies in human sera by using a single immunoassay experiment in which 18 different antigens were spotted onto aldehyde-coated slides. After activation of a 96-well teflon-coated glass plate (Erie Scientific) with aminopropylsilane, protein antigens were attached to the pre-defined spots to form a high-throughput microarray-based enzyme-linked immunosorbent assay (ELISA) [11]. Inside the wells, the authors spotted arrays of 144 (4 _ 36) elements each using a 36-capillary based print head. This array format thus appears as a compromise between high-density

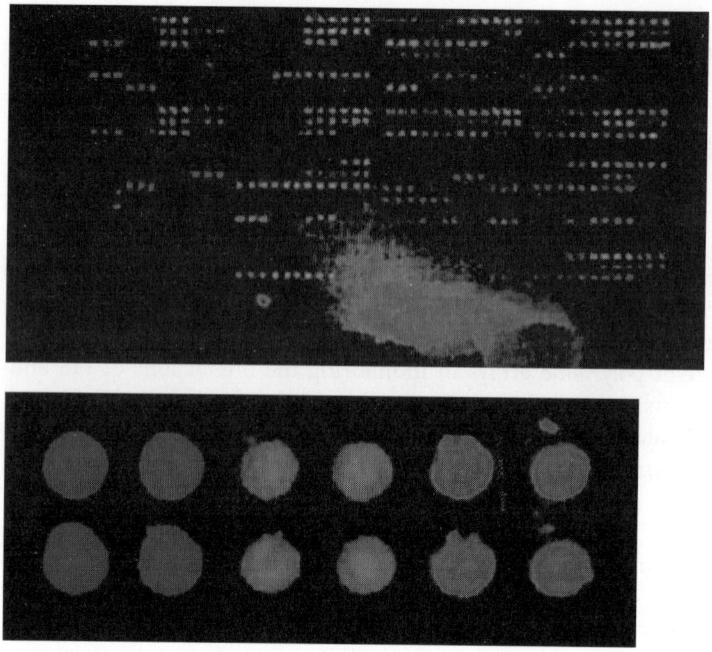

Figure 2.1 (a) Smearing effects on antibody-immobilized poly-lysine arrays [8]. (b) Snapshot of microarray spots with poor morphology and sur face chemistry (within the individual spots). Differences in spot homogeneity can result in varying signal intensities within the same spot [9]. (see Colour Plate p. XVII).

microarrays and conventional micro-titerplates. Antibodies can also be arrayed on aminosilated glass surfaces using a micro-molded hydrogel stamper [12]. Many approaches use micro-contact printing with PDMS stamps [13] or photoreactions in combination with masks [14] for the patterning of glass surfaces with proteins. The profiling of cancer cells from a cellular lysate was carried out with aldehyde- or poly-lysine-coated glass microarray containing 146 distinct antibodies. The target proteins were labeled with Cy5 or Cy3 dyes and no amplification methods such as ELISA were needed for monitoring the interactions [15]. In another interesting study, Schweitzer *et al.* designed microarrays containing 75 antibodies against cytokines on chemically-derivatized glass slides to study cytokine secretion from human dendritic cells induced with lipopolysaccharide or tumor necrosis factor-α [16].

The superiority of nickel complex-coated surfaces over conventional aldehyde slides was reported by Snyder in his study on a proteome chip containing 6566 GST yeast proteins spotted onto a glass slide (Figure 2.2). In this particular case, the fusion proteins attach via their His-Tags and it is believed that the binding sites orient away from the surface [17]. This example shows that the use of tags can help to control the orientation of the immobilized biomolecules on the surface.

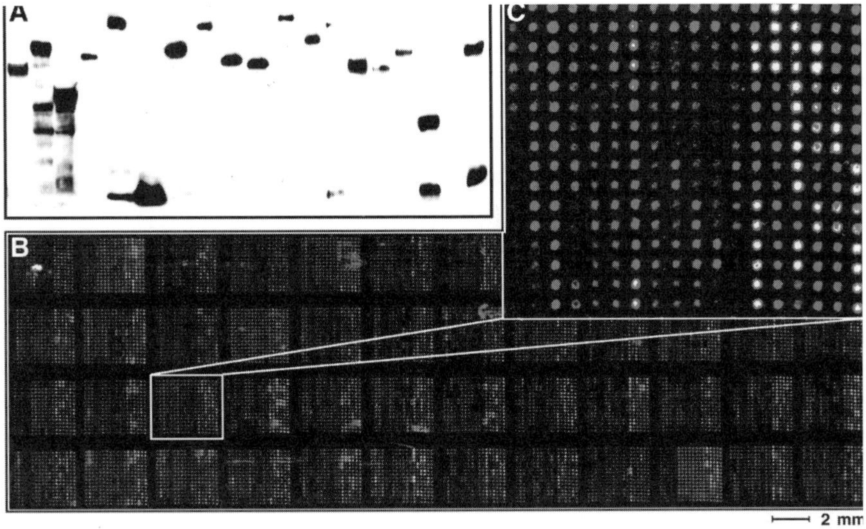

Figure 2.2 GST yeast protein analysis. (A) 60 samples were examined by immunoblot analysis using anti-GST; 19 representative samples are shown. (B) 6566 protein samples representing 5800 unique proteins were spotted in duplicate onto a single nickel-coated microscope slide which was then probed with anti-GST. (C) Enlarged image of one of the 48 blocks. (see Colour Plate p. XVIII).

2.3 Microarrays based on Gold Substrates

The use of gold-based substrates for monitoring protein interactions has a relatively long history. Instead of using polycrystalline gold surfaces, thin gold films (10–200 nm) can easily be deposited on solid supports similar to glass or silicon wafers by evaporation or sputtering in an ultrahigh vacuum (UHV) [18]. Since the adhesion of gold on glass or silicon is poor, pre-coating of a thin layer of Cr or Ti (1–10 nm) is generally required prior to gold deposition.

Since the work of Allara [19] and Whitesides [20] in the mid-1980s, the spontaneous adsorption of organosulfur compounds onto gold has been a widely used method for the preparation of self-assembled monolayers (SAMs). If the alkyl chain is of sufficient length, the resulting molecular architecture is very stable and orients itself nearly perpendicular to the surface [21]. The close proximity of the adsorbed molecules causes van der Waals forces to play an important role in intermolecular stabilization. The thickness of these monolayers is typically in the nanometer range in the vertical direction and as a result, they can be termed nanoassemblies. It is important to note however, that in the horizontal plane, i.e. over the substrate surface, these monolayers can extend to a macroscopic length.
Although SAMs have been constructed on a wide range of surfaces, monolayers based on alkanethiols on gold have been investigated extensively for the following reasons: (1) ease of fabrication, (2) degree of perfection, (3) chemical stability under laboratory conditions, (4) availability of materials, and (5) flexibility in introducing

various chemical functional groups to give different surface properties. They provide excellent control over the surface properties by individually tailoring/engineering the building blocks. Using these principles, customized surfaces with good biocompatibility properties can be developed to meet the complex needs of proteomic studies. Unlike glass- or polymer-based arrays, SAMs on gold also have the added advantage of the surface properties being efficiently characterized by optical, mechanical and electrochemical analytical tools.

Many commercial suppliers offer gold-coated microscope slides with sufficiently high surface quality. In spite of the above advantages, gold-based microarrays still are not in widespread use due to their cost and stability (chemical and thermal). A major limitation to the use of SAMs as an immobilization matrix is their low temperature stability due to the weak bonding (interactions) between sulfur and gold. While this is not a concern in the case of protein arrays (where experiments are conducted at low temperatures), such microarrays are found to be unsuitable for DNA array applications where high stringent wash and temperature conditions are employed. Because of the excellent electronic properties of gold, these surfaces can be used in more advanced and label-free detection technologies such as Surface Plasmon Resonance (SPR, Graffnity, Table 2.3. A reference to Biacore technology is provided in Section 2.4.1) and mass spectrometry (refer to Section 2.5).

Table 2.3 Examples of commercially available gold-based microarray platforms.

Coating	*Supplier*	*Advantages*	*Disadvantages*
Functionalized alkanethiols	Zyomyx	Good control over surface quality, good biocompatibility	More expensive than glass or polymer arrays
	Graffinity		

2.3.1
Surface Modifications

Natural proteins contain a wide range of functional groups on the side chains of their polypeptide backbone (refer to Table 2.1). In principle, all of them can be used during the direct chemical coupling reaction on specially prepared SAM surfaces with appropriate terminal reactive groups. Scheme 2.3 shows the various terminal functional groups of alkanethiols that are potentially available for biosensor applications.

Hydroxy terminal alkanethiols (oligo ethylene glycol units [22] and mannitol groups [23]) undergo low non-specific interactions with proteins and enzymes, and as a result are excellent candidates for making mixed monolayer-based protein array platforms. For other surface modifications which resist protein adsorption, refer to the extensive review by Whitesides [24].

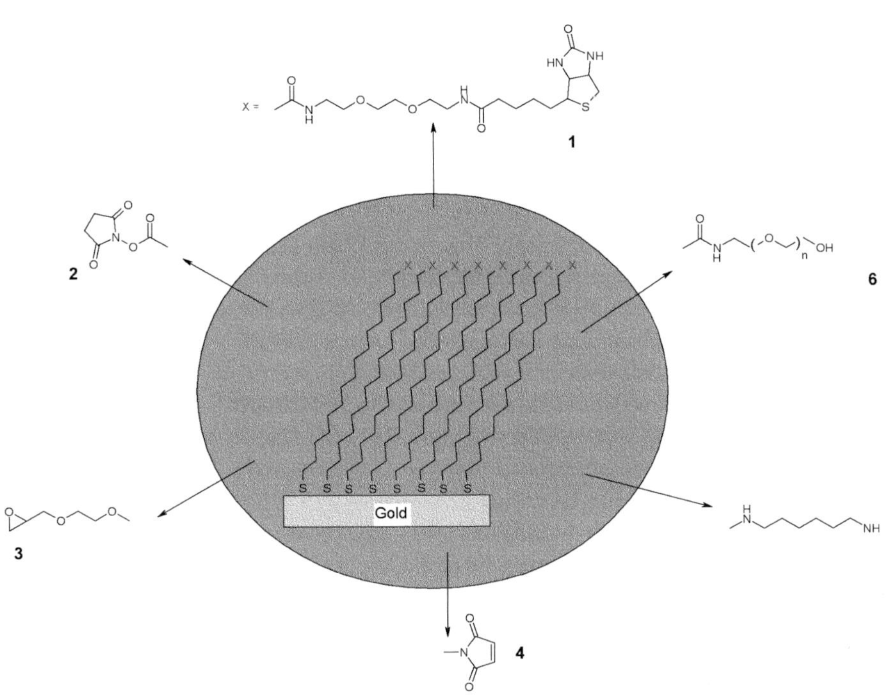

Scheme 2.3 Schematic respresentation of long-chain alkanethiol monolayers (e.g. 16 mercapto-hexadecanoic acid) on gold with different terminal functional groups. **1**, biotin; **2**, NHS-ester (NHS, N-hydroxysuccin-imide); **3**, epoxy-ethylene glycol; **4**, maleimide; **5**, diaminohexane; **6** oligo (ethylene glycol).

2.3.2
Current State of Gold-based Protein Microarrays

The most widely used gold-based protein microarrays are based on biotin–streptavidin technology. As a result, the following discussion will focus primarily on this technology. One of the reasons for the rapid growth in the application of biotin–streptavidin technology is the ease with which it is possible to biotinylate various biomolecules. The following two premises are, however, inherent when we talk about biotin-conjugated molecules: (a) the biochemical and physical properties of the biomolecules are not affected significantly by the biotinylation step and (b) the properties of the biotin moiety are not affected by the derivatized biomolecule. The relative ease of attachment of biotin to organic molecules substituted with a thiol makes it possible to use gold substrates as the underlying platform for constructing protein arrays. The streptavidin surface (formed on top of the underlying biotin-terminated SAM) can be used for immobilizing a wide range of biotinylated molecules (nucleic acids, proteins, saccharides, etc.).

Single component SAMs containing biotin tethered at the terminal end are known to generate poor streptavidin monolayers due to steric constraints imposed by the tight biotin layer. The close-packed biotin layer shields the individual biotin units and also prevents the effective docking of the bulky streptavidin molecules. This problem can be obviated by using either spacer molecules or via surface coupling schemes that involve lower yields.

The research groups of Knoll [25] and others [26] have actively investigated the interaction of biotin and streptavidin monolayers on a variety of sensor formats. By exploring many different combinations of biotinylated thiols and various spacer molecules, Knoll and co-workers have shown that an optimum monolayer of streptavidin can be generated by using a 1 : 9 mole ratio of biotin and hydroxyl-terminated thiols. The latter have a dual function: first, they act as efficient spacers between the biotinylated components and second, they minimize the non-specific interaction of streptavidin with the sensor substrate. Other research groups [27], using various streptavidin mutants, have indicated that an optimum streptavidin monolayer can be obtained by using approximately 10–25 % biotin on the surface.

Another method of generating a well-packed streptavidin surface is by using a surface reaction between a hydrophobic interface (Figure 2.3) comprising carboxy groups [28] and biotin-terminated linker units attached to amino groups. A densely-packed streptavidin surface is obtained when the surface coupling yield of biotin is between 25 and 30 %. This value is in the same range as that of the mixed SAM layers. An application of this surface in screening patients affected with Lyme Borreliosis is shown in Figure 2.4. No blocking solutions were required during this assay and highly reproducible results with high signal-to-noise ratios were observed because the assay surface was quite homogeneous (i.e. uniform probe packing density) [29].

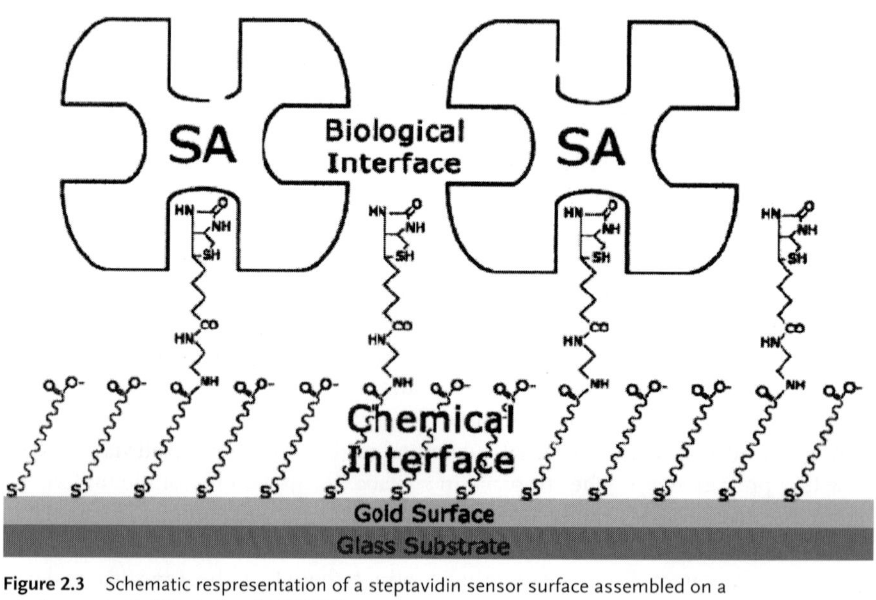

Figure 2.3 Schematic respresentation of a steptavidin sensor surface assembled on a reaction-controlled biotinylated SAM [28].

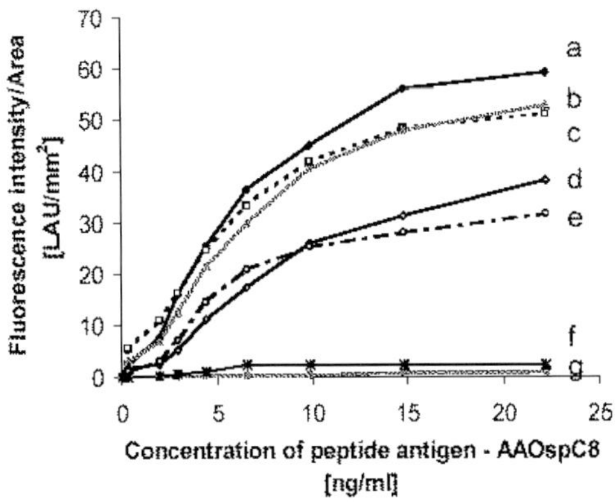

Figure 2.4 Ap[placation of stretavidin sensor platform (refer to Figure 2.3) for the diagnosis of Lyme Borreliosis. a-e, different patients suffering from the disease; f and g, control experiments using healthy subjects. The difference in the above biomolecular interaction curves can be attribued to patients at various stages of infection. Irrespective of the stage of infection, good signal-to-noise ratios can be observed between infacted patients and healthy subjects using the streptavidin microarray platform [29].

While both the above methods can be used in assembling a streptavidin surface, the mixed SAM approach is relatively easy to construct and can thus accelerate the production process. The drawback is that these molecules are not readily available and as a result, need to be specially synthesized. The synthesis of these molecules is not trivial and the synthesis steps usually have poor success (yield) rates. These problems can be overcome by using the readily available carboxy-terminated thiols for constructing the streptavidin sensor surface. It is important to note that both these techniques result in a high quality streptavidin surface. The biocompatibility of the SAM interface is likely to make such sensor surfaces interesting candidates for the screening of antibodies, patient sera and other complex proteins.

Several reports dealing with surface patterning of proteins on gold-coated surfaces using particle lithography [30], photoimmobilization [31], chemical selectivity of mixed monolayers [32] and microcontact printing have been published [33]. Atomic Force Microscopy (AFM) [33,26b] or Scanning Tunneling Microscopy (STM) [34] techniques can also be used for direct visualization of such protein patterns on gold surfaces. In the latter case, a thiolated protein was directly adsorbed onto the surface. Issues pertaining to protein denaturation still remain as the study involved the direct contact of the proteins with the metal surface.

Gold-coated electrodes can also serve as amperometric biosensors in the form of enzyme electrodes. The enzymes on such biosensors are immobilized via functionalized SAMs [35], conducting polymers such as polypyrrole or polythiophenes [36] or by direct absorption onto the electrode surface [37]. Some prominent examples include the glucose sensor based on the electrochemical detection of enzymatically-produced hydrogen peroxide [38] or the screening of neurotoxins with immobilized acetylcholinesterase in the form of competitive assays (using acetyl-thio-choline) [39]. However, none of these approaches is likely to be used in a microarray format for HTP studies in the near future.

2.4 Microarrays based on Polymer Substrates

Most of the above discussions have focused primarily on two-dimensional (2-D) surface architectures. Since the functional properties of a protein are highly dependent on its conformation and structure, it is possible that delicate proteins may not be able to survive the harsh conditions of 2-D surfaces. As a result, other formats based on three-dimensional (3-D) surfaces which provide a fluidic environment that closely mimics the biological environments of various proteins, have been proposed. Some prominent examples of 3-D-based systems include hydrogels and polymer membranes. The basic underlying tenet of such 3-D systems is that a fluidic environment preserves the conformation and, as a result, the activity of the protein. Additional benefits of 3-D architecture include larger surface area and greater probe density. Polymer membranes such as aminated nylon [40] or nitrocellulose [41] were the first commonly used substrates for the preparation of microarrays. For over two decades, these platforms have been widely employed in microbiology laboratories as blotting supports. Detection on these supports has been carried out mainly by radioimaging, although a combination of other detection technologies such as fluorescence is also possible.

Table 2.4 provides an overview of the polymer microarray platforms that are currently available on the market.

Table 2.4 Examples of common polymer and hydrogel substrates.

Polymer substrate	*Supplier*	*Overall advantages*	*General limitations*
Nylon (aminated) membrane	Pall, Macherey & Nagel	Cheap, versatile, high biocompatibility, applicable to blotting	Only electrostatic absorption, low spot homogeneity and reproducibility, mass transport limitations, high background smearing and non-specific interactions
Nitrocellulose membrane	Schleicher and Schuell Pierce, Bio World		
Polystyrene	Nunc		
Dextran-Hydrogels	BiaCore		
Polyacrylamide	Perkin Elmer		
Polymer brush	Prolinx		

Gel-based studies in combination with downstream operations such as mass spectrometry, dominate most proteomic investigations. Since it is possible to transfer proteins and antibodies directly onto the blotting membranes after electrophoresis, these platforms definitely have certain advantages when it comes to sample preparation and in terms of the overall cost. Hydrogel-based arrays based on gold substrates have dominated the field of real-time protein interaction studies for the past decade. This has been mainly possible due to the widespread application of SPR technology by Biacore. Polymer brushes and hydrogels on glass substrates (conventional microarray format) have also been developed (refer to Table 2.4), although their usage is still limited.

2.4.1 Surface Modifications

The immobilization of biomolecules on functionalized polymer membranes such as nylon or cellulose occurs via electrostatic absorption. This is also the case for hydrogels based on polyacrylamide platforms. As a result, it is difficult to obtain reliable, high quality data from such systems for potential application in microarray studies. During the past decade, self-assembled monolayers modified with dextran have become the system of choice for many biomolecular interaction studies. The widespread use of this 3-D architecture was mainly a result of the enormous success of the commercial SPR instrument maker, Biacore. Well-established protocols of self-assembly techniques on gold were utilized by researchers at Biacore to synthesize these dextran surfaces [42]. Epoxy-terminated thiol monolayers were used for coupling dextran polymers to the sensor surface. The immobilization of ligands was achieved by using carboxy-methyl-modified dextran containing extensive reactive handles for activation and covalent attachment of the molecules of interest. The resulting layer is a hybrid material composed of a 2-D monolayer on gold and a 3-D hydrogel which is highly flexible, non-crosslinked and extends 100–200 nm from the coupling surface under physiological buffer conditions (Scheme 2.4).

While such 3-D surfaces have numerous benefits, they are still plagued by problems relating to mass transport effects and high background signals emanating from non-specific interactions. The 3-D dextran surfaces are known to impose mass transfer limitations that can significantly alter the inherent kinetics of the interaction. Diffusion effects play an important role within these 3-D structures, and can lead to the formation of several local concentration gradients. During real-time kinetic studies using surface plasmon resonance measurements, such local gradients can result in the calculation of false kinetic rate constants. Protein immobilization with the help of functionalized polymer brushes or by photochemically-activated polymers will be discussed in Sections 2.4.2 and 2.6.

2.4.2 Current State of Polymer-based Microarrays

Numerous protein chip surfaces can be created by activating the above-mentioned dextran surface. Using the dextran as the base, surfaces incorporating thiol, amine, (strept)-avidin or aldehyde groups can be generated (Scheme 2.5). These surface modifications have resulted in the commercial success of numerous sensor chips [43] which are routinely used in a plethora of applications ranging from HIV analysis using various monoclonal antibody–antigen interactions to the area of small

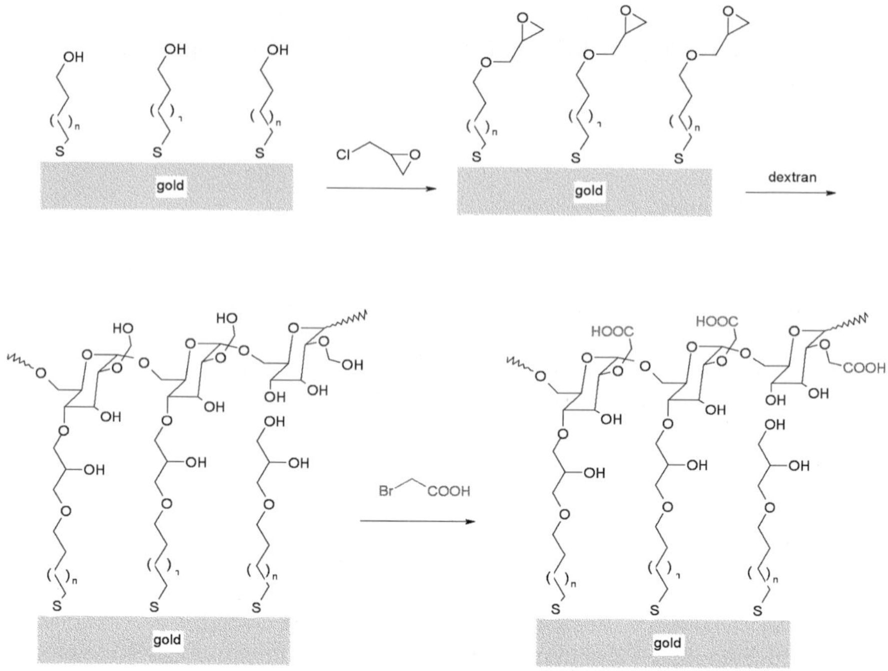

Scheme 2.4 Reaction schemes for preparing a hydrogel dextran matrix [41].

molecule drug discovery operations and quality control. Since these sensor chips are quite robust and can withstand stringent acidic or basic regeneration steps, they can be re-used several times without any substantial loss of the surface density/activity of the ligands.

Another new approach that combines self-assembled monolayer techniques with a 3-D microenvironment is the synthesis of functionalized dendrimers on solid supports such as glass. The activated dendritic structures are terminated by NHS ester- or isothiocyanate- groups [44]. Emerging technologies use the (co)polymerization of methacrylate, styrene or vinyl alcohol monomers on the surface of a glass chip followed by subsequent functionalization using photografting [45]. The optical properties of the chip surface can be tailored to the reflectance of light or scattering effects. The tuning of the refractive index of the surface can also be accomplished by effective implementation of high-refractive index materials (TiO_2) or low-refractive layers (polymers, SiO_2) [46]. This approach has been used to make highly reflective microarray surfaces at the applied wavelength (emission and excitation) for potential applications in highly sensitive luminescent-based assay systems.

A comparative study involving the screening of 11 different array surfaces on polymer gel and silane-coated glass slides was carried out by applying five different monoclonal antibodies in a 20 x 18 matrix. All the important aspects of the procedure are discussed in this work, including ease of chip preparation and handling,

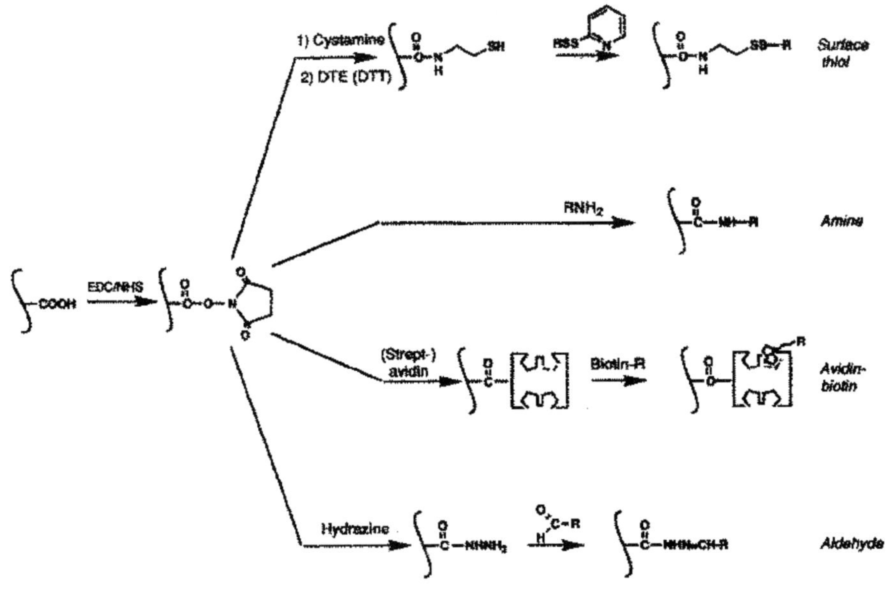

Scheme 2.5 Reaction schemes for generating different protein chip surfaces using the dextran matrix as the underlying platform.

storage, signal intensities, reproducibility and detection limits [47]. Based on these investigations, reflective slides with a mirror-like surface (3-aminopropyl-triethoxysilane, Amersham Biosciences [48]) were found to provide high levels of uniformity and reproducibility, whilst the polyacrylamide-coated slides were characterized by their high levels of sensitivity. The authors conclude by stating that the selection of the array support material strongly depends upon the special requirements of the experiment and the properties of the probe biomolecules.

Currently, many groups have reported on the fabrication of custom-made antibody arrays on nitrocellulose or nylon membranes for the detection of cytokines in conditioned media or human serum using a multiplexed sandwich assay [49].

2.5. Special Formats: Microfluidic Devices and Integrated Semiconductor Chips

The lab-on-chip technology involving microfluidics has evolved rapidly over the last decade. This technology allows for the parallelization and integration of multiple sample processing approaches in the field of diagnostics or proteomic research. These devices are however, more complex than conventional microarrays, and while they are suitable for small-scale sample preparation applications, their throughput is still relatively low. Nevertheless, as these systems continue to move from academic laboratories to commercialized technology, the user-friendliness of such microfluidic systems is slowly improving [50]. In spite of the above limitations, this technology has certain advantages such as the use of very small volumes of human sera or urine samples for conducting rapid diagnostic assays. This technology is poised to make rapid advances in the area of nucleic acid-based testing, a field that is set to explode during this decade. The lab-on-a-chip SNP genotyping systems by Caliper [51] and Nanogen [52] are clear illustrations of the scope of such technologies. Whilst most of the early microfluidic formats have focussed primarily on DNA interactions, such systems can also be used for the detection of protein–protein or antibody–antigen interactions if appropriate surface immobilization strategies are employed.

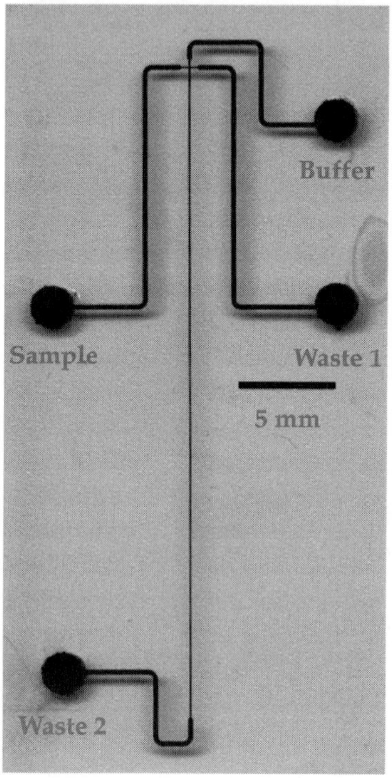

Figure 2.5 Schematic diagram of a microfluide system for the analysis of DNA samples. Image by courtesy of US DOE Genomes to Life Program,

While Nanogen uses a plastic cartridge in their silicon wafer biochip unit, many microfluidic systems are based on glass or silicon using classical etching and processing technologies of the semiconductor industry for the structuring of chips with microchannel systems. The planar waveguide glass chips from Zeptosens use integrated microfluidics combined with high refractive index material (Ta_2O_5) deposited on the glass support for monitoring biomolecular interactions using highly sensitive fluorescence detection methods [53]. The Flow-thru Chip™ from MetriGenix is also based on a three-dimensional microchannel glass chip [54]. CombiMatrix [55] attaches a proprietary porous three-dimensional reaction layer to the active side of semiconductor surfaces for the immobilization of biomolecules. Yet another microfluidic protein biochip platform based on silicon is currently under development at Zyomyx [56].

Microchannels in polymeric materials such as poly-(methyl methacrylate) (PMMA) can also be used to make microfluidic devices and at the same time the methyl ester groups on the surface of the PMMA can be chemically derivatized with reactive groups to immobilize biomolecules [57].

Some of the potential advantages of miniaturization can also be combined with other detection technologies for effective analysis of protein purification and interactions. Two examples of this are the coupling of microfluidics with ionization mass spectroscopy (ESI-MS) [58] or with micro-capillary electrophoresis [59].

Another technology that has been borrowed from the semiconductor industry in the field of integrated microarrays is the fabrication of microelectrode structures on silicon wafers. These gold, silver or platinum electrodes can be coated with functionalized monolayers or conducting polymers that are biocompatible and reactive towards biomolecules. On such electrodes, a voltage can be applied for the selective immobilization of the physiologically charged biomolecules. The microarray system from Nanogen is one example of such an approach, while the chip read-out is still undertaken using conventional optical techniques [60]. Other approaches are one step ahead and use electrochemical detection of DNA hybridization in combination with redox-active labels [61] (such as Motorola's eSensor™ DNA detection system) or intercalators [62]. Currently these technologies are primarily focused on DNA arrays although in future, additional solutions for monitoring protein–protein or antibody–antigen interactions can also be envisaged by using such techniques. Technologies such as impedance spectroscopy or surface capacitance measurements [63] can potentially be the key to these application (e.g. at BIONCHIP [64]). In analogy to luminescence-based assays, electrochemical detection methods can also be amplified by ELISA techniques. Enzymes such as horseradish peroxidase or alkaline phospatase can be conjugated to the target molecules, and the reaction products can be measured by impedance spectroscopy [65] or with amperometic methods [66] if a suitable substrate has been added. A fully integrated electronic CMOS (Complementary Metal Oxide Silicon) biochip (Figure 2.6) based on enzyme-linked sensor techniques has been recently launched by Infineon [67].

2.6 Chemical Immobilization Techniques for Proteins

Generally, all the functional surface coatings described above can act as biocompatible interfaces during the chemical coupling of proteins to any microarray substrate, either via covalent binding, complex coordination chemistry or supramolecular interactions. In addition to the above methods, proteins also have a tendency to interact with the surface via electrostatic interactions. This principle is used in the case of simple electrostatic adsorption of proteins onto nylon, nitrocellulose,

Figure 2.6 CMOS-based biochip platform with integrated electronic microstructure (gold on silicon) suitable for an electrochemical or optical chip read-out by Infineon (top) and Bionchip (bottom). Images by coutesy of Infineon and Bionchip BV. (see Colour Plate p. XIX).

poly-L-lysine or aminopropylsilane-coated surfaces. These interactions constitute what is widely known as non-specific interactions. Special SAMs based on polyethylene glycols [68] or poly(L-lysine)-g-poly(ethylene glycol) [69] are known to resist such interactions and are thus widely used in the design of most biosensor and microarrays. In the following sections, various strategies for coupling proteins using covalent, photochemical tags and site-directed immobilization approaches will be addressed.

2.6.1
Covalent Chemical Coupling

In order to achieve covalent chemical coupling to microarray substrates, proteins should have the following chemical functionalities (see also Table 2.1) in the side chains of their polypeptide backbone: –SH (Cysteine), $-NH_2$ (Lysine), –COOH (aspartic acid, glutamic acid), –OH (Serine), Ph-OH (Ph = phenyl, Tyrosine), imidazole (Histidine). In principle, all of them can be used during the direct chemical coupling reaction on specially prepared SAM surfaces. NHS (N-hydroxy-succinimide ester), aldehyde or epoxy groups form stable amide bonds with the free amino groups of the peptide (Scheme 2.6). The activation of carboxyl groups with EDC (*N*-3-dimethylaminopropyl-N-ethylcarbodiimide) and NHS, as well as further coupling schemes with NHS-modified surfaces are shown in Scheme 2.5. In another approach, the carboxylic acid groups in the side chains of the polypeptide can be activated by carbodiimides or NHS and can be covalently coupled to free amino groups on the sensor surface.

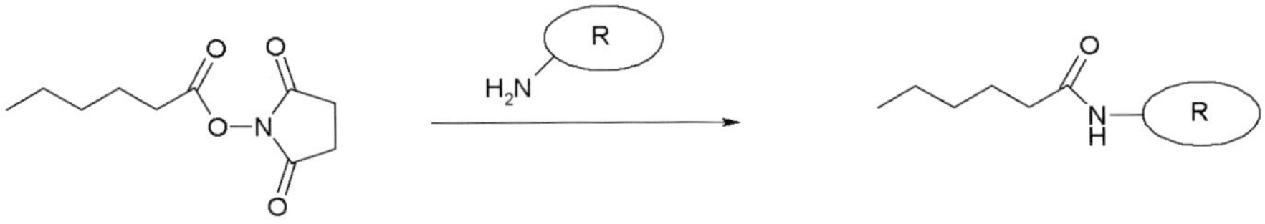

Scheme 2.6 Surface coupling reaction of NHS-esters with the amino residues of the side-chains of polypeptides (lysine units). R, protein.

Other chemicals which are currently available for the covalent coupling of biomolecules to microarray surfaces include maleimides, pyridyl disulfide and vinyl sulfone, all of which will react with thiol groups [70] (Scheme 2.7). Epoxy-modified surfaces can also be used for coupling hydroxy groups (see Scheme 2.4) and isothiocyanates for binding with amino groups [71] (Scheme 2.8).

2.6.2
Photochemical Cross-Coupling

In principle, there are many approaches available for the photochemical cross-coupling of biomolecules to specially functionalized surfaces. For example, the binding of biomolecules to epoxy- or aldehyde-coated slides can also be induced by irradiation with UV light. But there are also some typical photoreactive groups that are currently available. A prominent example is the azidophenyl group. Photoinduced crosslinking of proteins with azidophenyl-functionalized layers can be achieved by irradiation with UV or visible light. The excitation wavelength of this reaction can be changed by the use of additional substituents on the azidophenyl ring [72] (Scheme 2.9).

Another strategy for the photochemical attachment of biomolecules has been developed by researchers at the commercial firm, Exiqon. They have taken advantage of an anthrachinone linker that can be coupled to a wide variety of solid polymeric supports (e.g. polystyrene, polyethylene, polypropylene, PMMA, nylon) by irradiation (Scheme 2.10) and can be conjugated with either biomolecules, ligands or reactive groups (e.g. carboxylic acids or amines) [73].

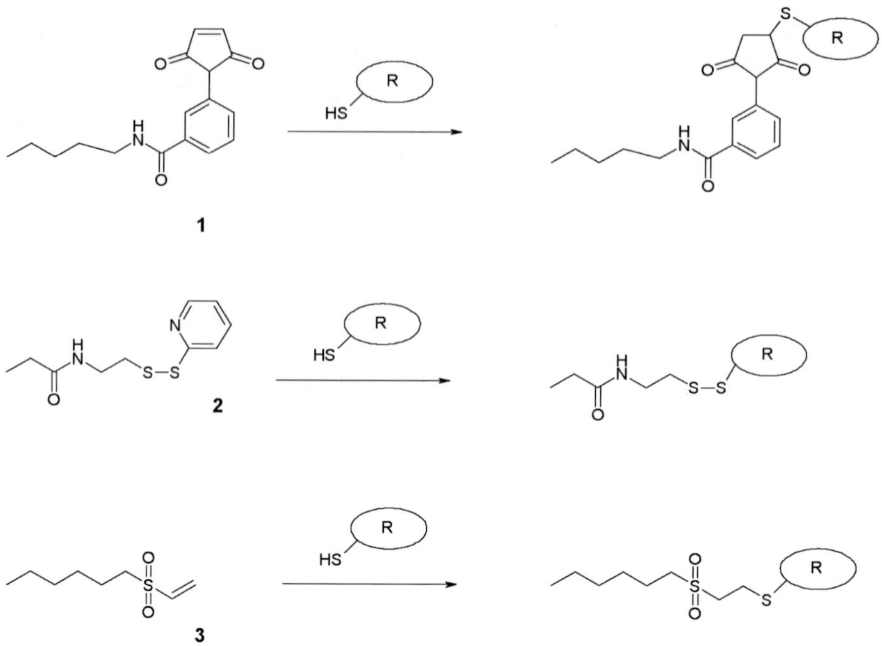

Scheme 2.7 Surface modifications for the attachment of thiol residues (which are present in the side chains) of polypeptides (cysteine units). R, protein, **1**, maleimide; **2**, disulfides; **3**, vinyl sulfone.

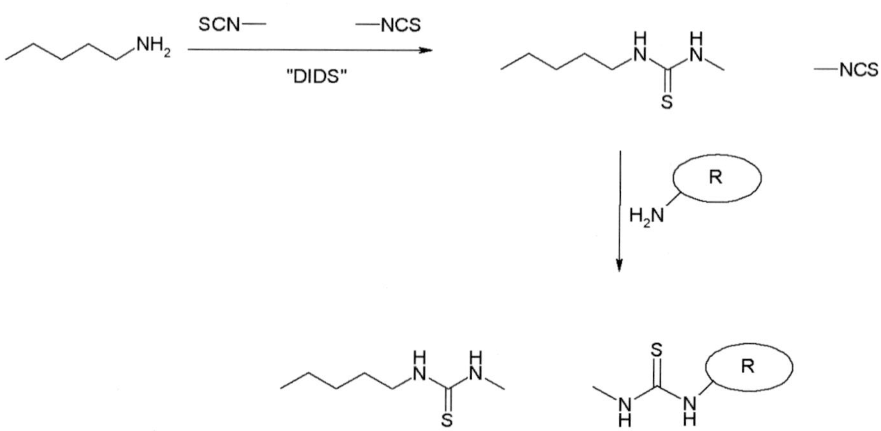

Scheme 2.8 Immobilization of proteins and enzymes on amino-terminated monolayers using the bifunctional reagent DIDS (4,4 diisothiocyanato-*trans*-2,2-disulfonic acid disodium salt).

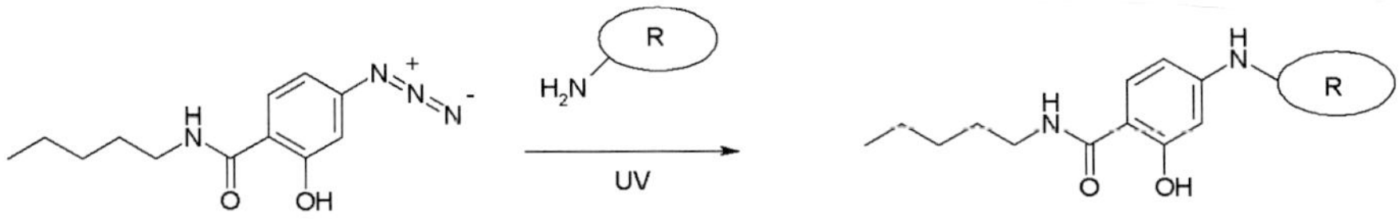

Scheme 2.9 Photo-induced cross-linking of proteins R and azidophenyl-functionalized layers (λ = 265-275 mm; if the hydroxyl substituent is replaced with a nitro group, then λ = 300-460 mm).

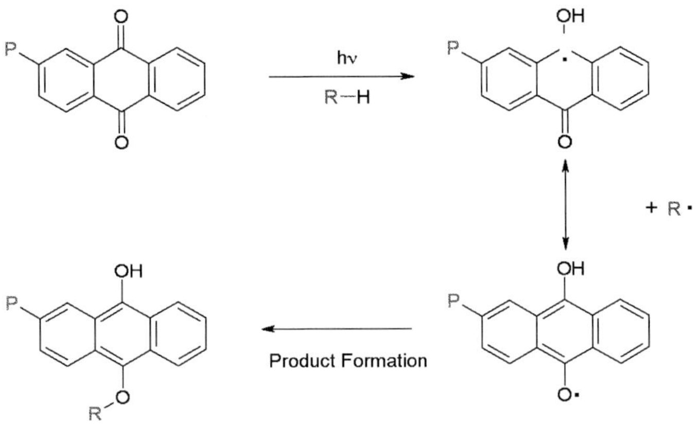

Scheme 2.10 Exiqon's patented anthrachinone (AQ) covalent attachment photochemical method. R, polymeric solid; P, specific ligand (biomolecule) or reactive group.

Yet another versatile method for the photochemical cross-linking of proteins to surfaces is the photoaptamer technique used by SomaLogic. However, in contrast to the above methods where the capture probes are photochemically immobilized, here the molecular recognition event is stabilized by consecutive photocrosslinking. Photoaptamers are chemically synthesized single-stranded DNA molecules that can be labeled at their 5-end with acrydite, amine, thiol or biotin. These labels facilitate the covalent attachment of the photoaptamers to solid supports including beads, multi-well plates and activated slides. Photoaptamers derived from the PhotoSELEX™ process which substitutes a brominated deoxyuridine (BrdU) for the thymidine base (T) normally found in DNA, are high-affinity ssDNA compounds that can function as highly specific capture agents for proteins. The photocross-linking of the BrdU of the photoaptamer to an electron-rich amino acid side chain of the protein adds a second dimension of specificity to the normal protein binding event. Since photoaptamers can be covalently bound to their target analytes before signal detection, such arrays can be vigorously washed to remove non-specifically bound proteins, yielding superior signal-to-noise ratios [74].

The Versalinx Protein Microarray Technology by Prolinx utilizes a synthetic chemical affinity (Scheme 2.11) based on the reversible complexation of phenyldiboronic acid (PDBA) with salicylhydroxamic acid (SHA). The microarray surface is then coated with 3-D polymer brushes which have SHA as their terminal units [75]. This surface permits the reversible binding of PDBA-conjugated proteins.

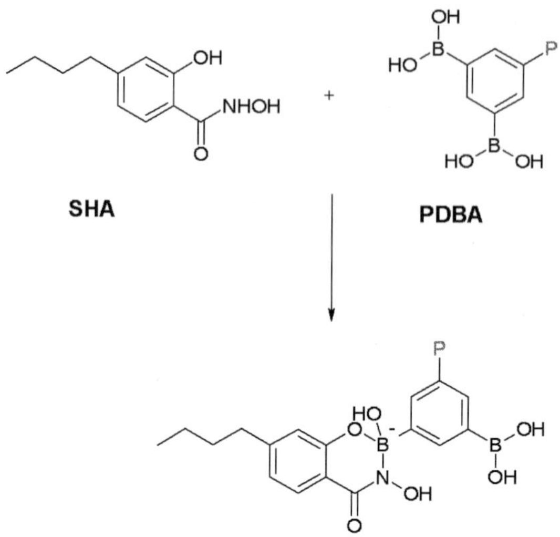

Scheme 2.11 Reversible formation of a complex of PDBA (phenyldiboronic acid)-conjugated proteins with SHA (salicyl hydroxymid acid)-terminated polymer brushes.

2.6.3
Tagged Proteins

To prevent protein denaturation, it is preferable to design surface modifications that target protein functional groups which are furthest from their CDRs. It is important that the epitope mapping domains of the proteins are not affected or modified during the surface immobilization step. By understanding the chemical nature of the proteins of interest, smart surfaces can be constructed for use in high-quality protein expression studies. Several protein or antibody samples currently exist in conjugated forms having so-called tags such as biotin, His-Tag, Strep-Tag, Avi-Tag, etc. Tags (Table 2.5) enhance the efficiency of the surface-coupling step but at the same time have a tendency to alter the properties of pristine proteins. Most research laboratories have tools that are adapted to these tags and if protein chips have to be integrated into routine test procedures, then they also have to be compatible with such tagged molecules.

The use of biotin–streptavidin technology in generating biocompatible surfaces has been highlighted in a large number of research articles [76]. Streptavidin surfaces can be used for immobilizing a wide range of biotinylated molecules (nucleic acids, proteins, saccharides, etc.) and biotin–streptavidin architectures have been

Table 2.5 Examples of commercially available tagged proteins and their corresponding immobilization systems.

Protein tag	*Manufacturer of tagged proteins or antibodies*	*Immobilization matrix*
Biotin	Roche, MorphoSys; Biotinylation kits by Molecular Probes or Avidity	Streptavidin (Greiner, Perkin Elmer, Xenopore)
Strep-Tag	IBA	Strep-Tactin (IBA)
His-Tag	Novagen, ClonTech	Ni-NTA complex (Greiner, Qiagen, Xenopore)

widely employed as sensor supports for studying numerous biomolecular interactions. One of the reasons for the rapid growth in the application of biotin–streptavidin technology is the ease with which it is possible to biotinylate various biomolecules. In addition to conventional procedures, new developments are continuously being established in the area of biotinylation of biomolecules. In this area, it is worth mentioning a new method for labeling proteins with biotin by using the so-called AviTag™ technology[77]. Many commercial suppliers offer both biotinylated microwell plates/microarray slides and biotin-conjugated biomolecules for biomolecular screening experiments. Another interesting tag for coupling to streptavidin-related surfaces is the Strep-Tag. The Strep-Tag consists of eight amino acids and has a high specific affinity for Strep-Tactin, a synthetically-engineered streptavidin derivative. Recombinant proteins or enzymes can be labeled with such a Strep-Tag. This technology is currently used for protein purification, although microplates and microarrays coated with Strep-Tactin can also be used as a platform for high-throughput screening assays [78].

A well-known label in protein chemistry is the widely used His-Tag. Proteins with six histidine amino acid units at the C- or N-terminus of the peptide chain can be expressed in E. coli. By analogy to affinity chromatography columns, nickel chelate complexes [79] can be used as immobilization matrices on biochip surfaces for His-tagged proteins. An example of a chelate complex is Ni-NTA (NTA = nitrilotriacetic acid). For this purpose, N-(5-amino-1-carboxypentyl)iminodiacetic acid can be synthesized by methods similar to those reported in the literature [80] or by solid phase peptide synthesis coupled to an activated carboxylic acid monolayer and subsequent treatment with a Ni2+ solution. The two free binding sites of the central nickel metal can be coordinated to two histidine units of the tag (Scheme 2.12). One of the main advantages of such systems is their re-usability. The immobilized proteins can be removed after the screening experiments by the addition of a stong nickel-complexing agent such as EDTA. Although the binding of a histidine tag to nickel is relatively weak compared to the biotin–streptavidin system, Snyder *et al.* [17] have obtained relatively good results from their nickel-coated protein chips.

There are several commercial suppliers who provide Ni-based products for biomolecular screening. Nickel chelate-coated glass slides are, for example, provided by Greiner [80]. As part of their LiquiChip System, Qiagen also offers Ni-NTA-coat-

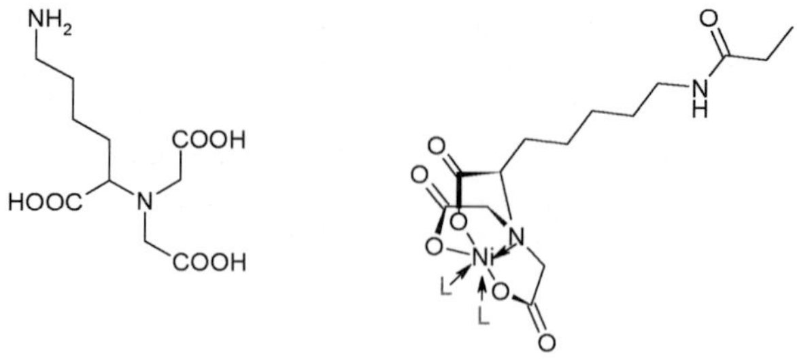

A B

Scheme 2.12 (A) Structure of N-(5-amino-1-carboxypentyl)iminodiacetic acid. (B) The quadridentate nitrilotriacetic acid (NTA) ligand forms a complex with four binding sites on the nickel metal which is present in the center. The two remaining binding sites can be coordinated with histidine ligands (L).

ed beads. This bead-based microplate platform can be used for a wide range of protein-based assays for proteomics and drug discovery [81]. Whitesides and others have also been granted a patent that covers the use of alkanethiol SAMs on gold surfaces which are terminated with Ni-chelate complexes for the coordination of polyamino acid-tagged biomolecules [82].

2.6.4
Site-Specific Immobilization of Antibodies

Antibodies are known to bind reversibly to protein A-coated surfaces. Antibodies of the IgG type can be attached to protein A surfaces for high-throughput immunoassay applications by means of their Fc-domains of the heavy chains. Under such conditions, the Fab-domain (fragment of antigen binding) is oriented away from the surface and is easily available for the antigens in the sample solution. In situations where monoclonal IgG antibodies do not bind to protein A surfaces, protein G can be used for antibody immobilization. A review of various currently used immunoassay strategies for protein interactions has been recently reported by MacBeath [83].

Another covalent-linking approach involves the use of hydrazide-terminated monolayers. These functionalities bind exclusively to aldehyde groups that appear only in the glycosylated part of the heavy chains of antibodies, but not in the protein chains of the Fab-domains (Scheme 2.13).

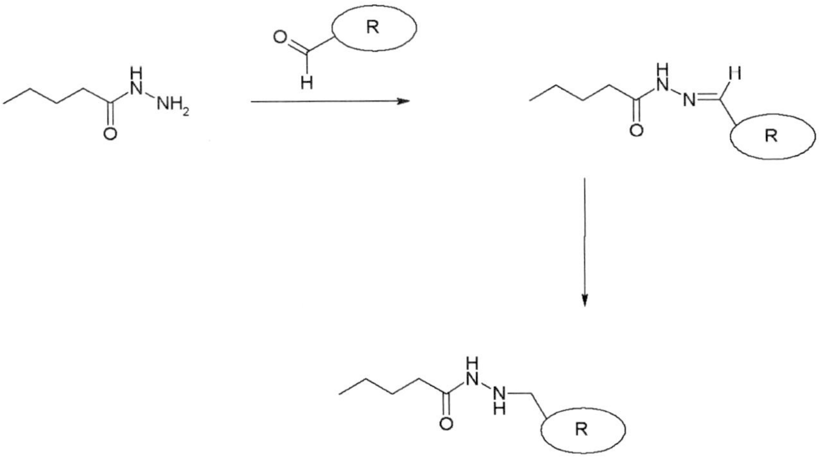

Scheme 2.13 Coupling of aldehyde residues of glycoproteins to hydrazide-terminated monolayers.

2.7
Conclusions

This chapter has reviewed the numerous strategies which are currently available for immobilizing proteins and related biomolecule families. Since proteins are not as homogeneous as nucleic acids, a wide range of challenges will continue to arise depending upon the nature of the end application. Protein denaturation will continue to be the rate-limiting step in most protein microarray platforms. However, selection of smart chemistries coupled with a sound knowledge of the structural and chemical properties of the probes (which are to be immobilized) will ultimately result in the development of reliable strategies which can be used in the near future to develop functional screening platforms for biomolecules.

References

1. [x1] A. Ulman, *Chem. Rev.* **1996,** 96, 1533-1554.
2. N. Zammatteo, L. Jeanmart, S. Hamels, S. Courtois, P. Lhevesi, J. Remacle, *Anal. Biochem.* 2000, **280,** 143–150.
3. M.C. Pirrung, *Angew. Chem.* Int. Ed. 2002, **41,** 1276–1289.
4. A. Ulmann, *Chem. Rev.* 1996, 1533–1554.
5. J. Gun, J. Sagiv, *J. Colloid. Interface Sci.* 1986, **112,** 457.
6. K. Bierbaum, M. Grunze, *Adhes. Soc.* 1994, 213.
7. R. Cohen, N. Zenou, D. Cahen, S. Yitzchaik, *Chem. Phys. Lett.* 1997, **279,** 270–274.
8. B.B. Haab, M.J. Dunham, P.O. Brown, *Genome Biol.* 2001, 2, 4.1–13.
9. G. MacBeath, S.L. Schreiber, *Science* 2000, **289,** 1760–1763.
10. T.O. Joos, M. Schrenk, P. Höpfl, K. Kröger, U. Chowdhury, D. Stoll, D. Schörner, M. Dürr, K. Herrick, S. Rupp, K. Sohn, H. Hämmerle *Electrophoresis* 2000, **21,** 2641–2650.
11. L.G. Mendoza, P. McQuary, A. Mongan, R. Ganghadharan, S. Brignac, M. Eggers, *BioTechniques* 1999, **27,** 778–788.
12. B.D. Martin, B.P. Gaber, C.H. Patterson, D.C. Turner, *Langmuir,* 1998, **14,** 3971–3975.
13.] A. Bernard. E. Delamarche, H. Schmid, B. Michel, H.R. Bosshard, H. Biebuyck, *Langmuir* 1998, **14,** 2225–2229.
14. S.K. Bhatia, J.L. Teixeira, M. Anderson, L.C. Shriver-Lake, J.M. Calvert, J.H. Georger, J.J. Hickman, C.S. Dulcey, P.E. Schoen, F.S. Ligler, *Anal. Biochem.* 1993, **208,** 197–205.
15. A. Sreekumar, M.K. Nyati, S. Varambally, T.R. Barrette, D. Ghosh, T.S. Lawrence, A.M. Chinnaiyan, *Cancer Res.* 2001, **61,** 7585–7593.
16. B. Schweitzer, S. Roberts, B. Grimwade, W. Shao, M. Wang, Q. Fu, Q. Shu, I. Laroche, Z. Zhou, V. T. Tchernev, J. Christiansen, M. Velleca, S.F. Kingsmore, *Nature Biotechnol.* 2002, 20, 359–365.
17. H. Zhu, M. Bilgin, R. Bangham, D. Hall, A. Casamayor, P. Bertone, N. Lan, R. Jansen, S. Bidlingmaier, T. Houfek, T. Mitchell, P. Miller, R.A. Dean, M. Gerstein, M. Snyder, *Science* 2001, **293,** 2101–2105.
18. L. Bertilsson, B. Liedberg, *Langmuir* 1993, **9,** 141–149.
19. a) R.G. Nuzzo, D.L. Allara, J. *Am. Chem. Soc.* 1983, **105,** 4481–4483.
    b) M.D. Porter, T.B. Bright, D.L. Allara, C.D. Chidsey, J. *Am. Chem.* Soc. 1987, 109, 3559–3568.
20. a) G.M. Whitesides, C.D. Bain, *Science* 1988, **240,** 62–63.
    b) L. Strong, G.M. Whitesides, *Langmuir* 1988, 4, 546–558.
21. a) L.H. Dubois, B.R. Zegarski, R.G. Nuzzo, *Proc. Natl. Acad. Sci. USA* 1987, **84,** 4739–4742.
    b) C.D. Bain, G.M. Whitesides, J. *Am. Chem. Soc.* 1988, **110,** 3665–3666.
22. K.L. Prime, G.M. Whitesides, *Science* 1991, **252,** 1164–1167.
23. Y.Y. Luk, M. Kato, M. Mrksich, *Langmuir* 2000, **16,** 9604–9608.
24. E. Ostuni, R.G. Chapman, R.E. Holmlin, S. Takayama, G.M. Whitesides, *Langmuir,* 2001, **17,** 5605–5620.
25. W. Knoll, M. Liley, D. Piscevic, J. Spinke, J., M.J. Tarlov, *Adv. Biophys.* 1997, **34,** 1–15.
26. a) L.S. Jung, K.E., Nelson, C.T. Campbell, P.S. Stayton, S.S. Yee, V. Perez-Luna, G. Lopez, *Sensors Actuators B* 1998, **54,** 137–144.
    b) M. Schäferling, M. Kruschina, F. Ortigao, M. Riepl, K. Enander, B. Liedberg, *Langmuir* 2002, **18,** 7016–7023.
27. L.S. Jung, K.E. Nelson, P.S. Stayton, C.T. Campbell, *Langmuir* 2000, **16,** 9421–9432.
28. M. Mecklenburg, B. Danielsson, F. Winqvist, Patent SE, PCT/EP 97/03317.
29. P. Pavlickova, N. Mejlhede Jensen, H. Paul, M. Schäferling, C. Giammasi, M. Kruschina, W.-D. Du, M. Theisen, M. Ibba, F. Ortigao, D. Kambhampati, *J. Prot. Res.* 2002, **1,** 227–231.
30. J.C. Garno, N.A. Wadu-Mesthrige, G.-Y. Liu, *Langmuir* 2002, **18,** 8186–8192.
31. E. Delamarche, G. Sundarababu, H. Biebuyck, B. Michel, C. Gerber, H. Sigrist, H. Wolf, H. Ringsdorf, N. Xanthopoulos, H.J. Mathieu, *Langmuir* 1996, **12,** 1997–2006.
32. M. Veiseh, M.H. Zareie, M. Zhang, *Langmuir* 2002, **18,** 6671–6678.
33. F. Morhard, J. Pipper, R. Dahint, M. Grunze, *Sensors Actuators B* 2000, **70,** 232–242.

34. G.J. Leggett, C.J. Roberts, P.M. Williams, M.C. Davies, D.E. Jackson, S.J.B. Tendler, *Langmuir* 1993, **9,** 2356–2362.

35. I. Willner, E. Katz, *Angew. Chem.* Int. Ed. 2000, **39,** 1180–1218.

36. W. Schuhmann, *Mikrochim. Acta* 1995, **121,** 1–29.

37. a) T. Tatsuma, T. Watanabe, Anal. Chem. 1991, 63, 1580–1585. b) M. McRipley, R.A. Linsenmeier, *J. Electroanal.Chem.* 1996, **414,** 235–246.

38. A. Riklin, I. Willner, *Anal. Chem.* 1995, **67,** 4118–4126.

39. a) S. Andreescu, L. Barthelmbs, J.-L. Marty, *Anal. Chim. Acta* 2002, **464,** 171–180.
b) G. Jeanty, A. Wojciechowska, J.-L. Marty, M. Trojanowicz, *Anal. Bioanal. Chem.* 2002, 373, 691–695.

40. J. Gershoni, G. Palade, *Anal. Biochem.* 1982, **124,** 396–405.

41. H. Towbin, T. Staehelin, J. Gordin, *Proc. Natl. Acad. Sci.,* USA 1979, **76,** 4350–4354.

42. B. Johnsson, S. Lofas, G. Lindquist, *Anal. Biochem.* 1991, **198,** 268–277.

43. http://www.biacore.com

44. R. Benters, C.M. Niemeyer, D. Wöhrle, *ChemBioChem* 2001, **2,** 686–694.

45. B. de Boer, H.K. Simon, M.P.L. Werts, E.W. van der Vegte, G. Hadziioannou, *Macromolecules* 2000, **33,** 349–356.
b) C. Preininger, H. Clausen-Schaumann, A. Ahluwalia, D. DeRossi, *Talanta* 2000, **52,** 921–930.

46. a) M. Mennig, P.W. Olivera, H. Schmidt, Thin Solid Films 1999, **351,** 99–102.
b) W.E. Vargas, *J. Appl. Phys.* 2000, **88,** 4079–4084.
c) W. Que, Z. Sun, Y.L. Chan, C.H. Kam, *Thin Solid Films* 2000, **359,** 177–183.

47. P. Angenendt, J. Glökler, D. Murphy, H. Lehrach, D.J. Cahill, Anal. Biochem. 2002, **309,** 253–260.

48. http://www.apbiotec.com

49. a) R.P. Huang, R. Huang, Y. Fan, Y. Lin, *Anal. Biochem.* 2001, **294,** 55–62.
b) R. Wiese, Y. Belosludtsev, T. Powdrill, P. Thompson, M. Hogan, *Clin. Chem.* 2001, **47,** 1451–1457.
c) S.W. Tam, R. Wiese, S. Lee, J. Gilmore, K.D. Kumble, *J. Immunol. Methods* 2002, **261,** 157–165.

50. D. Figeys, *Proteomics,* 2002, **2,** 373–382.

51. http://www.calipertech.com

52. http://www.nanogen.com

53. M. Pawlak, E. Schick, M.A. Bopp, M.J. Schneider, P. Oroszlan, M. Ehrat, *Proteomics,* 2002, **2,** 383–393.

54. V. Benoit, A. Steel, M. Torres, Y.-Y. Yu, H. Yang, *J. Cooper, Anal. Chem.* 2001, **73,** 2412–2420.

55. http://www.combimatrix.com

56. http://.www.zyomyx.com

57. S.A. Soper, A.C. Henry, B. Vaidya, M. Galloway, M. Wabuyele, R.L. McCarley, *Anal. Chim. Acta* 2002, **470,** 87–99.

58. a) D. Figeys, S.P. Gygi, G. McKinnon, R. Aebersold, *Anal. Chem.* 1998, **70,** 3728–3734.
b) C.-H. Chiou, G.-B. Lee, H.-T. Hsu, P.-W. Chen, P.-C. Liao, *Sensors and Actuators B* 2002, **86,** 280–286.

59. a) H. Becker, M. Arundell, A. Harnisch, D. Hülsenberg, *Sensors and Actuators B* 2002, **86,** 271–279.
b) D. Figeys, Y. Zhang, R. Aebersold, *Electrophoresis* 1998, **19,** 2338–2347.

60. P. Swanson, R. Gelbart, E. Atlas, L. Yang, T. Grogan, W.F. Butler, D.E. Ackley, E. Sheldon, *Sensors Actuators B* 2000, **64,** 22–30.

61. C.J. Yu, Y. Wan, H. Yowanto, J. Li, C. Tao, M.D. James, C. L. Tan, G. F. Blackburn, T. J. Meade *J. Am. Chem.* Soc. 2001, **123,** 11155–11161.

62. E.M. Boon, D.M. Ceres, T.G. Drummond, M.G. Hill, J.K. Barton, *Nature Biotech.* 2000, **18,** 1096–1100.

63. a) M. Dijksma, B. Kamp, J.C. Hooglvliet, W.P. Bennekom, *Anal. Chem.* 2001, **73,** 901–907.
b) M. Riepl, V.M. Mirsky, I. Novotny, V. Tvarozek, V. Rehacek, O.S. Wolfbeis, *Anal. Chim. Acta* 1999, **392,** 77–84.
c) C. Berggren, B. Bjarnason, G. Johansson, *Biosens. Bioelectron.* 1998, **13,** 1061–1068.

64. http://www.bionchip.com

65. a) F. Patolsky, E. Katz, A. Bardea, I. Willner, *Langmuir* 1999, 15, 3703–3706. b) C. Ruan, L. Yang, Y. Li, *Anal. Chem.* 2002, **74,** 4814–4820.

66. C.N. Campbell, D. Gal, N. Cristler, C. Banditrat, A. Heller, *Anal. Chem.* 2002, **74,** 158–162.

67. http://www.infineon.com

68. A. G. Frutos, J. M. Brockman and R. M. Corn, *Langmuir* 2000, **16,** 2192–2197.

69. L. A. Ruiz-Taylor, T. L. Martin, F. G. Zaugg, K. Witte, P. Indermuhle, S. Nock, and P. Wagner *Proc. Natl Acad. Sci. USA.* 2001, **98,** 852–857.

70. http://www.piercenet.com

71. a) I. Willner, A. Riklin, B. Shoham, D. Rivenzon, E. Katz, *Adv. Mater.* 1993, **5,** 912–915.
b) K. Lindroos, U. Liljedahl, M. Raitio, A.-C. Syvänen, *Nucl. Acids Res.* 2001, **29,** e69.

72. a) J. Rinke, M. Meinke, R. Brimacombe, G. Fink, W. Rommel, H. Fasold, *J. Mol. Biol.* 1980, **137,** 301–304.
b) J. Ciesiolka, P. Gornicki, J. Ofengand, *Biochemistry* 1985, **24,** 4931–4938.

73. http://www.exiqon.com

74. http://www.somalogic.com

75. http://www.prolinx.com

76. M. Schäferling, S. Schiller, H. Paul, M. Kruschina, P. Pavlickova, M. Meerkamp, C. Giammasi, D. Kambhampati, *Electrophoresis* 2002, **23,** 3097–3105.

77. http://www.avidity.com

78. http://www.iba-go.de

79. E. Hochuli, H. Döbeli, A. Schacher, *J. Chromat.* 1987, **411,** 177–184.

80. http://www.greinerbioone.com

81. http://qiagen.com

82. C.C. Bamden, G.B. Sigal, J.L. Strominger, G.M. Whitesides, USP 6,197,515, March 2001.

83. G. MacBeath, *Nature Genet.* 2002, 32, 526–532.

3
Optimization of a Protein Microarray Platform Based on a Small-molecule Chemical Affinity System

Karin A. Hughes, Lisa R. Booth, Robert J. Kaiser,
Kevin P. Lund and Douglas A. Spicer

Abstract

We describe here the optimization of a platform for protein microarray preparation based on a small-molecule, chemical affinity system. We compare the properties of glass slides containing 2-D or 3-D salicylhydroxamic acid (SHA) surface coatings as array substrates, and present the results from a protein microarray-based immunoassay using phenyldiboronic acid-modified capture proteins immobilized on 3-D SHA-coated slides.

3.1
Introduction

Proteomics, the systematic study of protein structure and function, is a natural extension of genomics. As with genomic research, proteomic studies require high-throughput platforms that enable screening of multiple proteins in single assays. The protein microarray holds promise as one such a platform. Indeed, the utility of protein microarrays has been demonstrated in applications such as comparative protein expression analysis [1], high-throughput protein function determination [2–4], miniaturized high-sensitivity multi-analyte immunoassays [5–7], and immunodiagnostics of disease [8–10]. However, in order for protein microarrays to be broadly applied to studying proteomics questions, significant improvements are required in the tools for protein microarray generation.

Microarrays were originally developed for the high-throughput analysis of nucleic acids. However, tools and technologies designed for DNA microarrays have found limited success when applied to protein microarrays. Whereas the synthesis and immobilization of the correct linear DNA sequence is critical to nucleic acid microarrays, reliable maintenance of protein three-dimensional (3-D) structure and activity post-spotting as well as optimal presentation of active regions are critical to the successful generation of protein microarrays. Issues inherent in the fundamental nature of proteins – their fragility and diversity – present additional challenges to the adaptation of DNA microarray tools for the analysis of proteins [10–14]. One of the keys to the successful implementation of protein microarrays is both the

Protein Microarray Technology. Edited by Dev Kambhampati

ISBN: 3-527-30597-1

approach used to attach proteins to a substrate surface as well as the characteristics of the substrate surface itself.

An ideal methodology for the generation of protein microarrays is one that immobilizes the proteins of interest in a fully functional and active orientation. In addition, the method should be robust and reproducible, and utilize substrates that exhibit excellent stability before, during and after spotting. Such a method should be compatible with a wide variety of proteins which exhibit diverse characteristics (e.g., low and high molecular weights; hydrophobic and hydrophilic surface properties), and produce low non-specific binding that may otherwise affect accuracy or dynamic range. There should be little to no increase in the intrinsic background of the substrate in the detection method of choice, and the surface should afford arrayed elements with predictable and regular morphology (i.e. shape and signal uniformity).

Several methods exist for the immobilization of proteins on surfaces [15], including passive adsorption, covalent coupling, and affinity-based binding. Passive adsorption relies on non-covalent interactions between a protein and a hydrophobic or charged surface [10,12,16–18]. While this strategy requires no specialized chemistry and is easy to use, limitations include frequent partial protein denaturation and the inability to control protein orientation on the surface. In addition, this type of linkage can be unstable and protein may be lost during wash steps. Covalent coupling uses a variety of chemically-activated surfaces (e.g., aldehyde, epoxy, activated esters) to react with specific side-chains on proteins or peptides [2,4,6,9,19–21]. Such chemistries can allow for oriented immobilization of proteins on surfaces, are moderately reproducible, and result in a stable linkage of proteins to solid supports. However, not all activated substrates are stable to the conditions required for protein array printing. Most require a step in which unreacted active groups are rendered inactive prior to further processing of the assay. Furthermore, reduced protein activity is often observed using this strategy, and multiple surface chemistries are often required to specifically orient proteins using different protein functional groups. Use of biological affinity-based binding (e.g., (strept)avidin/biotin, recombinant affinity-tagged proteins, protein A) [2,20,22,23] is specific by virtue of the affinity interaction used. These methods also allow for proteins to be attached to surfaces in a defined orientation, but linkages resulting from the use of recombinant tags are often weak and may not be stable over the course of the assay [18,24]. Use of streptavidin/biotin results in a highly stable protein-surface linkage, but requires the use of a large protein, limiting capacity and requiring special handling [24].

To address the above issues, we developed an alternative strategy for the preparation of protein microarrays using a non-biological, chemical affinity system based on phenyldiboronic acid (PDBA) and salicylhydroxamic acid (SHA; [25–29] Figure 3.1). In order to prepare a protein microarray, a PDBA reagent is selected (Table 3.1) to modify solvent-accessible functional groups present on the capture protein while avoiding the active site. Conjugation of proteins with PDBA reagents occurs in solution under mild reaction conditions separate from immobilization, resulting in a more uniform distribution of affinity moieties while insuring a high retention of protein activity relative to other methods. After conjugation, and without the need for purification, the PDBA-modified protein is immobilized on a 3 x 1-inch glass microarray slide coated with SHA. We have developed multiple methods of coating surfaces with SHA. For protein microarrays, we have developed methodologies for both 2-D and 3-D presentations of SHA on glass microarray slides. For 2-D SHA-coated slides, the slides are first amine-functionalized and then reacted with an NHS-ester derivative of SHA. For the 3-D SHA-coated slides, SHA is incorporated into a polymer grown from the surface under living polymerization conditions resulting in a brush polymer with high graft density and no cross-linking. Using both methods, we evaluated 2-D and 3-D presentations of SHA on slides for protein microarray substrates. Based on the results of these experiments, we selected the 3-D SHA-coated surface and demonstrated its utility in conjunction with PDBA-conjugated protein capture ligands for the generation of protein microarrays.

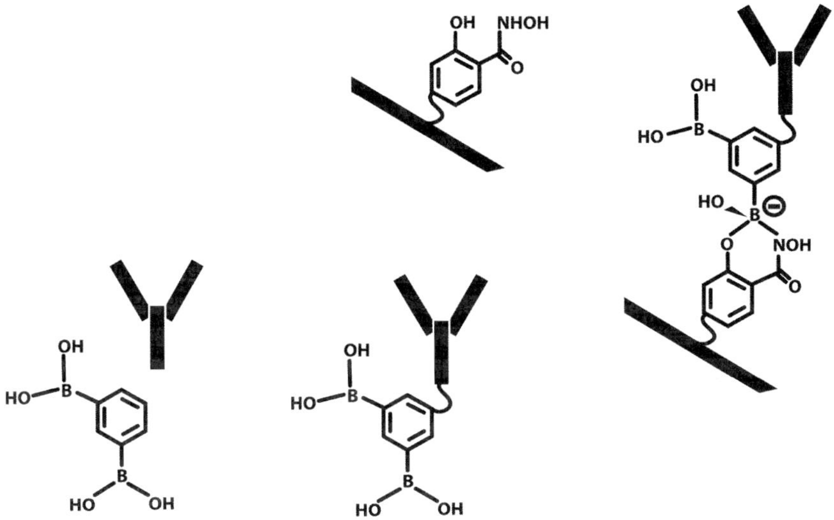

Figure 3.1 Strategy for use of phenyldiboronic acid (PDBA) and salicylhydroxamic acid (SHA) for the generation of protein microarrays. PDBA is covalently bound to proteins using known chemistries. SHA is incorporoated into a 3-D matrix bound to a 3 x 1-inch glass microarray slide. Proteins are immobilized on the surface by the specific complex formation of PDBA with SHA.

Table 3.1 Reagents used for protein conjugation with PDBA.

Reagent	*Reagent reactive group*	*Targeted protein site*
Amine modifying reagent	NHS ester	e-amino group of lysine and N-terminal amino group
Sulfhydryl alkylating reagent	Maleimide	Sulfhydryl of cysteine
Carbohydrate modifying reagent	Hydrazide	Oxidized carbohydrates

3.2 Experimental

3.2.1 Reagents and Materials

3-Aminotriethoxysilane was obtained from Acros Organics, and was distilled under dry nitrogen prior to use. It was stored in a rubber septum-sealed bottle under dry nitrogen in a desiccating chamber. Anhydrous N,N-dimethylformamide, N,N-diisopropylethylamine (redistilled, 99.5%) and p-toluenesulfonic acid monohydrate were purchased from Aldrich. Methanol (HPLC grade), toluene (ACS reagent), concentrated sulfuric acid, 30% hydrogen peroxide, and glycerol (ACS reagent) were obtained from Fisher Scientific. Glass slides (1 by 3 inches) were a product of Corning and were obtained from Fisher Scientific. Versalinx™ Protein Microarray Technology slides (3-D SHA-coated glass slides), spotting solution, wash buffers 1, 2, and 3, slide wash jars, Versalinx Amine Modifying Reagent (PDBA-X-NHS), and 4-[3-hydroxy-4-(tetrahydro-pyran-2-yloxycarbamoyl)-benzylcarbamoyl]-butyric acid 2,5-dioxo-pyrrolidin-1-yl ester [SHA(THP)-X-NHS ester] were from Prolinx Inc. Human IgG (ChromPure, whole molecule, 11.6 mg mL^{-1}), goat anti-human IgG [H + L] (AffiniPure), goat anti-rabbit Fc-specific IgG (min X Hu Sr Prot, 1.8 mg mL^{-1}), goat anti-rabbit Fc specific $F(ab)_2$ fragment (affinity purified, min X Hu Sr Prot, 1.3 mg mL^{-1}), goat anti-human $F(ab)_2$ fragment (affinity purified, 1.8 mg mL^{-1}), goat anti-mouse Fcg $F(ab)_2$ fragment (affinity purified, min X Hu Bov, Hrs, Sr Pro, 1.3 mg mL^{-1}), Cy5-conjugated goat anti-human IgG [H + L] (AffiniPure), Cy3-conjugated mouse IgG (ChromPure, whole molecule, 1 mg mL^{-1}), Cy5-conjugated rabbit IgG (ChromPure, whole molecule, 1 mg mL^{-1}), and Cy3-conjugated human IgG (ChromPure, whole molecule, 1 mg mL^{-1}) were obtained from Jackson Immunoresearch Laboratories. Bovine serum albumin (BSA) was obtained from Sigma Chemical and used as received or was dialyzed against several changes of distilled deionized water at 4 °C before use. Phosphate buffered saline (PBS; 0.01 M sodium phosphate, 0.138 M sodium chloride, 0.0027 M potassium chloride, pH 7.4) was obtained as dried powder in foil packs from Sigma Chemical and was reconstituted according to the manufacturer's instructions. 3-[N-morpholino] propanesulfonic acid (MOPS) hemi-sodium salt was also obtained from Sigma Chemical. Jacquard Gutta resist clear polymer (Rupert Gibbon and Spider Inc., Headsburg, CA) was purchased from Daniel Smith Art Supply store (Seattle, WA) and is available at most craft stores.

3.2.2 Comparison of the Intrinsic Fluorescence and Non-Specific Protein Binding of 2-D and 3-D SHA-Coated Glass Slides

3.2.2.1 Preparation of 2-D SHA-Coated Glass Surfaces

Thirty standard glass slides were arranged in a metal slide holder (Wheaton Sciences) and placed in a glass staining dish (Wheaton Sciences). The slides were soaked overnight in “piranha solution” (70 : 30 (v/v) concentrated sulfuric acid : 30% hydrogen peroxide), then transferred to a clean staining dish and rinsed well under a stream of distilled deionized water. Finally, slides were placed in a staining dish containing sufficient methanol to cover the slides completely, and were sonicated for 15 min. The slides (still in the rack) were removed from the solvent and allowed to dry at room temperature.

The air-dried, clean slides were amine-functionalized using a chemical vapor deposition process. The rack of slides was placed in a 2000-mL glass reaction flask (Corning) such that the slides were held almost horizontally in the flask. The flask was then fitted with a glass head with three joints; a Teflon gasket was used between ground glass joints on the head and the flask. A hose-adapter with a Teflon stopcock was fitted to one of the outer joints, while the other outer joint held a thermometer/adapter assembly. The center joint was fitted with a rubber septum. An oil vacuum pump was attached to the hose-adapter, and the flask was evacuated. An aliquot of 1.0 mL of 3-aminopropyltriethoxysilane was transferred under anhydrous conditions to the bottom of the reaction flask using a syringe and an 18-inch stainless steel needle. Care was taken not to let any of the liquid silane drop onto the glass slides. The vacuum was then shut off from the flask by means of the stopcock, and the flask was placed in a 2000-mL heating mantle (Glas-Col) designed to cover the entire body of the flask. The flask was heated until the thermometer read 120 °C, and was then maintained at this temperature overnight (14–16 h). The next day, the flask was removed from the mantle and allowed to cool to room temperature. Air was introduced into the flask, and the rack of slides was removed. The slides were immersed in toluene in a glass staining dish and sonicated for 15 min. The rack was transferred to a dish containing methanol, and the slides were again sonicated for 15 min. Finally, the slides (still in the rack) were removed from the solvent, placed on a lint-free tissue, covered lightly with another lint-free tissue, and allowed to dry at room temperature.

The air-dried aminated slides were checked for any visible deposits of white polymeric silane; slides showing such deposits were discarded. The slides were immersed in a 5-mM solution of SHA(THP)-X-NHS ester in 98 : 2 (v/v) anhydrous N,N-dimethylformamide:N,N-diisopropylethylamine, covered with plastic film and a glass lid, and allowed to react overnight at room temperature. The rack of slides was then removed from the reaction mixture and washed by sequential immersions in N,N-dimethylformamide (2 x) followed by methanol (2 x). Finally, the slides (still in the rack) were removed from the solvent, placed on a lint-free tissue, covered lightly with another lint-free tissue, and allowed to dry at room temperature.

The tetrahydropyranyl protecting group was removed from the SHA-modified slides by incubating the slides in a 2% w/v solution of p-toluenesulfonic acid in methanol for 4 h at room temperature. The slides were then washed with three immersions in methanol, placed on a lint-free tissue, covered lightly with another lint-free tissue, and allowed to dry at room temperature. The air-dried SHA-modified slides were stored in a plastic slide holder (Fisher Scientific) in a desiccator at room temperature.

3.2.2.2 Preparation of PDBA-modified Bovine Serum Albumin

A total of 3.0 g (44 µmoles) of BSA was dissolved in 30 mL of distilled deionized water in a 50-mL plastic centrifuge tube and 3.3 g of MOPS hemi-sodium salt was added to give a buffered solution, pH 7.4. A solution of 0.55 g (1.1 mmoles; 25-fold molar excess over protein) of PDBA-X-NHS in 5 mL of anhydrous methyl sulfoxide was prepared and added to the protein solution. The mixture was vortexed well and allowed to sit overnight at room temperature. Next day, the reaction mixture was briefly centrifuged to remove a small amount of insoluble material. The supernatant was then dialyzed overnight against 5 L of water at room temperature. The dialysis water was changed, and dialysis continued overnight. Finally, the dialysate was lyophilized and the PDBA-modified BSA stored in an amber bottle at 4 °C.

3.2.2.3 Preparation of PDBA-modified Human IgG

Of the stock solution of human IgG 86 µL was diluted into 914 µL of ice-cold 0.1 M sodium bicarbonate buffer, pH 9.0, to yield a 1 mg mL^{-1} solution (approximately 6.7 µM). A 10 mg mL^{-1} solution of PDBA-X-NHS in anhydrous methyl sulfoxide was prepared, and a 5-µL aliquot (1 x 10^{-7} moles, 15-fold molar excess over protein) was added to the IgG solution. The mixture was vortexed and allowed to sit at 4 °C, overnight. The reaction mixture was then dialyzed overnight against 1 L of PBS at 4 °C. The stock solution of PDBA-modified human IgG was stored at 4 °C.

3.2.2.4 Printing and Development of 2-D and 3-D SHA-coated Glass Slides

Stock solutions of PDBA-modified and unmodified human IgG were diluted into spotting solution (Prolinx, Inc.) to give solutions of 10, 5 and 2 µg mL^{-1}. A 2-D and a 3-D SHA-coated glass slide were each spotted manually using a variable volume pipettor, applying 2-µL aliquots of each protein dilution just above the slide surface to avoid physical damage to the SHA capture layer. The slides were then placed in a closed humid chamber and incubated for 1 h at room temperature. The slides were rinsed by immersion in PBS for 5 min. The 2-D SHA-coated slide was then blocked by soaking in a solution of 1 mg mL^{-1} PDBA-BSA in PBS.Both slides were then washed by immersion in PBST for 10 min. The slides were developed by soaking in a solution of 0.1 µg mL^{-1} Cy5-labeled goat anti-human IgG in wash buffer 3 for 30 min, and rinsed well by immersion and agitation in wash buffers 2 (10 min), and 3 (5 min) and water (2 min). Finally, the slides were dried under a stream of nitrogen prior to fluorescence scanning.

3.2.2.5 Comparison of the Intrinsic Fluorescence of Unmodified, 2-D and 3-D SHA-coated Glass Slides

Surface fluorescence was measured using a Packard BioChip Technologies ScanArray LITE slide scanner. Scanning was performed under the following conditions for all slides: area scanned, 1.5 x 1.5 cm, near the center of the slide; 100% laser power; 100% PMT gain; 5 µm resolution; both Cy3 and Cy5 fluorophore channels. The instrument software was used to determine the number of counts in the scanned area for each channel.

3.2.2.6 Determination of the Non-Specific Protein Binding of 2-D and 3-D SHA-coated Glass Slides

Surface fluorescence was measured using a Packard BioChip Technologies ScanArray LITE slide scanner. Scanning was performed under the following conditions for both slides: 100% laser power; 100% PMT gain; 5 µm resolution; Cy5 fluorophore channel.

3.2.3 Immunoassay Using a 3-D SHA-coated Glass Slide

3.2.3.1 Preparation of PDBA-Cy3-modified Bovine Serum Albumin, PDBA-Human IgG, PDBA-Goat anti-Human IgG

BSA (16.1 mg) was dissolved in 10.0 mL of 0.1 N $NaHCO_3$, pH 8.3. A total of 620 µL of the BSA solution (1 mg BSA) was added to a microfuge tube which contained the contents of a tube of Cy3 monofunctional reactive dye (Amersham Pharmacia), enough to label 1.0 mg of protein. The solution was incubated at 0 °C for 1 h and then passed through an NAP-25 column (Amersham Pharmacia) equilibrated with 10 mL of PBS. The faster running band was collected in 1.25 mL of PBS. A 25-mM solution of PDBA-X-NHS (12.5 mg mL^{-1}) was prepared in anhydrous N,N-dimethylformamide and 4.5 µL of the PDBA solution (15 : 1 molar input ratio) was added to 625 µL of the Cy3-BSA solution. The solution was incubated at 0 °C for 1 h before use.

A 25-mM solution of PDBA-X-NHS (12.5 mg mL^{-1}) was prepared in anhydrous N,N-dimethylformamide. Human IgG (11.3 mg mL^{-1}, 500 µL) was diluted with 500 µL of 0.1 N $NaHCO_3$, pH 8.3. A total of 22.6 µL of the PDBA solution (15 : 1 molar input ratio) was added to the human IgG and the solution incubated at 0 °C for 1 h. The solution was diluted with 1.0 mL of PBS and stored at 4 °C. In a separate reaction, goat anti-human IgG (2.4 mg mL^{-1}, 750 mL) was diluted with 250 µL of 0.1 N NaHCO3, pH 8.3 and 7.2 µL of the PDBA solution (15 : 1 molar input ratio) was added and the solution incubated at 0 °C for 1 h. The solution was diluted with 0.5 mL of PBS and stored at 0 °C.

3.2.3.2 Printing the Array and Analyzing the Data

PDBA-BSA-Cy3 (30 µL, 800 µg mL^{-1}) was diluted to 1.0 mL (25 µg mL^{-1} final concentration) with spotting solution (Prolinx, Inc.). Of the diluted PDBA-BSA-Cy3 60 µL was transferred to a 96-well polypropylene microtiter plate. PDBA-human IgG stock solution (20 µL, 2825 mg mL^{-1}) was diluted to 1.0 mL (50 µg/mL^{-1} final concentration) with spotting solution and 60 µL of the diluted PDBA-human IgG was transferred to the 96-well plate. PDBA-goat anti-human IgG stock solution (40 µL, 1200 µg mL^{-1}) was diluted to 1.0 mL with spotting buffer (50 µg mL^{-1} final concentration) and 60 µL of the diluted PDBA-goat anti-human IgG was transferred to the 96-well plate.

The 96-well plate was transferred to the deck of a non-contact Packard BioChip Arrayer and PDBA-conjugates were printed in 10 x 10 arrays on a Versalinx Protein Microarray Technology slide (350 pL $spot^{-1}$, 180 µm diameter, 500 µm pitch, one-tip arrays). The slide was incubated at 23 °C and 54 % humidity for 1 h. The slide was slowly dipped (upside-down and at a 45° angle) into approximately 30 mL of wash buffer 1 in a slide wash jar and incubated for 2 min. The slide was transferred to a solution of wash buffer 2 (30 mL) which contained Cy3-human IgG (0.4 μg mL^{-1}) and incubated on ice at 0 °C for 30 min. The slide was transferred to a solution of wash buffer 2 (30 mL) and incubated for 5 min. The slide was transferred to a solution of wash buffer 2 (30 mL) which contained Cy5-goat anti-human (0.4 μg mL^{-1}) and incubated at 0 °C for 30 min. The slide was then transferred into 30 mL of wash buffer 3 and incubated for 5 minutes. Wash buffer 3 was decanted and the slide was washed five times with ddH_2O. The slide was dried under a gentle stream of N_2. Surface fluorescence was measured using a Packard BioChip Technologies ScanArray LITE slide scanner. Scanning was performed under the following conditions: 100 % laser power, 95 % PMT gain, 5 μm resolution for the Cy3 fluorophore channel and 90 % laser power, 90% PMT gain, 5 μm resolution for the Cy5 fluorophore channel. The Cy3 and Cy5 images were imported into QuantArray Microarray Analysis Software (version 2.1) to give the composite image. Each image was imported individually and quantified using the above analysis software.

3.2.4 Stability of the PDBA–Protein Conjugates Immobilized on 3-D-SHA-coated Glass Slides

3.2.4.1 Preparation of PDBA-modified Goat Anti-rabbit Fc-specific IgG, PDBA-modified Goat Anti-rabbit Fc-specific $F(ab)_2$, PDBA-modified Goat Anti-human $F(ab)_2$, and PDBA-modified Goat Anti-mouse Fcγ-specific $F(ab)_2$

A 10-mM solution of PDBA-X-NHS was prepared in anhydrous N,N-dimethylformamide from a 100-mM stock solution in the same solvent. Of the goat anti-rabbit Fc-specific IgG (1.8 mg mL^{-1}) a 500-μL aliquot was placed into a microcentrifuge tube, 6.0 μL of 10 mM PDBA-X-NHS (10 : 1 molar input ratio) was added and the solution incubated for 30 min at room temperature. The solution was stored at 4 °C. Of the goat anti-rabbit Fc-specific $F(ab)_2$ fragment (1.3 mg mL^{-1}) a 500-μL aliquot was placed into a microcentrifuge tube, 6.9 μL of 10 mM PDBA-X-NHS (12 : 1 molar input ratio) was added and the solution incubated for 30 min at room temperature. The solution was then stored at 4 °C. A 500-μL aliquot of goat anti-human $F(ab)_2$ fragment (1.8 mg mL^{-1}) was placed into a microcentrifuge tube, 9.6 μL of 10 mM PDBA-X-NHS (12 : 1 molar input ratio) was added and the solution incubated for 30 min at room temperature. The solution was stored at 4 °C. A 500-μL aliquot of goat anti-mouse Fcγ-specific $F(ab)_2$ fragment (1.3 mg mL^{-1}) was placed into a microcentrifuge tube, 6.9 μL of 10 mM PDBA-X-NHS (12 : 1 molar input ratio) was added and the solution incubated for 30 min at room temperature. The solution was stored at 4 °C.

3.2.4.2 **Printing and Reading the Array**

Three hours before printing the array, Gutta resist polymer was applied to the microarray slide to demarcate the areas for the desired number of sub-arrays. The polymer created a hydrophobic barrier and allowed for the use of reduced volumes of reagents in subsequent development steps.

Each of the above PDBA-modified proteins and unmodified proteins were diluted to 200 µL (50 µg mL^{-1} final concentration) with spotting solution. Of each protein solution 50 µL was transferred into four separate wells of a 96-well polypropylene microtiter plate.

The 96-well plate was transferred to the deck of a non-contact Packard BioChip Arrayer and the proteins were printed in two 8 x 24 arrays, on 10 Versalinx Protein Microarray Technology slides (350 pL $spot^{-1}$, 180 µm diameter, 500 µm pitch, one-tip array). The slides were incubated at 19 °C and 55 % humidity for 1 h. Slides 2–9 were placed into slide wash jars (two slides per jar) and stored horizontally (spots upward) at – 20 °C. The remaining two slides (1, 10) were processed immediately as follows. Each slide was slowly dipped (upside-down and at an angle of 45°) into approximately 30 mL of wash buffer 1 in a slide wash jar and left for 2 min. The slides were removed from the slide wash jar and each array was washed with 3 x 1 mL of wash buffer 2. Each array was then incubated with 100 µL of wash buffer 2 containing the appropriate detection protein (see Figure 3.4) at 10 µg mL^{-1} for 30 min at room temperature in a humid chamber to prevent the array drying out. Next, each array was washed three times with 1 mL of wash buffer 2, three times with 1 mL of wash buffer 3, rinsed with ddH_2O and dried under a gentle stream of N_2. After 14 days slides 4 and 5 were removed from – 20 °C storage and processed as described above.

Surface fluorescence was measured using a Packard BioChip Technologies ScanArray LITE slide scanner. Scanning was performed under the following conditions: 100 % laser power, 75 % PMT gain, 5 µm resolution in the Cy3 fluorophore channel and 90 % laser power, 75 % PMT gain, 5 µm resolution in the Cy5 fluorophore channel. The Cy3 and Cy5 images were imported into QuantArray Microarray Analysis Software (version 2.1) for image analysis and quantification. Signal quantification for each spot was accomplished by centering a 160-µm diameter circle within each 180-µm protein spot and recording the average signal. Background quantification for each spot was accomplished by centering a 250- and a 350-µm diameter circle on each 180-µm protein spot and recording the average signal between the outer circles.

3.3
Results and Discussion

3.3.1
Comparison of the Intrinsic Fluorescence and Non-specific Protein Binding of 2-D and 3-D SHA-coated Glass Slides

Protein microarray applications require high detection sensitivity (i.e. the ability to differentiate a (potentially very) low signal from the background), making it necessary to ensure that the surface coating contributes minimally to the intrinsic background of the substrate. While a number of detection strategies are being used for imaging protein microarrays (e.g., surface plasmon resonance, atomic force microscopy, resonance light scattering, etc.), the majority of laboratories currently rely on fluorescence-based slide scanners. We wished to compare the contribution of the 2-D and 3-D SHA surface coatings to the background fluorescence of unmodified glass. Individual slides coated with 2-D or 3-D SHA surface chemistries were scanned in both the Cy3 and Cy5 fluorescence channels. Both the 2-D and 3-D SHA-coated surfaces exhibited little to no increase in intrinsic fluorescence over unmodified glass at either wavelength (Table 3.2). This result suggests that both surfaces are compatible with the use of fluorescence-based detection and that neither surface would adversely affect assay sensitivity.

Table 3.2 Comparison of the intrinsic fluorescence exhibited for unmodified and SHA-coated glass slides.

Surface	*Cy3 channel*	*Cy5 channel*
Clean glass slide	289	2130
2-D coated SHA slide	272	1873
3-D coated SHA slide	263	1835

Another desirable feature of protein microarray surfaces is low non-specific protein binding. In order to evaluate the two surfaces with respect to non-specific binding of unmodified proteins, 2-D and 3-D SHA-coated slides were spotted by hand with PDBA-modified human IgG and unmodified IgG. After washing and prior to development, only the 2-D SHA-coated slide was blocked with a solution of PDBA-modified BSA. Both slides were developed with Cy5-goat anti-human IgG, washed and scanned (Figure 3.3). Despite being blocked, the 2-D SHA-coated slide displayed very high non-specific binding of unmodified IgG, presumably due to protein adsorption to poorly-coated regions of glass. In contrast, the 3-D SHA-coated slide exhibited very low non-specific binding of unmodified human IgG, without the use of a blocking step. The three-dimensional nature of the polymer matrix coating is effective in preventing proteins from contacting the glass substrate, thus eliminating non-specific protein adsorption and obviating the need for blocking. Blocking of non-specific protein binding to surfaces is often accomplished by using proteins such as BSA or by addition of detergents. Such reagents can result in reduced activity of immobilized proteins and may obscure detection of desired protein binding events [4,8,18]. Thus, use of the 3-D SHA-coated surface not only reduces the time required for array processing, but may also afford increased assay sensitivity and reduced numbers of false positive results.

3.3.2
Immunoassay on a 3-D SHA-coated Slide

To further investigate the utility of 3-D SHA-coated slides in conjugation with PDBA-modified proteins as a suitable protein microarray platform, we spotted three 10 x 10 arrays onto a single 3-D SHA-coated slide. Cy3-modified BSA, a human IgG antigen and a goat anti-human IgG antibody were conjugated separately with PDBA using PDBA-X-NHS. The resulting PDBA-conjugates were printed without purification on a 3-D SHA-coated slide using a non-contact arrayer. After incubation of the slide for 1 h under conditions optimized to prevent evaporation of the spot, the slide was washed and then sequentially developed in solutions containing Cy3-labeled human IgG followed by Cy5-labeled goat anti-human IgG. The resulting Cy3 and Cy5 images were quantified (Table 3.4) and used to generate a composite false-color image (Figure 3.3).

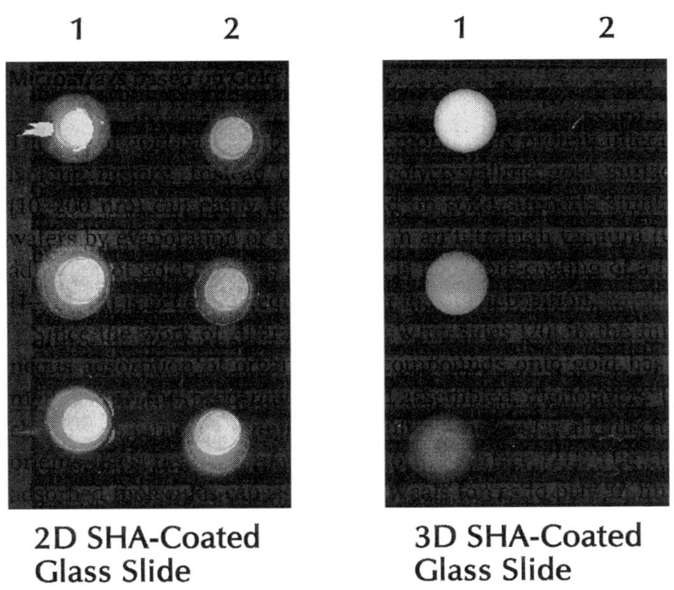

Figure 3.2 Effect of surface chemistry on non-specific protein binding. 2-D and 3-D SHA-coated slides were hand spotted in triplicate with PDBA-modified human IgG (column 1) and unmodified IgG (column 2). The 2-D SHA-coated slide was blocked with a solution of PDBA-modified BSA prior to development. Despite the blocking step, the 2-D SHA-coated slide displayed very high non-specific binding of unmodified IgG, presumably due to protein adsorption to non-coated portions of glass.

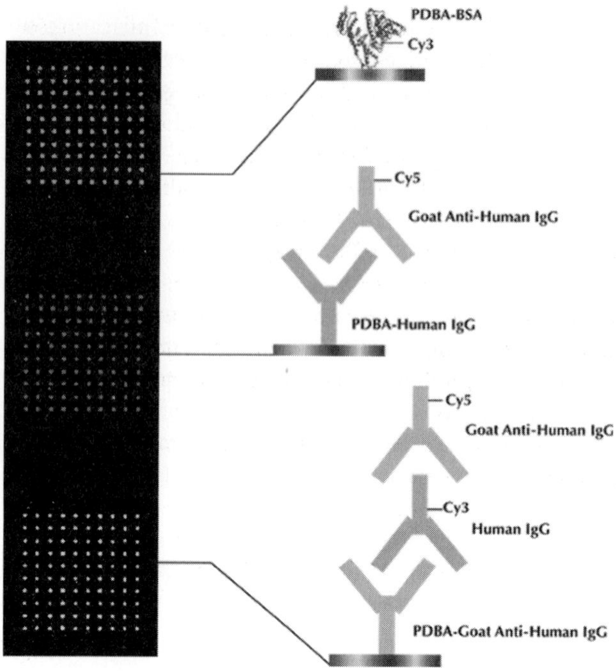

Figure 3.3 Protein microarray immunoassay on a 3-D SHA-coated slide. Fluorescence false-color composite image of 3–10 x 10 sub-arrays corresponding to PDBA-BSA-Cy3, PDBA-human IgG antigen detected with Cy5-labeled goat anti-human IgG and PDBA-goat anti-human IgG detected with Cy3-labeled human IgG antigen and Cy5-labeled goat anti-human IgG. Green indicates signal detected in the Cy3 channel, red in the Cy5 channel and yellow in both the Cy3 and Cy5 channels. (see Colour Plate p. XX).

Table 3.4 Quantitative analysis of the signal averages (n = 24) for the arrays in Figure 3.4.

	Mean FIU	*Standard deviation*	*CV (%)*
Day 0, slide 1, array 1	49697	1794	4
Day 0, slide 10, array 1	51419	1090	2
Day 14, slide 4, array 1	47563	3003	6
Day 14, slide 5, array 1	47036	1881	4
Day 0, slide 1, array 2	38796	3287	8
Day 0, slide 10, array 2	34154	3445	10
Day 14, slide 4, array 2	28740	3592	12
Day 14, slide 5, array 2	27175	5272	19

All three proteins were readily detected in the expected fluorescence channels. Cy3-BSA was directly detected in the Cy3 channel (green). Human IgG antigen exhibited signals in the Cy5-channel (red), concordant with specific recognition by the Cy5-anti-human IgG. The anti-human IgG antibody exhibited fluorescence signals in both the Cy3 and Cy5 channels, consistent with specific recognition by the Cy3-human IgG followed by the Cy5-anti-human IgG in a sandwich format. The intense fluorescence signals afforded by the human IgG antigen and anti-human IgG antibody arrays suggest that PDBA modification and subsequent conjugate immobilization on the 3-D SHA-coated slide did not result in loss of protein activity as defined by specific recognition. Quantitative analysis of signal strength averages for each of the arrays (Table 3.4) revealed little deviation from the mean, consistent with good spot morphology and signal uniformity as well as little variability in binding capacity across the slide surface. The low background (average background counts 501 FIU for the Cy3 channel and 406 FIU for the Cy5 channel) was consistent with low non-specific protein binding and provided excellent signal-to-noise ratios.

3.3.3 Stability of PDBA–Protein Conjugates Immobilized on 3-D SHA-coated Glass Slides

DNA is robust and able to withstand harsh conditions. Oligonucleotides, for example, can be absorbed onto substrates, dried and re-hydrated without loss of activity. This inherent stability facilitates DNA microarray production by allowing for long array print times and long-term storage of spotted and dried arrays. While stability is somewhat protein dependent, many proteins undergo irreversible denaturation upon drying, making the production and subsequent storage of printed protein microarrays difficult. Results from our group and others have demonstrated the importance of continued hydration of spotted proteins to prevent denaturation and subsequent loss of signal intensities indicative of loss of protein activity [4,30]. In one study we compared the stability of two antibodies spotted onto a slide: PDBA-anti rabbit IgG and PDBA-human IgG [31]. If slides were developed immediately after printing while the spots were hydrated, a good signal was observed for both proteins. If the slides were dried for 1 min prior to development, little loss in signal intensity was observed for the human IgG, whereas nearly half of the initial signal intensity was lost for the anti-rabbit antibody. Similarly, if slides were stored at 4 °C after being dried for 18 h prior to development, 25 % of the signal intensity was lost for the human IgG and 75 % of the signal intensity was lost for the anti-rabbit antibody.

As a control, these results were compared to those for peptides spotted onto a slide; petides, like oligonucleotides, are much more stable to drying. Slides that were spotted with peptides, dried and stored for 18 days at room temperature prior to development displayed signal intensities comparable to those obtained from slides which were developed immediately (results not shown). These results confirm that stability is protein dependent, but that in general, dry storage of protein microarrays is limited.

Recently, Snyder et al. reported some success with storage of hydrated arrayed protein slides [30]. Slides were printed using a spotting solution containing 40 % (v/v) glycerol. After printing, slides were used immediately or stored at – 20 °C until use. The use of glycerol in the spotting solution served to keep the spots hydrated over time while preventing freezing at reduced temperature, lending enhanced protein stability.

We wished to examine the storage stability of PDBA-modified proteins arrayed on 3-D SHA-coated slides under hydrated conditions. We spotted two 8 x 24 arrays on each of 10 3-D SHA-coated glass slides in the pattern shown in Figure 3.4. After printing, two slides (slides 1 and 10) were developed immediately (day 0) and the remaining slides were stored at 20 °C without washing. After 14 days, two additional randomly selected slides (slides 4 and 5) were removed from – 20 °C storage. These slides were developed using the same procedure as for the first two slides. The resulting Cy3 scans were used to generate a composite image (Figure 3.4). Quantitative analysis of the averages of signal strength for each of the arrays is shown in Table 3.4. The remaining slides were left at – 20 °C.

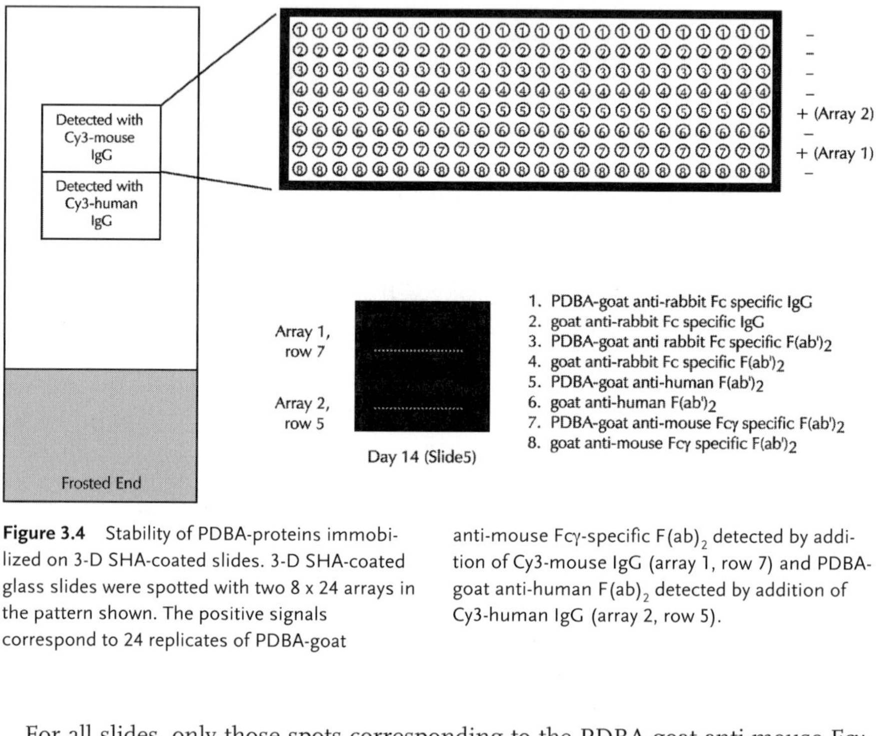

Figure 3.4 Stability of PDBA-proteins immobilized on 3-D SHA-coated slides. 3-D SHA-coated glass slides were spotted with two 8 x 24 arrays in the pattern shown. The positive signals correspond to 24 replicates of PDBA-goat anti-mouse Fcγ-specific $F(ab)_2$ detected by addition of Cy3-mouse IgG (array 1, row 7) and PDBA-goat anti-human $F(ab)_2$ detected by addition of Cy3-human IgG (array 2, row 5).

For all slides, only those spots corresponding to the PDBA-goat anti-mouse Fcγ-specific $F(ab)_2$ in array 1 (row 7) and the PDBA-goat anti-human $F(ab)_2$ in array 2 (row 5) for which the appropriate detection system had been added were detected. Average background counts were very low: 363 and 397 FIU for non-PDBA-labeled goat anti-mouse Fcγ-specific $F(ab)_2$ (array 1, day 14, slides 4 and 5 respectively) and 124 and 11 FIU for non-PDBA-labeled goat anti-human $F(ab)_2$ (array 2, day 14, slides 4 and 5 respectively). The specifically bound PDBA-modified proteins afforded average signal intensities of > 45,000 FIU (array 1, day 14) and > 25,000 FIU (array 2, day 14) at similar concentrations. In addition, no signal over background was observed for any of the other PDBA-modified proteins that lacked appropriate detection proteins. These results are again consistent with the 3-D SHA-coated slides affording excellent signal-to-noise ratios and low non-specific binding without the need for a blocking step.

Intra-slide variability, determined by comparing the mean signal intensities for 24 replicate spots for each of the arrays on the four slides from day 0, had coefficients of variation (CV) below 10 %. As with the data shown in Figure 3.3, these results are consistent with good spot morphology and signal uniformity as well as little variability in binding capacity across the slide surface. By day 14, there were no observed changes in gross spot morphology prior to slide development. The intra-slide variability was still good for spots corresponding to PDBA-labeled goat anti-mouse Fcγ-specific $F(ab)_2$ (CV < 10 %, array 1, slides 4 and 5), but had dropped precipitously for the goat anti-human $F(ab)_2$ (CV of 12 and 19 %, array 2, slides 4 and 5 respectively). While differences across the slide surface cannot be ruled out, it is more likely, based on previous results, that the observed increase in the CV is due to instability of the goat anti-human $F(ab)_2$ protein.

Comparison of the average signal intensities for array 1 on day 0 versus day 14 of PDBA-labeled goat anti-mouse Fcγ-specific $F(ab)_2$ detected with Cy3 mouse IgG (50,558 and 47,300 FIUs respectively), corresponded to a 6 % loss of signal consistent with little to no loss in protein activity. Comparison of the average signal intensities for array 2 on day 0 versus Day 14 for PDBA-labeled goat anti-human $F(ab)_2$ detected with Cy3 human IgG (36,475 and 27,958 FIUs respectively), corresponded to a 41 % loss of signal intensity. This loss of intensity may suggest loss in protein activity due to instability of the anti-human : human antibody pair.

3.4 Conclusions

We have demonstrated the feasibility of producing protein microarrays using reagents based on a small-molecule, chemical affinity system. The technology utilizes stable, specific, easy-to-use chemistry involving conjugation of capture proteins with PDBA followed by immobilization of PDBA-labeled proteins simply by spotting unpurified reactions onto SHA-coated slides. Protein conjugation methods use simple and familiar protocols, enabling reproducible modification of proteins with no evidence of interference with biological activity. Comparison of performance of 2-D and 3-D SHA-coated glass slides revealed that, while both surfaces contribute little to the intrinsic fluorescence of the underlying glass substrate, the 3D-SHA coated slides exhibited superior performance with respect to low non-specific protein adsorption without the need for blocking steps. In addition, the 3-D SHA-coated slides produced spots with predictable and regular morphology (i.e. shape and signal uniformity) facilitating data analysis. The 3-D SHA-coated slides are easy to use, requiring no preparation before use and are suitable for use with normal protein handling conditions without exhibiting degradation. The methodology is compatible with direct, indirect and sandwich fluorescence detection. Results from studies examining stability of printed protein microarrays suggest that hydration is important to maintaining protein activity, but that even under hydrated conditions, protein activity degrades over time. Additional work will be required to stabilize printed protein microarrays. Alternatively, use of antibody fragments or protein structure mimics (e.g., affibodies, aptamers, etc.) may impart greater stability and supersede the use of antibodies as capture ligands. We conclude that Versalinx Microarray Technology is a convenient platform for the production of protein microarrays.

Acknowledgments

The authors would like to gratefully acknowledge Mary Wang for preparation of figures and Leslie Linkkila for her insight and critical review of this manuscript.

References

1. Wang, C. C., Huang, R.-P., Sommer, M., Lisoukov, H., Huang, R., Lin, Y., Miller, Y., Miller, T., and Burke, J. Array-based multiplexed screening and quantification of human cytokines and chemokines. *J. Proteome Res.* 2002, **1**, 337–343.
2. Zhu, H., Bilgen, M., Bangham, R., Hall, D., Casamayor, A., Bertone, P., Lan, N., Jansen, R., Bidlingmaier, S., Houfek, T., Mitchell, T., Miller, P., Dean, R. A., Gerstein, M., and Snyder, M. Global analysis of protein activities using proteome chips. *Science* 2001, **293,** 2101–2105.
3. Zhu, H., Klemic, J. F., Chang, S., Bertone, P., Casamayor, A., Klemic, K.G., Smith, D., Gerstein, M., Reed, M. A., and Snyder, M. Analysis of yeast protein kinases using protein chips. *Nature Genet.* 2000, **26,** 282–289.
4. MacBeath, G., and Schreiber, S.L. Printing proteins as microarrays for high-throughput function determination. *Science* 2000, **289,** 1760–1763.
5. de Wildt, R., Mundy, C., Gorick, B., and Tomlinson, I. Antibody arrays for high-throughput screening of antibody-antigen interactions. *Nature Biotechnol.* 2000, **18,** 989–994.
6. Arenkov, P., Kukhtin, A., Gemmell, A., Voloshchuk, S., Chupeeva, V., and Mizabekov, A. Protein microchips: use for immumoassay and enzymatic Reactions. *Anal. Biochem.* 2000, **278,** 123–131.
7. Haab, B. B., Dunham, M. J., and Brown, P. O. Protein microarrays for highly parallel detection and quantitation of specific proteins and antibodies in complex solutions. *Genome Biol.* 2001, **2,** 1–13.
8. Mezzasoma, L., Bacarese-Hamilton, T., Di Cristina, M., Rossi, R., Bistoni, F., and Cristani, A., Antigen microarrays for serodiagnosis of infectious diseases. *Clin. Chem.* 2002, **48,** 121–130.
9. Robinson, W. H., DiGennaro, C., Hueber, W., Haab, B. B., Kamachi, M., Dean, E. J., Fournel, S., Fong, D., Genovese, M. C., Neuman de Vegvar, N., Skriner, K., Hirschberg, D. L., Morris, R. I., Muller, S., Prujin, G. J., van Venrooij, W. J., Smolen, J. S., Brown, P. O., Steinman, L., and Utz, P. J. Autoantigen microarrays for multiplex characterization of autoantibody responses. *Nature Med.* 2002, **8,** 295–301.
10. Miller, J. C., Bulter, E. B., Teh, B. S., and Haab, B. B. The application of protein microarrays to serum diagnostics: prostate cancer as a test case. *Disease Markers* 2001, **17,** 225–234.
11. Service, R. F. Searching for recipes for protein chips. *Science* 2001, **294,** 2080–2082.
12. Wilson, D. S., and Nock, S. Functional protein microarrays. *Curr. Opin. Chem. Biol.* 2001, **6,** 81–85.
13. Boguslavsky, J. Protein chips still growing. *Drug Discovery* March 2001, S25–S26.

14. Haab, B. B. Advances in protein microarray technology for protein expression and interaction profiling. *Curr. Opin. Drug Discovery Develop.* 2001, **4,** 116–123.
15. Rao, S. V., Anderson, K. W., and Bachas, L. G. Oriented immobilization of proteins. *Mikrochim. Acta* 1998, **128,** 127–143.
16. Butler, J. E., Ni, L., Nessler, R., Joshi, K. S., Suter, M., Rosenberg, B., Chang, J., Brown, W. R., and Cantarero, L. A. The physical and functional behavior of capture antibodies absorbed on polystyrene. *J. Immunol. Methods* 1992, **150,** 77–90.
17. Butler, J. E., Ni, L., Brown, W. R., Joshi, K. S., Chang, J., Rosenberg, B., and Voss, E. W. The immunochemistry of sandwich ELISAs. VI. Greater than 90 % of monoclonal and 75 % of polyclonal anti-fluorescyl capture antibodies (Cabs) are denatured by passive adsorption. *Mol. Immunol.* 1993, **30,** 1165–1175.
18. Houseman, B. T., and Mrksich, M. Towards quantitative assays with peptide chips: a surface engineering approach. *Trends Biotechnol.* 2002, **20,** 279–281.
19. Seong, S.-Y. Microimmunoassay using a protein chip: optimizing conditions for protein immobilization. *Clin. Diagn. Lab. Immunol.* 2002, **9,** 927–930.
20. Turkova, J. Oriented immobilization of biologically active proteins as a tool for revealing protein interactions and functions. *J. Chromatogr. B* 1999, 777, 11–31.
21. Williams, R. A., and Blanch, H. W. Covalent immobilization of protein monolayers for biosensor applications. *Biosens. Bioelectron* 1994, **9,** 159–167.
22. Butterfield, D. A., Bhattacharyya, D., Daunert, S., and Bachas, L. Catalytic biofunctional membranes containing site-specifically immobilized enzyme arrays: A review. *J. Membr. Sci.* 2001, **181,** 29–37.
23. Ruiz-Taylor, L. A., Martin, T. L., Zaugg, F. G., Witte, K., Indermuhle, P., Nock, S., and Wagner, P. Monolayers of derivatized poly(L-lysine)-grafted poly(ethylene glycol) on metal oxides as a class of biomolecular interfaces. *Proc. Natl. Acad. Sci. USA* 2001, **98,** 852–857.
24. Hodneland, C. D., Lee, Y.-S., Min, D.-H., and Mrksich, M. Selective immobilization of proteins to self-assembled monolayers presenting active site-directed capture ligands. *Proc. Natl. Acad. Sci. USA* 2001, **99,** 5048–5052.
25 Stolowitz, M. L., Ahlem, C., Hughes, K. A., Kaiser, R. J., Kesicki, E. A., Li, G., Lund, K. P., Torkelson, S. M., and Wiley, J. P. Phenylboronic acid-salicylhydroxamic acid bioconjugates I: A novel boronic acid complex for protein immobilization. *Bioconjugate Chem.* 2001, **12,** 229–239.
26. Hughes, K., and Wiley, J. Use of small-molecule affinity-based matrices for rapid protein purification. In: Vaillancourt, P. E. (Ed.): *E. Coli* Gene Expression Protocols, Methods in Molecular Biology. Humana Press: Totowa, NJ, 2002, 215–223.
27. Wiley, J. P., Hughes, K. A., Kaiser, R. J., Kesicki, E. A., Lund, K. P., and Stolowitz, M. L. Phenylboronic acid-salicylhydroxamic acid bioconjugates II: Polyvalent immobilization of protein ligands for affinity chromatography. *Bioconjugate Chem.* 2001, **12,** 240–250.
28. Hughes, K. A., Lucas, D. L., Stolowitz, M. L., and Wiley, J. P. Novel affinity tools for protein immobilization: Implications for proteomics. *Am. Biotechnol. Lab.* 2001, 1, 36–38.
29. Booth, L. R., Clary, S. T., Gall, A. S., Hughes, K. A., Kaiser, R. J., Lund, K. P., and Spicer, D. A. A fabrication of peptide microarrays utilizing small-molecule affinity technology. *Am. Biotechnol. Lab.* 2002, **20,** 76–78.
30. Zhu, H., Bilgin, M., Bangham, R., Hall, D., Casamayor, A., Bidlingmaier, S., and Snyder, M. Global analysis of biochemical activities using protein chips. Presented at Macroresults Through Microarrays 3, Boston, MA, April 30, 2002.
31. Hughes, K. A. Surface chemistry for optimized protein and peptide microarrays. Presented at Surface Chemistry for Protein Microarrays pre-conference Symposium, Hamburg, Germany, September 23, 2002.

4
Seeing Beneath the Surface of Biomolecular Interactions: Real-time Characterization of Label-free Binding Interactions using Biacore's Optical Biosensors

Gary Franklin and Alan McWhirter

4.1
Introduction

The phenomenon of SPR (surface plasmon resonance) was first shown to be applicable to biological molecules as long ago as 1983 [1] and has since played an increasingly important role in the study of biomolecular interactions. Biacore, the pioneering company in the optical biosensor field, combined SPR detection with an advanced system of microfluidics and specially developed sensor surface coupling chemistries to generate a unique technology for generating high quality, information-rich data on molecular binding events. The technology is label-free, highly flexible in terms of the range of biomolecules that it can accommodate and can deliver a wide range of data types, including analyses of concentration, specificity, affinity and kinetics.

The technology has been used very successfully in a number of areas, of which the three major areas have been academic life science research, commercial pharmaceutical and biotechnology applications and within the food industry. The instruments available cover a wide assortment of user requirements, ranging from manual systems designed for simple, flexible basic research applications, to automated, high-throughput systems aimed at the drug development industry, as well as systems designed to function within the GxP (Good Laboratory Practice, Good Manufacturing Practice, Good Clinical Practice etc.) environment now required by many regulatory authorities.

Biacore users have taken advantage of the inherent flexibility of the technology to study many diverse types of molecules, although proteins have consistently represented the principle focus. As the genomics era begins to give way to the age of proteomics, the need for high quality data on the molecular interactions of proteins has never been greater. While Biacore, along with many other technologies, was making an important contribution towards the understanding of protein interactions long before the term "proteomics" ever existed, the widely anticipated proteomics explosion will certainly present new challenges and opportunities to all working within this field.

Protein Microarray Technology. Edited by Dev Kambhampati

ISBN: 3-527-30597-1

4.1.1 The Proteomics Revolution

During the last decade of the 20th century, biomedical science was characterized by the technological advances that powered the genomics revolution. The race to complete the sequencing of the human genome was probably the most absorbing scientific development of its time. We are now in possession of a staggering amount of information regarding the nucleic acid sequence of our own, and many other species. The total number of human genes (which has not yet been totally covered by the "completion" of the human genome project) is currently estimated at around 27 000 [2]. When compared to genomic sequencing projects carried out in other species however, it is clear that there is no real correlation between gene number and biological complexity among different organisms and that higher levels of genetic organization and control of expression of the genetic material must be involved.

The primary challenge now is to relate this vast store of sequence information to its biological functions. Since the vast majority of genes are currently thought to function at the protein level, the next logical step is to re-examine the genome in terms of the proteins it produces. This will require an even larger-scale research effort however, since it is becoming increasingly clear that the number of distinctly functional human proteins produced is much higher than the number of genes that exist. This provides a likely explanation for the "gene number versus complexity" paradox referred to above. The genes of higher eukaryotes are complex in structure, often containing multiple promoters that direct alternative transcripts, and recent evidence suggests that around 50 % of all genes may be subject to alternative splicing [3]. In addition to multiple transcripts and splice variants, it has also been estimated that on average, each human gene may produce three different post-translationally modified protein variants, all with distinct functions [4]. Taken together, these variables mean that there are multiple proteins produced from each gene, with current estimates ranging anywhere between 3 and 20 proteins per gene.

The term "proteome" was first coined in the mid-1990s to introduce the concept of a protein-level equivalent of "genome", and can be loosely defined as "the entire protein complement that is encoded by a genome". The proteome concept is considerably more complex than its genomic predecessor however, since the protein complement varies enormously between different cells within the organism and can change from minute to minute within an individual cell, depending on the internal and external signals received. Another factor is that the functional roles of proteins in the proteome involve interaction on a much greater scale of complexity than for nucleic acids. In addition, identification of a protein on the basis of its amino acid composition alone may not provide the crucial information in all cases, since the post-translational modifications alluded to previously can play a critical role in physiological and pathological functions (reviewed in [4]).

Consequently, there are multiple challenges for the much-vaunted "new era" of proteomics. Very high throughput discovery programs will be required to identify all the proteins produced by the genome and to determine where and when these proteins are actually expressed in very complex, dynamic combinations. In addition, entirely different, lower throughput approaches will be required to provide the detailed characterization of interactions between functionally connected biomolecules. This last aspect in particular (embodied in the branch known as "interaction proteomics" or "functional proteomics") illustrates the magnitude of the problems ahead. It should be clear from the application examples presented here (see Section 4.3) that understanding biomolecular interactions in the context of biological function requires a far higher level of detail than "do they bind?" and that the ability to describe interactions in terms of affinity and (particularly) rate constants will be a prerequisite for real progress in the future. Biomolecular interactions are transient events and their kinetic characteristics are vitally important for biological function. A full understanding of how proteins function on a mechanistic level cannot, therefore, rely on steady-state measurements alone [5–8]. This temporal aspect of functional proteomic studies adds yet another level of complexity. Many different technologies will doubtless play important roles in the development of proteomics, and Biacore is ideally placed to make a significant contribution. In fact, even before the term "proteomics" existed, Biacore has always been used in approaches that fundamentally belong to the proteomics field. The ability of Biacore to measure protein concentrations, assess binding specificity, to characterize binding interactions in terms of affinity and rate constants and (as described in Section 4.4.1) provide identification by linking SPR to MS, make this a key proteomics technology. Moving from the study of isolated proteins or individual binding partners into more industrialized proteomics applications does not, therefore, represent a fundamental paradigm shift but is merely an issue of scale, as will be discussed later.

This chapter will now describe how Biacore's approach to SPR technology functions on a technical level and how these features contribute to the study of biomolecular interactions. This will be followed by selected examples of how the technology has been successfully applied to a number of important questions, with relevance to both academic biomedical research and the pharmaceutical and biotechnology industries. Finally, current and future developments will be discussed in the context of the evolving areas of proteomics and protein microarrays.

4.2. Biacore Technology

Biacore's optical biosensors are built around three principal components:

- an *optical system* responsible for the generation and detection of the SPR signal that is triggered by biomolecular interaction events
- a *sensor chip* with a biospecific surface where biomolecular interactions take place
- a liquid handling system with precision pumps and an *integrated microfluidic cartridge* (IFC) for controlled flow of sample over the sensor surface

These three cornerstones of the technology will be described in detail, along with a discussion of their practical and theoretical consequences. In the interests of clarity, some specialized terminology relating to the interactants used in a Biacore assay should also be pointed out. The interactant immobilized on the sensor surface is referred to as the *ligand* and should not be confused with the use of this term in the biological context of ligands/receptors. The ligand can be attached either directly to the sensor surface, or indirectly via an immobilized *capturing molecule*, and the interactant in free solution is referred to as the *analyte*. This is shown schematically in Figure 4.1.

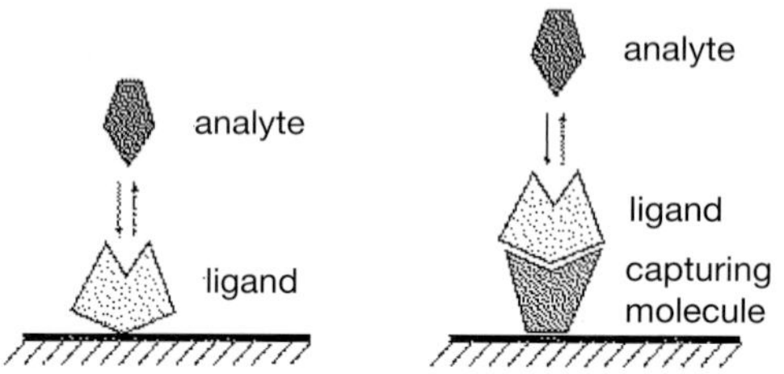

Figure 4.1 Ligand, analyte and capturing molecule in relation to the sensor surface, as defined in a Biacore assay.

4.2.1
Surface Plasmon Resonance

The detection principle of the technology depends upon the phenomenon of surface plasmon resonance (SPR), which occurs in thin conducting films at an interface between media of different refractive index [9]. In Biacore systems, the media are the glass of the sensor chip and the sample solution, and the conducting film is a gold layer on the sensor chip surface.

Under conditions of total internal reflection, the light leaks an electric field intensity called an *evanescent wave field* across the interface into the medium of lower refractive index, without actually losing net energy. The amplitude of the evanescent field wave decreases exponentially with distance from the surface, and the effective penetration depth is about half the wavelength of the incident light. In Biacore instruments, the light source is a near-infrared LED of peak intensity wavelength 760 nm and the penetration distance of the evanescent wave is about 300 nm.

At a certain combination of angle of incidence and energy (wavelength), the incident light excites *plasmons* (electron charge density waves) in the gold film. As a result, a characteristic absorption of energy via the evanescent wave field occurs and SPR is seen as a drop in the intensity of the reflected light at a specific incident angle (the *SPR angle*). These principles are illustrated in Figure 4.2.

The reduced intensity of reflected light is not caused by light absorption in the sample in the conventional (transmission spectroscopy) sense. The light is totally

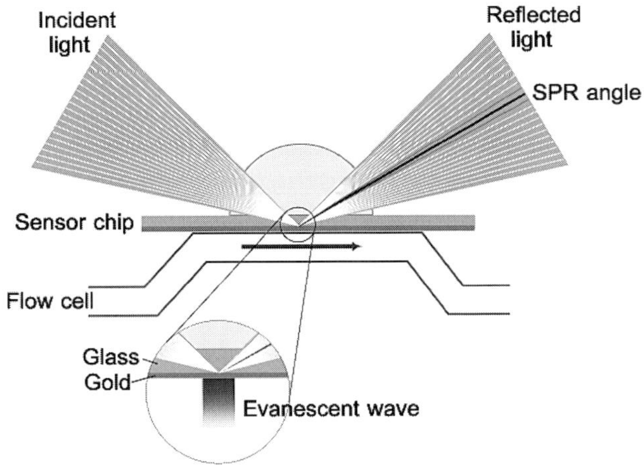

Figure 4.2 The principle of SPR detection in Biacore instruments. Schematic drawing showing the optical interface, sensor chip and microfluidic components that comprise the main technology cornerstones in Biacore's approach to SPR biosensors. As drawn here, the ligand is immobilized on the lower (dark grey) surface of the sensor chip and the analyte is passed over the surface under continuous flow conditions, via the integrated microfluidics system. An evanescent wave is created at the chip surface by the action of total internal reflection of the incident light, and this penetrates the sample side of the interface. As indicated in the enlarged view, the amplitude of this wave decays exponentially with distance from the surface. Polarized components of the evanescent wave field excite electromagnentic surface plasmon waves within the gold film, enhancing the evanescent wave and causing a characteristic drop in the reflected light intensity. Changes in refractive index at the flow cell side of the interface caused by changes in mass concentration at the sensor surface, result in a shift in the angle of incidence required to create surface plasmon resonance. This SPR angle is - measured as a change in the detector position for the reflected intensity dip that results from the SPR phenomenon.

internally reflected inside the optical unit, and it is the evanescent wave that penetrates the sample. The significance of this feature is that measurements can be made on turbid or even opaque solutions, without interference from conventional light absorption or scattering by the sample.

When the detecting molecule is attached to the sensor chip or when analyte binds to the detecting molecule, the concentration at the sensor chip surface increases leading to an SPR response, observed as a change in the SPR angle. The specific refractive index contribution (i.e. the change in refractive index produced by a unit change in concentration) is very similar for different proteins regardless of composition, and values for other macromolecules are of the same order of magnitude. Consequently, the response measured is related to the mass of analyte bound and is largely independent of the nature of the analyte. Refractive index contributions for different molecules are additive, so that the amount of detecting molecule attached and the amount of analyte bound can both be measured with the same detection principle.

4.2.2
The Sensor Surface and Immobilization Chemistry

As outlined in the previous section, SPR measurements with Biacore are based on the interaction of an analyte in solution with a ligand attached to the sensor surface. At the heart of this process is the sensor chip, which provides both the physical conditions necessary to generate the SPR signal, and the surface on which the interaction being studied occurs.

The specificity of an SPR analysis is determined through the nature and properties of the molecule attached to the sensor surface. Broadly speaking, biomolecules may be attached to the surface of the sensor chip using three different approaches:

1. Covalent immobilization, where the molecule is attached to the surface through a covalent chemical link.
2. High affinity capture, where the molecule of interest is attached by non-covalent interaction with another molecule (which in turn is usually attached using covalent immobilization).
3. Hydrophobic adsorption, which exploits non-specific hydrophobic interactions between the molecule of interest (or a hydrophobic carrier such as a lipid monolayer or bilayer) and the sensor chip surface.

4.2.2.1 Sensor chips

The sensor chip has two essential features:

- A glass surface covered with a thin gold layer, creating the conditions required for generating an SPR signal.

- A coating of some sort on the gold layer, providing a means for attaching the ligand and an environment where the interaction being studied will occur. The coating does not contribute directly to the SPR effect, and varies between different types of chips.

Although SPR can be generated in thin films made from different materials, all Biacore chips are coated with a uniform thin layer of gold. Gold gives a well-defined reflectance minimum with easily handled visible light wavelengths, and is also amenable to covalent bonding of surface matrix layers, and at the same time the metal is largely inert in physiological buffer conditions. Unmodified gold is not generally a suitable surface environment for biomolecular interactions however, and is covered with a covalently bonded monolayer of alkanethiol molecules. This serves to protect the biological samples from contact with the gold and at the same time provides a means of attachment of a surface matrix, which further enhances the usefulness of the surface in biomolecular contexts.

On most sensor chips, the surface is covered with a "hydrogel" matrix of carboxymethylated dextran, a flexible unbranched carbohydrate polymer forming a surface layer of approximately 25–100 nm thick. The dextran is covalently attached to an epoxide-modified alkanethiol monolayer. Surfaces for various applications have been tailored by the size of the dextran, differing in the range of 10 kDa to 1

MDa [10]. The dextran matrix imparts several important properties to the sensor surface:

- It provides a hydrophilic environment favorable to most solution-based biomolecular interactions.
- It provides a defined chemical basis for covalent attachment of biomolecules to the surface using a wide range of well-defined chemistries (as will be described shortly).
- It increases the surface capacity many-fold in comparison with a flat surface.
- It extends the surface detection principle to encompass a thin layer with an extension of the same order of magnitude as the penetration depth of the evanescent wave responsible for the detection principle. The sensitivity of the detection is thereby increased considerably in comparison with a flat surface.

Dextran is also an inert molecule in the context of biomolecular interactions, and the flexibility of the unbranched polymer chains allows the "surface-attached" biomolecules to move with relative freedom within the surface layer. Dextran-based surface matrices therefore provide an excellent environment for biomolecular interactions [10]. Furthermore, the most frequently utilized carboxymethylated derivative of dextran improves the hydrophilicity and serves as an excellent starting point for various immobilization alternatives. The negatively charged polymer may give rise to unwanted electrostatic interactions with the immobilized ligands and thus to a risk of denaturation. These potential effects are normally eliminated by working under physiological salt conditions, where the electrostatic effects are effectively shielded. The basic structure of a sensor chip and its surface is shown in Figure 4.3.

While this is not the forum for a detailed description of specific sensor chips manufactured by Biacore, a number of variations exist on the basic structure described which are designed for a range of specific application areas. Some of these specific application-related formats will be discussed later, where appropriate.

4.2.2.2 **Immobilization Chemistry**

As mentioned previously, ligands can be attached to the sensor chip using a variety of chemical methods, allowing a broad range of different types of biomolecules to be analyzed by SPR. The gold layer and dextran matrix on the sensor surface are stable under a wide range of conditions, including extremes of pH and many organic solvents. Once the ligand has been immobilized, the stability of the sensor surface is determined primarily by the stability of the attached ligand itself. The carboxymethylated dextran matrix on the sensor surface is amenable to a range of chemistries for ligand immobilization which exploit different groups on the ligand molecule. Immobilization approaches may be directed towards amine, carboxyl,

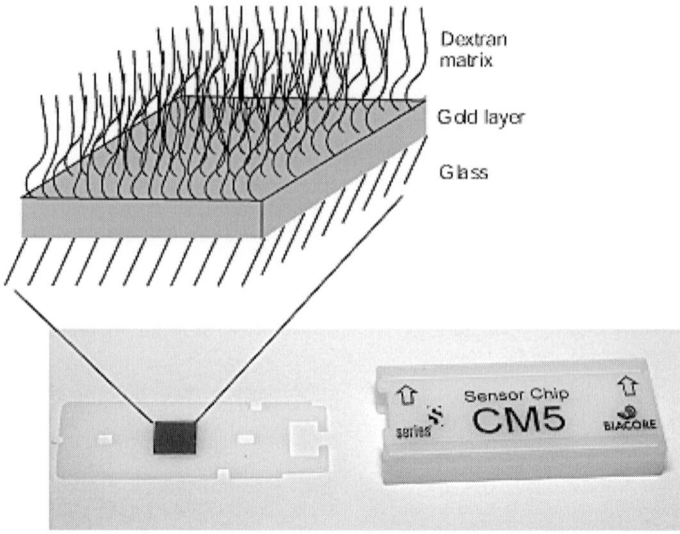

Figure 4.3 Schematic illustration of the structure of the sensor chip surface.

thiol or hydroxyl groups on the ligand, or may use specific tags introduced into the ligand by either chemical modification or recombinant techniques.

Amine coupling chemistry is the most widely applicable approach for attaching biomolecules covalently to the sensor surface. With this method, the dextran matrix on the sensor chip surface is first activated with a mixture of 1-ethyl-3-(3-dimethylaminopropyl) carbodiimide (EDC and N-hydroxysuccinimide (NHS) to give reactive succinimide esters [11]. The ligand is then passed over the surface and the esters react spontaneously with amino groups or other nucleophilic groups to link the ligand covalently to the dextran (see Figure 4.4).

Most proteins contain several amine groups so that efficient attachment can be achieved without seriously affecting the biological activity of the ligand. In some instances, however, amine coupling may involve groups at or near the active site or binding site of the ligand, with the result that attachment is accompanied by loss of activity. In such cases, the ligand can be attached using alternative coupling chemistry or a capturing approach.

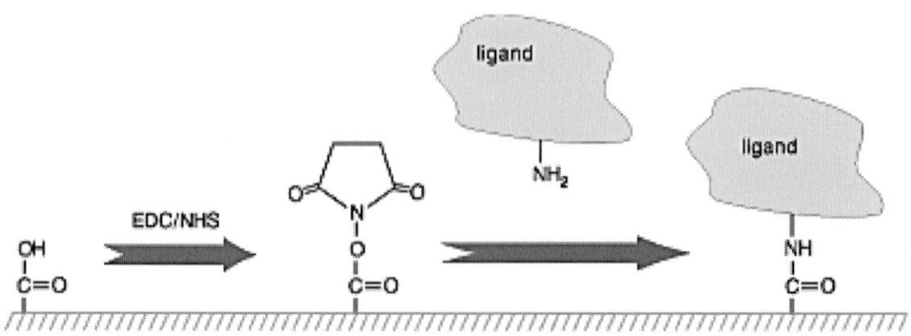

Figure 4.4 Amine coupling of ligand to the sensor chip surface.

Thiol coupling [12] utilizes exchange reactions between thiol and active disulfide groups. The active disulfide moiety may be introduced either on the dextran matrix (to exchange with a thiol group on the ligand, referred to as the *ligand thiol* approach) or on the ligand molecule (to exchange with a thiol group introduced on the dextran matrix, referred to as the *surface thiol* approach). The reagent 2-(2-pyridinyldithio) ethaneamine (PDEA) can be used for introducing active disulfide groups. The amine group in PDEA can be used to attach the molecule to activated carboxyl groups on either the surface or the ligand. The thiol approach can help to immobilize ligands in a defined orientation, since the number of potential attachment sites is often less than with amine coupling, and in many cases is reduced to one single site. Surface thiol coupling is also valuable for acidic proteins, since the substitution with PDEA raises the isoelectric point of the protein, improving the electrostatic interaction of ligand and surface matrix prior to coupling. The two different thiol coupling approaches are illustrated in Figure 4.5.

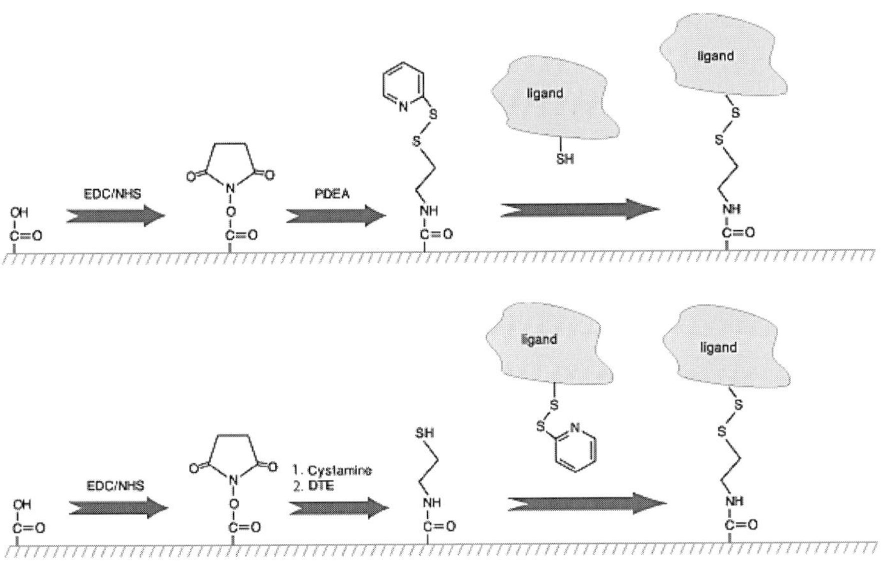

Figure 4.5 Ligand thiol (A) and surface thiol (B) coupling of the ligand to the sensor surface.

Aldehyde coupling is a third chemical immobilization method [12], and is useful for ligands containing aldehyde groups, which can be either native or introduced by oxidation of *cis*-diols. In this case, immobilization requires activation of the surface with hydrazine or carbohydrazide. Aldehyde coupling provides an alternative approach for immobilizing glycoproteins and other glycoconjugates. The method is particularly suitable for ligands containing sialic acid, since these residues are very easily oxidized to aldehydes. The chemistry of aldehyde coupling is summarized in Figure 4.6.

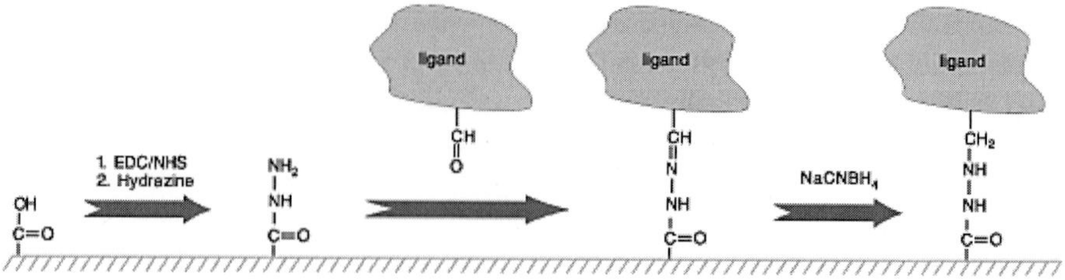

Figure 4.6 Aldehyde coupling of ligand to the sensor surface.

4.2.2.3 **General Capture Methods**

Immobilization of ligand by capturing involves high affinity binding of the ligand to an immobilized capturing molecule. In general, the regeneration step in the assay procedure removes ligand along with any remaining bound analyte, so that new ligand needs to be captured for each assay cycle. The capturing molecule is attached to the surface using one of the covalent chemical approaches described previously.

Capturing approaches can provide an alternative to covalent immobilization in situations where it is difficult to find chemical methods that give satisfactory results, or where the assay requires that the ligand on the surface can be changed. Capturing can also provide a ligand purification step, for example, when tagged recombinant ligands are captured from partially purified material through binding of the tag to a specific capturing molecule. The basic requirement for successful capturing is a robust high affinity interaction between the capturing molecule and the ligand. Monoclonal antibodies are frequently used as capturing agents.

4.2.2.4 **Specialized Capture Methods**

Although one of the major advantages of SPR technology is label-free detection of binding interactions, commonly used labeling techniques employed for purification or other purposes can also be taken advantage of for the purposes of ligand immobilization. Biotin-labeled molecules for example, can be captured directly onto sensor chips pre-immobilized with streptavidin [12]: while this is technically a form of capture, the affinity of the streptavidin–biotin interaction (10^{-15} M) makes this a permanent immobilization of the ligand in practice. This option is particularly useful when using nucleic acid ligands, but is also used for other applications, such as biotinylated liposomes [13] and biotinylated glycosaminoglycans such as heparin [14].

Histidine (his)-tagged recombinant proteins may also be conveniently captured onto a specialized sensor chip surface on which nitrilotriacetic acid (NTA) has been covalently linked to the carboxymethylated dextran matrix. This surface is activated by a simple pulse of $NiCl_2$, which forms a chelating complex with NTA that binds polyhistidine peptides [15]. This is a useful generic method for working with his-tagged proteins and the captured ligands are easily removed by a pulse of EDTA.

4.2.2.5 **Lipids and Membrane Proteins**

Hydrophobic molecules such as lipids present specialized requirements for SPR analysis, specifically in terms of sensor surface attachment, since they cannot be directly coupled to a standard carboxymethylated dextran surface for interaction studies with other molecules. This is an important issue, particularly since membrane-bound receptors are also a very important class of molecules for biomedical research and these are most appropriately studied in the context of a lipid environment. To facilitate these types of applications, specific sensor chips are available. These either have a flat hydrophobic surface, suitable for adsorbed lipid monolayers, or a dextran surface modified with lipophilic compounds, which enables liposomes to maintain a membrane-like lipid bilayer structure on the sensor surface. In addition to these specific sensor chips, several other inventive approaches to lipid-related applications have also been demonstrated [13,16–19].

Taken together, the range of available surface chemistry options enables the study of almost any conceivable ligand, from low molecular weight substances to whole cells. In the event of immobilization difficulties with a particular ligand, the option of using a capture molecule is also available (see Figure 4.1).

4.2.3
The Microfluidics System

The method of delivery of samples to the sensor surface is also a critical function and can vary considerably among the different commercial variants of SPR technology. The solution employed by Biacore involves a flexible microfluidics system based around an *integrated microfluidic cartridge,* or IFC [20]. This consists of a series of micro-channels and membrane valves encased in a plastic housing, and serves to control delivery of liquid to the sensor chip surface. Samples are transferred through a needle and into the IFC, which connects with the detector flow cells. These are formed by the sensor chip pressing against molded channels in the IFC. The precise construction of the IFC and the number of flow cells available varies amongst the different instrument series, but the common purpose is to allow analyte to pass over the sensor surface in a continuous, pulse-free and controlled flow that maintains a constant analyte concentration at the sensor surface during sample injection. An example of a microfluidics system is presented in Figure 4.7.

In the first Biacore instruments, samples were addressed in separate flow cells formed by the IFC. More recently, a technique of hydrodynamic addressing has been introduced whereby solutions are directed over different detection spots in a single flow cell, allowing more precise comparisons between spots. The construction is also designed to provide an almost instantaneous switch from sample to buffer, with minimum dispersion between the injections. These are important features of the technology, as they permit the SPR measurement of rapid events and enable, for example, very high-resolution kinetic studies to be undertaken. A more detailed discussion of the microfluidics system from the assay perspective will be taken up in Section 4.2.5.

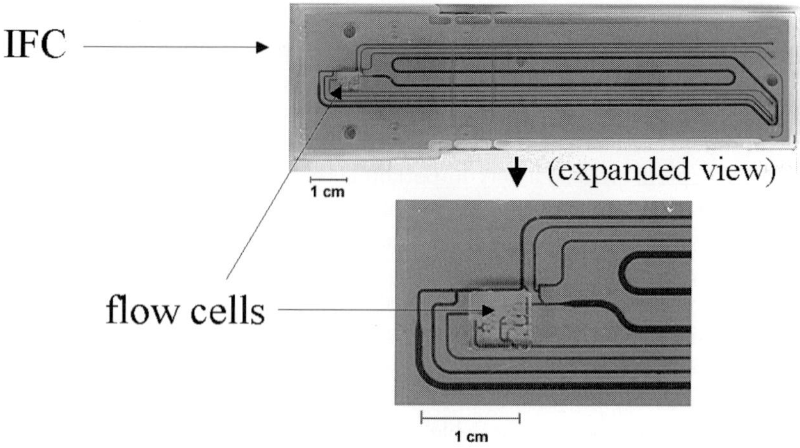

Figure 4.7 The microfluidics system and flow cells

4.2.4
Biacore Assay Basics

The definitions of basic assay formats and interaction partners are given at the beginning of Section 4.2, but it may also be useful to define some additional fundamental Biacore terminology at this point:

- The *resonance unit* (RU) is the unit of measurement of the SPR response and is directly proportional to the concentration of biomolecules on the sensor surface.

- The real-time SPR response is presented as a *sensorgram*, which plots response against time. Sensorgrams may be analyzed to provide information on the rates of interaction between molecules.

- *Report points* are sensorgram responses taken at specific times, averaged over a short time window, or taken as the slope of the sensorgram over a defined time window.

- *Regeneration* is the process of removing bound analyte from the surface after an analysis cycle, in preparation for a new cycle.

SPR studies, as with most advanced, high technology molecular techniques, require care and planning from the user in order to provide reliable, high quality results. Proper instrument maintenance, intelligent planning of assays and careful experimental technique all are important factors. As mentioned in several places within this chapter, Biacore systems are designed to enable all appropriate controls to be incorporated into assays. While care should always be exercised when carrying out assays, particular attention needs to be paid when carrying out affinity and kinetic assays. These approaches involve the simultaneous evaluation of multiple sensorgrams derived from concentration series of the analyte. It is very important here that the concentration dilutions are accurate and that the different solutions are carefully matched in order to calculate correct binding constants.

4.2.4.1 **Correlation of SPR Response with Surface Binding**

While the SPR response is proportional to the concentration of bound analyte at the sensor surface, changes in the refractive index caused by the bulk solution will also contribute to the measured response. In practice, the bulk response is usually irrelevant in terms of data evaluation, since the binding response can be processed to take account of this. The ***absolute response*** measures the total SPR signal (including bulk contribution) and is normally recorded for both ligand and reference spots. The *relative response* subtracts a specified baseline value from the total SPR signal obtained. Depending upon the particular system and assay set-up used, a *reference-subtracted response* corresponding to the difference between ligand and reference spots can also be presented. To a first approximation, this eliminates the contribution from the bulk solution and corresponds to the amount of ligand and analyte on the surface. A schematic sensorgram which indicates the bulk contribution effect is shown in Figure 4.8.

The precise relationship between SPR signal and the amount of surface-bound analyte depends to some degree on the nature of the molecule involved, but is linear for any given molecule over experimentally relevant ranges. Direct studies using various proteins, for example, have shown a linear response and established an approximate conversion factor of 1000 RU ng^{-1} protein mm^2, using a standard carboxymethylated dextran sensor chip [21]. Taking a matrix depth of 100 nm, a 1000-RU response approximates to an equivalent solution protein concentration of

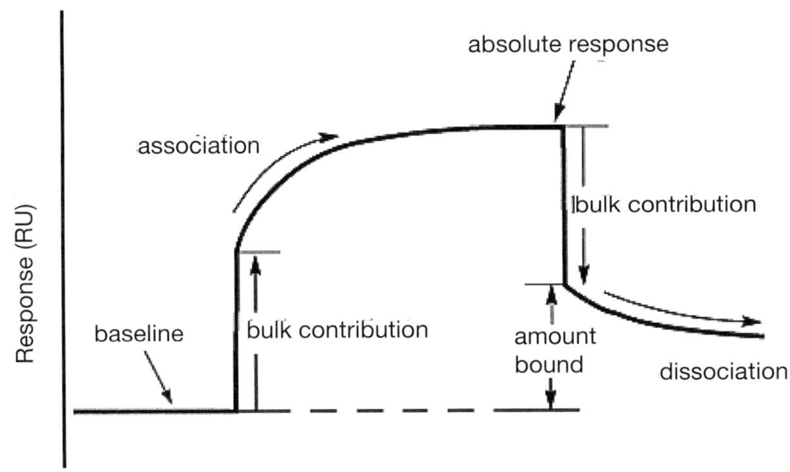

Figure 4.8 Schematic sensorgram, illustrating the contribution of the bulk refractive index to the SPR signal.

10 mg mL[1]. This relationship varies for other types of molecules and it is known for example that nucleic acids have a slightly higher specific refractive index increment than proteins. In terms of detection limits, these vary depending on the molecules involved and the assay format used (see below), but the specified limits quoted, based on antibody detection of myoglobin, vary between 50 pM to less than 1 pM depending on the specific type of Biacore instrument.

The sensorgram represents one of the key strengths of this technology, since it provides real-time, continuous monitoring throughout all phases of a biomolecular interaction. SPR responses are taken at a frequency of 1 to 10 s^{-1}, producing an information-rich data format than enables evaluation of complex interaction properties, but which also gives immediate visual feedback on each interaction. SPR responses at defined points during analysis cycles are taken from the sensorgram in the form of *report points* that are automatically assigned by the control software and/or assigned by the user.

4.2.4.2 **Assay Design: Assay Formats**

In addition to *direct binding assays* (whether directly to an immobilized or captured ligand), a number of other assay formats are possible. *Sandwich assays* (direct binding with enhancement) are an extension to the direct assay in which a secondary "enhancement molecule" is allowed to interact with the bound analyte in order to increase the level of the SPR response. This approach can also be used to confirm the identity of the analyte when working with complex mixtures for example, since two specific binding interactions must take place in order to see an enhanced response.

Inhibition assays (or solution competition assays) exploit the ability of the analyte to inhibit the binding of a high MW detecting molecule to the surface. Typically, the analyte or a derivative thereof is attached to the surface as the ligand, while the detecting molecule (added to the samples at a fixed, known concentration) is a macromolecule that binds specifically to the analyte. The mixture is incubated for a defined time, and then injected over the sensor surface to measure the remaining free detecting molecules. The amount of free detecting molecule is then inversely related to the concentration of analyte in the sample. Inhibition assays are most commonly used for low MW analytes. They may however, also be used as an alternative to direct binding level assays for macromolecular analytes, in situations where immobilization or regeneration of the detecting molecule on the sensor surface is not satisfactory.

In *surface competition assays*, a binding partner to the analyte is used as the ligand, and a high molecular weight analog to the analyte (typically analyte conjugated to a carrier protein) is added in constant amount to the samples to be measured. The basis of the assay is competition between analyte and the high MW analog for binding to the ligand. The measured response is the sum of the contributions from the analyte and the high MW molecule: as with the solution competition approach, the response is inversely related to the amount of analyte in the sample. The surface competition approach can have advantages over the more common inhibition assay format in situations where immobilization of the analyte on the sensor chip surface presents problems. The four basic assay strategies are shown in Figure 4.9. The sandwich and two competition-format assays are used widely for making concentration measurements, whereas affinity and kinetic measurements principally employ direct binding assays.

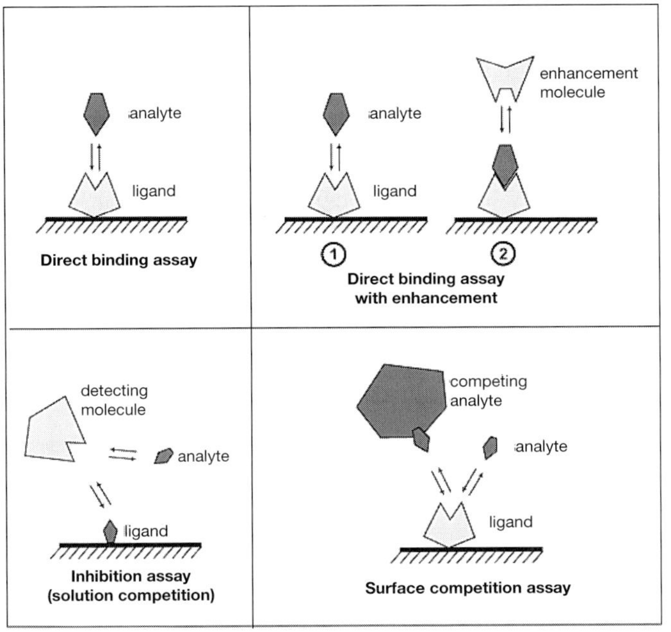

Figure 4.9 Schematic illustration of the four principle Biacore assay formats.

4.2.4.3 Assay Design: Surface Preparation

The range of surface coupling chemistry approaches supported by Biacore gives the user great freedom in designing an assay and choosing which interaction partner to immobilize on the sensor surface. In many cases, this choice can be made purely on the basis of experimental efficiency and convenience. In an ideal situation, the ligand and analyte identities of the biomolecules can be assayed in reciprocal experiments, confirming that surface coupling does not affect the observed binding behavior. As described in Section 4.2.2, certain types of ligand may necessitate using specific sensor chips, although capture-type approaches can always be considered. However, it may be preferable to use certain molecules as ligand or analyte specifically, such as in the case of RNA molecules, which are generally more likely to be used as ligands. Their strong negative charge can produce electrostatic repulsion effects with the surface matrix at physiological pH if used as analyte which might of course affect apparent binding and kinetic properties [22]. It is often convenient in any case to capture nucleic acid ligands onto a streptavidin surface, given the frequent and convenient use of 5' biotin labeling of oligonucleotides.

It is also worth remembering that the SPR signal is related to the mass concentration of bound molecule on the surface, so that the molecular weight (MW) of the interaction partners may also be an important factor. For a given number of interaction events, the SPR response is proportional to the MW of the analyte. Conversely, for a given level of ligand immobilization, a higher MW ligand will provide proportionately fewer binding sites. These issues are normally expressed in terms of R_{max}, which defines the theoretical maximum binding capacity of the surface for a defined pair of ligand and analyte molecules:

$$R_{max} = \frac{MW_{analyte}}{MW_{ligand}} \times R_{ligand} \times n \qquad (1)$$

where: R_{max} is the maximum analyte binding capacity in RU; $MW_{analyte}$ and MW_{ligand} are the molecular weights of analyte and ligand respectively; R_{ligand} is the amount of immobilized ligand in RU and n is the stoichiometry of analyte binding to ligand.

For example, with ligand molecular weight = 50,000 Da, analyte molecular weight = 10,000 Da and ligand immobilized level = 10,000 RU, the theoretical maximum binding capacity assuming 1 : 1 binding of analyte to ligand is given by

$$R_{max} = \frac{10{,}000}{50{,}000} \times 10{,}000 = 2{,}000 \text{ RU} \qquad (2)$$

In addition to aiming for immobilization levels in relation to the expected concentration and MW of the analyte, the optimal R_{max} can vary considerably depending on the particular assay. Concentration assays, for example often use a relatively high R_{max} of several hundred to several thousand RU, whereas for kinetic rate constant determinations, much lower R_{max} values of the order of 10–50 RU are recommended (reviewed in [23]).

4.2.4.4 **Regeneration**

The process of removing bound analyte from the sensor chip surface after analysis of a sample in preparation for the next analysis cycle (regeneration) is another important feature of Biacore's approach to SPR. This enables sensor surfaces to be prepared with immobilized ligand and re-used for multiple rounds of analysis. The number of times a sensor surface can be regenerated depends on the nature of the attached ligand, but normally ranges from fifty to several hundred. When the ligand is attached directly to the surface, the aim of regeneration is to remove the analyte from the ligand without destroying the ligand activity. When a capturing approach is used, regeneration generally removes both ligand and analyte from the capturing molecule, and the stability of the ligand under regeneration conditions is irrelevant. The optimal regeneration solution varies depending on the nature of the ligand–analyte interaction and the stability of the two molecules, but can often be achieved by simple exposure to acidic (dilute HCl or glycine-HCl) or basic (NaOH) solutions. Many other alternatives can be used, however, and it is possible to use sequential exposure to two different regeneration solutions and to vary the contact time for the solutions on the sensor surface.

Efficient regeneration is important for successful assays. Incomplete regeneration or loss of the binding activity from the surface impairs the performance of the assay and the useful lifetime of the sensor chip can be shortened. Sub-optimal regeneration can be tolerated in some simple types of yes/no binding assay, since a certain loss of ligand activity can be adjusted for using suitable controls. The demands on efficient regeneration are much stricter in the case of affinity or kinetic assays however, since it is important to maintain a constant surface activity between cycles within the concentration series for each sample.

4.2.5 Theoretical and Practical Implications of Technology Design for Biacore Assays

The technological approaches taken by Biacore are designed to optimize SPR analysis of biomolecular interactions. Some of the consequences of these technology implementations have been referred to in the previous sections, but attention will now be focused on the more detailed theoretical and practical implications for designing and running SPR assays with Biacore. David Myszka has also written a review that covers many of these issues [23].

4.2.5.1 Label-free Detection

The label-free detection principle that forms the basis of SPR technology represents one of its strongest advantages over more traditional methods. In addition to being inconvenient and time consuming, labeling procedures may also have adverse effects on the behavior of the labeled molecule, by changing the properties of, or occluding the binding site for example. The label itself can also introduce complications: fluorophores are usually hydrophobic and are frequently associated with background problems for example, and radiolabeling suffers from drawbacks connected with safety, isotopic decay and radiolysis-induced instability of labeled compounds. Labeling techniques are of course more sensitive in certain instances, but this must be weighed against the disadvantages of time-consuming and laborious procedures and high sample consumption.

One important aspect of label-free SPR technology however, is the generic nature of the detection principle. The SPR response from any given analysis is derived from changes in the total mass concentration at the sensor surface and does not distinguish between potentially different molecular identities. This raises an important issue with regard to specificity and experimental design. Non-specific binding of analytes to the native sensor surface can be easily controlled for by incorporating a "blank" flow cell in the analysis. This is rarely a problem however, and the most likely source of non-specific binding problems lies with the ligand chosen. For this reason, it is important to use very pure preparations of the intended ligand and to ensure that it is a sufficiently specific molecule for the intended assay, since this will largely determine the chances of success. Biacore's ongoing development of combining SPR with mass spectrometry (MS, see Section 4.4.1) represent an interesting approach to provide molecular identification along with binding information. This promises to be particularly useful in the case of *ligand fishing-type* experiments, where potentially unknown binding partners may be involved.

4.2.5.2 The Optical Detection System and Bulk Effects

As mentioned previously, the optical detection system in Biacore will detect all surface changes in refractive index, including those contributed by different bulk solutions. In most situations, using relative-responses and careful reference-subtraction will take care of these effects. Particular care is needed however, when working in conditions of low analyte responses, high surface ligand density and organic solvents with large bulk refractive index effects. These conditions most commonly arise when working with low molecular weight drug compounds, which often require 1–10 % dimethylsulphoxide (DMSO) in order to maintain solubility. Under these conditions, the excluded volume effect between ligand-bound and blank reference surfaces (allowing more of the solvent to access to the blank surface) can produce bulk subtraction errors that are of the same order of magnitude as the expected analyte responses. This requires great care when preparing samples, since small differences in DMSO concentration will produce a significant bulk response difference. Fortunately, this complication can be overcome by normalizing responses against a solvent correction curve prepared by running a concentration series of DMSO-containing buffer solutions over the ligand and reference surfaces [24].

4.2.5.3 Surface-based Versus Solution-based Analysis

SPR, as the name implies, is an intrinsically surface-based technique. Despite the many advantages that this provides, potential effects on molecular behavior caused by coupling one of the interaction partners to the sensor surface *per se* must also be considered. The possibility always exists for example, that chemical coupling of a ligand to the sensor surface may interfere with a critical analyte-binding region (labeling procedures also carry an analogous risk, as discussed previously). This possibility can usually be tested for by carrying out an activity test on the immobilized sensor surface, using a positive control sample. In the unlikely event of such problems occurring, the range of coupling chemistry options available provides alternative immobilization strategies, as well as the possibility of reversing the experimental design and using the molecule in question as the analyte.

Another question that has been raised is whether meaningful binding data can really be addressed by any surface-based method. The argument is that in contrast to solution-based assays, the restrictions imposed upon the interaction partner which is immobilized onto a surface will produce experimental artifacts. This is an over-simplification of course, and interactions in simple solution assays are unlikely to precisely reflect many biomolecular interactions that occur *in vivo*. Furthermore, the carboxymethyl dextran hydrogel on Biacore sensor chips presents a flexible matrix for ligand binding and probably allows a considerable degree of molecular movement [10]. Formal proof that surface-based methods are in fact highly appropriate for biomolecular interaction studies has come from studies designed to directly compare solution-based techniques with SPR. These have demonstrated that affinity, kinetic and thermodynamic constants obtained using Biacore are extremely consistent with those obtained using a range of solution-based techniques [25].

4.2.5.4 The Microfluidic Design and Mass Transport Effects

The liquid flow system in Biacore's approach to SPR lies at the heart of the design philosophy and contributes greatly to its powerful kinetics capabilities. As with all surface-based analysis methods however, there is one complicating factor that needs to be considered for both system and assay design, namely the issue of mass transport [26]. For the analyte to be able to bind to the sensor surface, the molecules must be transported from the bulk solution to the surface. This is a diffusion-controlled process and under the laminar flow conditions that exist in the IFC, the transport rate is proportional to the analyte concentration in the bulk solution, with a proportionality constant called the mass transport coefficient (k_m). This constant is proportional to the cube root of the liquid flow rate and depends on flow cell dimensions and the diffusion properties of the analyte in the sample solution [27].

If mass transport is much faster than the association of the analyte with the ligand, the observed binding will be determined by the interaction constants. Conversely, if mass transport is much slower than association, the binding will be wholly limited by the mass transport process and kinetic data for the interaction will not be obtained. In practice, many experimental systems will actually lie between these extremes, and the binding will be determined by contributions from both mass transport and interaction kinetics. This may be represented in a reaction scheme as:

$$A_{bulk} \underset{k_m}{\overset{k_m}{\rightleftharpoons}} A_{surface} + B \underset{k_d}{\overset{k_a}{\rightleftharpoons}} AB \tag{3}$$

where k_m is the rate constant for mass transport to and from the surface and k_a and k_d are the rate constants for the formation and dissociation of the complex.

It is important to understand the experimental parameters that determine whether the association rate is limited by mass transport or interaction. For a given analyte and flow cell dimensions, only two factors are significant:

1. The flow rate, which affects the mass transport rate constant km but has no effect on the interaction kinetics. Increasing the flow rate increases km, so that mass transport is less likely to be limiting. This effect is however relatively small, since km is proportional to the cube root of the flow rate.
2. The concentration of surface binding sites, which affects the absolute interaction rate but has no effect on mass transport. This is the most important parameter. Increasing the surface binding capacity increases the initial interaction rate, so that mass transport is more likely to be limiting.

As an interaction proceeds, the concentration of available surface binding sites falls and interactions which are limited by mass transport in the initial phase can therefore, become interaction-limited as the amount of free ligand sites decreases.

The mass transport rate is also influenced by the diffusion coefficient of the analyte, which in turn depends on the molecular weight. A large analyte has a lower diffusion coefficient, and will therefore be more subject to mass transport limitations than a small analyte with the same kinetic parameters. The very same parameters that govern the influence of mass transport during the association phase also apply during dissociation: released analyte can either diffuse into the bulk buffer flow and be removed, or may in theory re-bind to the ligand. In this context too, a high flow rate and low surface binding capacity will minimize any such effects.

It is important to realize that mass transport is not a "problem" in SPR analysis, but a physical property of the system which can be readily accounted for in data interpretation [28] and which can also be manipulated to some advantage during assay design. When carrying out concentration measurements, it can be desirable to perform assays under conditions where the interaction is entirely controlled by mass transport [29], since under such conditions, the binding rate is entirely dependent on the analyte concentration. This makes the assay independent of affinity effects and therefore more robust for analysis of multiple analytes. These conditions are best achieved by using low flow rates and immobilizing the ligand at a high surface-binding capacity.

For assays designed to measure kinetic rate constants on the other hand, the requirements are reversed. If the rate of mass transport is much faster than the association of analyte with ligand, the observed binding will be determined entirely by the interaction constants. The optimal conditions to achieve this are a combination of high flow rate and a low surface binding capacity [23]. The height of the flow cell also affects the phenomenon [20] and some Biacore systems are designed with a lower flow cell that helps to improve mass transport. Given that in practice assays can be governed by a combination of interaction constants and mass transport rate, the fitting algorithms used to determine rate constants from sensorgram data are designed to incorporate a mass transport component that ensures correct analysis [30]. In the extreme case of a completely mass transport-limited interaction however, the sensorgrams will not contain any kinetic information and the calculated rate constants will be meaningless. Fortunately, software tools are now designed to deal with this potential problem and can identify binding interactions that are entirely controlled by mass transport.

4.3 Biacore Applications in Basic and Clinical Research

4.3.1 Introduction

The publication of around 400 papers each year covering a wide range of biological and clinical research fields is testament to the versatility of Biacore's SPR technology, with examples of quantitative analyses of biomolecular interactions to be found in virtually all disciplines within the life sciences. New innovative applications are continually appearing in the literature and demonstrate that the technology has the potential to replace and often improve on the speed and data quality of established techniques. This section will review a selection of recently published applications with particular emphasis on protein : protein interactions and protein : nucleic acid interactions in cancer research and neuroscience, two areas within which Biacore has become firmly established in recent years.

4.3.2 Applications of Biacore in Cancer Research

The design of targeted therapeutic reagents to cellular signaling events that initiate oncogenic transformation or maintain a tumorigenic phenotype in a cell requires that the specificity and kinetics of the molecular interactions in the signaling cascades are accurately determined. Possibly the most accessible interactions against which intervention may be directed are those that occur at the cell surface between growth factors and constitutively active receptors. In one approach, peptide mimetics were designed based on a functionally inhibitory antibody to HER2, an oncogenic member of the EGF family of tyrosine kinase receptors which is commonly expressed in human breast, ovarian and colon cancer and which is associated with a poor prognosis [31]. A Biacore assay was used to screen and select peptide mimetics from a range of candidates that bound with high affinity and with favorable kinetics, most importantly those with slow dissociation rates (Figure 4.10).

Those with the slowest dissociation rates were also most effective at inducing functions such as reduced tyrosine phosphorylation of downstream signaling molecules in cell-based assays. Most encouragingly, cells treated with these mimetics were hypersensitive to radiation-induced apoptosis suggesting that the mimetics may be used clinically in conjunction with radiotherapy. Simultaneous competition studies using Biacore showed that the mimetics and the parent antibody were directed against the same epitopes on HER2 [32]. These studies also suggested that the receptor was disabled because the mimetics inhibited receptor dimerization [33], a prerequisite for the propagation of signals into the cytoplasm following engagement with a cognate ligand. A strategy of structure-based targeting of the dimerization interfaces of tyrosine kinase receptors may therefore be generally applicable to the design of antagonists of oncogenically transformed tyrosine kinase receptors. It is important to note that in this study, although the affinity of the mimetic for HER2 was lower than the parent antibody, the dissociation rate constant, kd, was similar, indicating that the complex was stable. Potential therapeutic reagents may therefore be missed during a screening process that relies on affinity data alone and both association and dissociation rate constants (ka and kd, respectively) may reveal effective binders (see [8] and [7] for detailed studies on the treatment of ka and kd data in Biacore assays to select effective binders from a large base of candidate drugs).

Although growth factors generally bind with high affinity to one favored cellular receptor, they may nevertheless also bind alternative receptors. Platelet-derived growth factor (PDGF), for example, exists both as a homodimer (AA or BB) and a

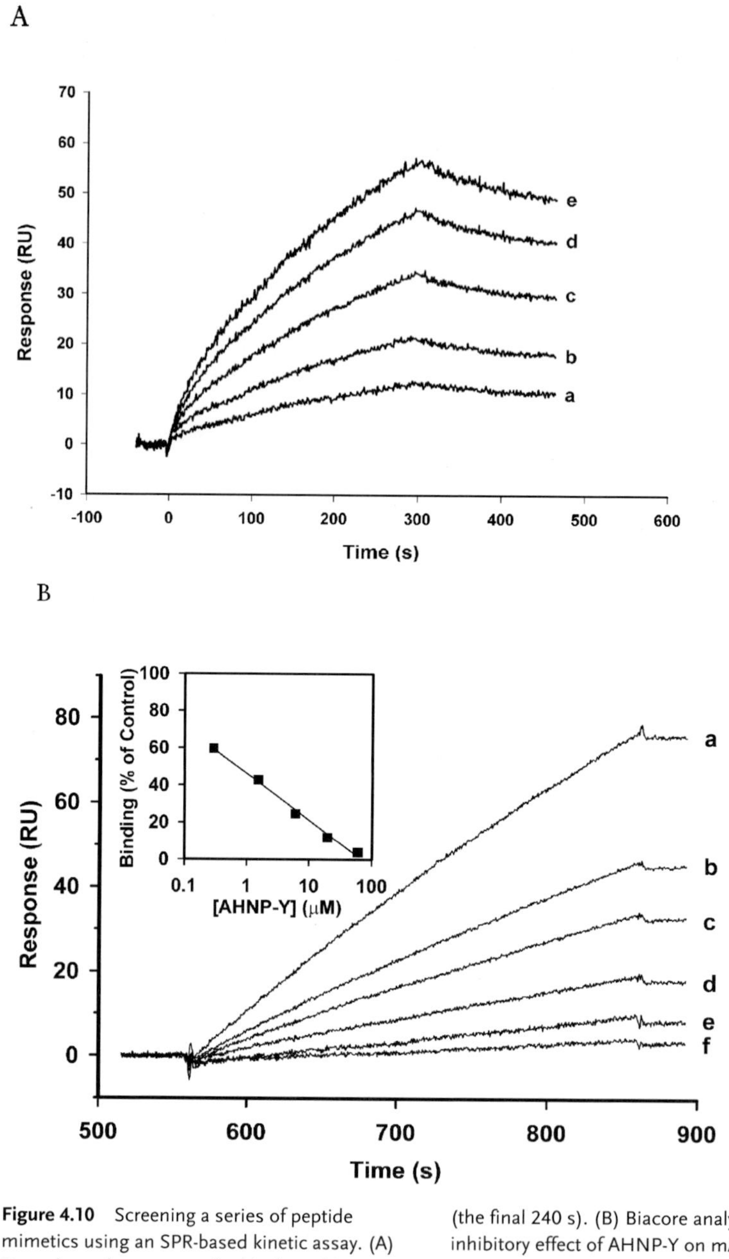

Figure 4.10 Screening a series of peptide mimetics using an SPR-based kinetic assay. (A) Dose dependence curves for binding of AHNP to the immobilized HER2 receptor. AHNP-Y was injected at concentrations of 0.5 (a), 1 (b), 2 (c), 4 (d) and 8 mM (e) at a flow rate of 20 μl min. The sensorgrams show binding of AHNP-Y to the immobilized HER2 (the first 300 s) followed by peptide dissociation from the receptor surface (the final 240 s). (B) Biacore analysis of the inhibitory effect of AHNP-Y on mAb 4D5 binding to immobilized HER2. Sensorgrams show binding of 1 nM mAb 4D5 to HER2 after pre-injection of 0 (a), 0.3 (b), 1.5 (c), 6 (d), 20 (e), and 60 mM (f) AHNP-Y. Inset: the correlation between the initial rate of 4D5 binding to HER2 and concentration of pre-injected AHNP-Y.

heterodimer (AB). While PDGF-AA binds only the PDGF-α receptor, PDGF-BB and PDGF-AB can bind both the PDGF-α and -β receptors. In addition, it was recently shown in a Biacore study that the pleiotropic scope of PDGF-BB is wider still and binds low-density lipoprotein receptor-related protein [34]. It does not always follow, however, that the functional consequences of binding alternative receptors are easily differentiated. The functions of two different cell membrane receptors, lymphotoxin-β receptor (LTβR) and herpes virus entry mediator (HveA), both of which bind a common ligand, have now been defined using Biacore. Peptides bearing selected point mutations were generated from LIGHT, a pro-apoptotic TNF ligand superfamily member. Mutants were selected that bound one, but not the other binding partner [35]. One selected mutant that bound HveA, but not LTβR in the Biacore assay, failed to induce apoptosis in target cells. The cell-based assays showed that a dominant negative mutant of TRAF3, an intracellular protein that binds the cytosolic domain of LTβR and transmits signals leading to cell death, blocked LIGHT-mediated apoptosis. These functional cell-based assays confirmed the Biacore binding data, indicating that LTβR but not HveA propagates LIGHT-mediated apoptosis.

Although mouse monoclonal antibodies have been raised to many human tumor-specific markers [36–39] their use in cancer treatment is greatly limited by the fact that the patient inevitably generates a potentially harmful immune response. Attempts to address these limitations have been made by the generation of single-chain Fv (scFv) molecules that retain specificity without presenting the highly antigenic constant regions of the whole antibody molecule to the host. In one variation of this strategy, a fully human, gene-fused bi-specific antibody (bs-Ab) was created with specificity both to IgG Fc receptors (CD16) on the surface of cytotoxic immune cells and to HER2 on tumor cells [40]. The hypothesis was that this bs-Ab might function as a bridge and actively recruit host cytotoxic immune cells to tumor cells. In separate Biacore experiments in which either CD16 or HER2 were immobilized on a sensor chip it was demonstrated that the gene fusion process did not greatly alter the binding kinetics of either of the component scFvs in the bs-Ab. Confirmation of "conservation-of-function" after conjugation is an important stage in the development of any carrier system in which an active molecule is conjugated to a carrier. The function of the bs-Ab was also tested *in vitro*, where it increased lysis of human ovarian cancer cells which over-expressed HER2 when presented together with peripheral blood lymphocytes.

While whole monoclonal antibodies are potentially useful reagents for diagnostic applications, human antibody fragments obtained via phage antibody technology are a more attractive alternative for human cancer therapy since they might be expected to penetrate tumors more efficiently than whole antibodies [41–43]. Fibronectin is a ubiquitous extracellular matrix protein, one variant of which contains a novel splice-generated domain (ED-B). This isoform is found in the stroma of fetal and neoplastic tissues and in the walls of developing neoplastic blood vessels, but is absent from normal adult tissues. Reasoning that this protein is a suitably generic tumor-specific target, a number of single chain antibody fragments and dimeric variants to ED-B were generated and an optimal binding fragment was selected for clinical trials [44]. Biacore's SPR technology was used to screen antibody fragment libraries and to compare detailed binding properties to immobilized ED-B. The affinity dissociation constants (K_D) of the antibody fragments for ED-B varied widely but those that bound ED-B with higher affinity than the parent fragment were shown in most cases to do so as a result of slower dissociation rates. When fragments were assessed for tumor targeting *in vivo* using real-time photodetection of fluorescently-labeled antibodies, the efficiency of tumor homing correlated closely with the slower off-rates of ED-B binding detected in the Biacore assays. Biacore's SPR technology was therefore used to select optimal antibody fragments for diagnostic imaging and may therefore be useful in the preclinical stage of selecting reagents for therapeutic strategies targeted at the tumor vasculature.

Cell surface receptors bind to their cognate ligands and to recruited intracellular second messengers at a hydrophobic interface: the external and internal faces of the plasma membrane, respectively. Proteins assume markedly different structures in hydrophobic and aqueous microenvironments. Even if the transmembrane domain of the receptor is the only part of the molecule that sits directly within a lipid bilayer, structural disturbances in this region of the protein may alter the protruding intra- and extracellular domains. It is thus reasonable to assume that interactions between proteins may yield a very different picture depending on the microenvironment in which the interaction is studied. A lipid microenvironment may be vital to allow immobilized receptors to assume a configuration that reasonably mimics that found *in situ* and thereby present a functional binding interface for interacting molecules.

For example, in order to interact with the serine/threonine kinase, *Raf*, the protooncogene, *Ras*, must insert into a biological membrane to transduce signals from activated receptor tyrosine kinases to the intracellular MAP-kinase pathway (Figure 4.11).

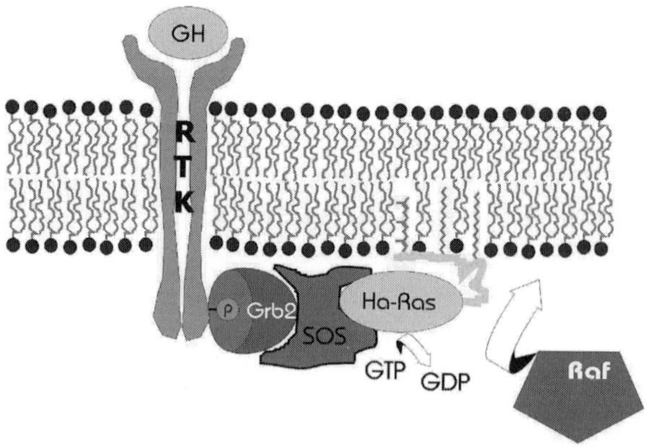

Figure 4.11 Signal transduction occurs in the plasma membrane. Ras is anchored in the plasma membrane via one farnesyl and one or two palmitoyl modifications at its C-terminus, where activation by GEF-proteins (here Sos) and recruitment of downstream effectors such as Raf-kinase occurs. (see Colour Plate p. XXI).

In order to be able to make kinetic measurements of the interaction between *Raf* and *Ras*, the plasma membrane was thus simulated *in vitro* on a hydrophobic sensor chip by loading the chip with liposomes prior to molecular interaction analysis [45]. Biacore experiments showed that GTP-bound *Ras* bound specifically to *Raf* only if *Ras* was immobilized on the liposome-loaded chip by prior fusion with a synthetic lipid tail (Figure 4.12). In functional assays, cells that were microinjected with nominally oncogenic truncated *Ras* without a lipid tail did not transform into tumor cells. These cells, in contrast and in full agreement with the Biacore data, only transformed if truncated *Ras* was co-injected along with a lipid anchor.

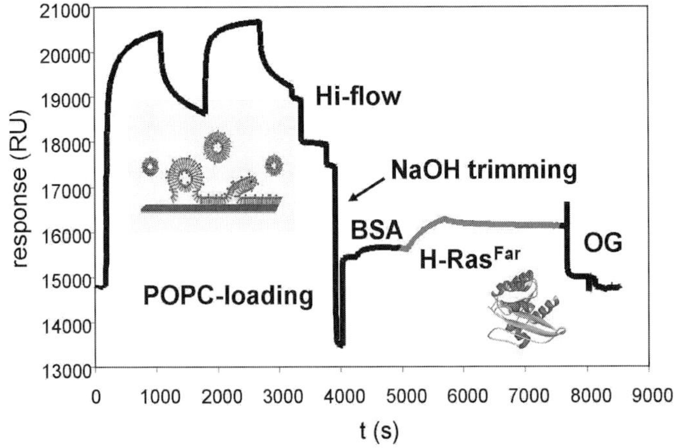

Figure 4.12 Insertion of lipid-modified Ras protein into an artificial membrane on Sensor Chip HPA. Sensor Chip HPA was treated with POPC (palmitoyl oleoyl phosphatidyl choline)-containing lipid vesicles and conditioned by treatment with NaOH and bovine serum albumin at high flow rates. Farnesylated Ras is inserted into the artificial membrane, while all non-covalently attached molecules are easily displaced from the lipid surface by the detergent octylglycoside. (see Colour Plate p. XXI).

To be able to grow beyond a volume of about 1 mm^3, a tumor, once initiated, requires its own microvasculature. Inhibition of angiogenesis, or the process of neovascularization, is an attractive strategy in the intervention of tumor spreading. Members of the vascular endothelial cell growth factor (VEGF) family are principal effectors of angiogenesis as they influence the division and migration of endothelial cells by activating cognate transmembrane receptors, among them VEGFR-1 and VEGFR-2. Recombinant, soluble variants of these receptors that lack transmembrane domains can be used to sequester VEGF in solution and block binding to native cell-bound receptors. However, while cell-based approaches indicate that heparin is necessary to facilitate the functional presentation of VEGF to the receptors, cell-free assays suggest that this is not the case for the isolated soluble receptors. Kinetic analysis using Biacore showed that heparin actually inhibited the interaction with soluble receptors [46] and that this was due to a reduction in association rate. In this way, Biacore analysis therefore enabled baseline parameters to be established for the binding of soluble VEGFR-2 to VEGF and to eliminate the potential problem of a heparin requirement in therapeutic applications.

Intervention in the deregulated cell cycle is another potential therapeutic anti-cancer strategy. Progression from the G1 (gap) phase to the S phase (DNA synthesis) of the cell division cycle is regulated by a family of cyclin-dependent protein kinases (CDKs) which are regulated in their turn by many other proteins including members of the cyclin family. In one study, binding between PCNA (proliferating cell nuclear antigen) and mutants of p21, a CDK inhibitor, was measured using both a yeast two-hybrid assay and Biacore [47]. Although these two approaches gave similar qualitative data, only Biacore provided information on binding affinity and further showed that trimerization of PCNA is a prerequisite for p21 binding. The technology was also used to screen for short peptides that inhibited the p21–PCNA interaction in competition assays. One p21-like peptide was found that abolished this interaction beyond the level of detection.

Cyclin G is regulated by p53 in response to growth factor stimulation and its over-expression has been shown to inhibit the proliferation of human sarcoma cells. Although this implies a role for cyclins in cell cycle control, the physiological functions of cyclin G have not been characterized. A cDNA clone encoding a potential cyclin G binding partner has been isolated [48]. This novel serine/threonine protein kinase was named cyclin G-associated kinase (GAK). Immunoprecipitation/Western blotting experiments corroborated GAK binding to cyclin G and indicated that both proteins interact weakly with CDK2. Since such cell-based experiments do not rule out indirect association of proteins within a complex, Biacore's SPR technology was used to confirm that GAK binds directly to the N-terminus of cyclin G. The kinetic data provided by the Biacore analysis also showed that this association was stable. When the weak interaction between cyclin G and CDK2 suggested by immunoprecipitation was also investigated by SPR using recombinant proteins, no binding was detected, suggesting that any *in vivo* association between these two proteins either requires post-translational modifications, or that it is indirect. Biacore's SPR technology therefore identified the primary molecular targets of cyclin G, providing the basis for future investigations into this system and its role in cancer.

Cell cycle arrest and apoptosis are often demonstrably co-regulated. For example, survivin, a factor expressed during the G2/M phase of the cell cycle, co-localizes in centrosomes with the apoptotic protein, caspase-3. Although available data suggest that survivin allows cells to progress from G2 to M by binding caspase-3 and thus diverting cells from apoptosis, this has been difficult to prove. Biacore was applied to find out if survivin binds directly to caspase proteins and if this inhibits caspase function [49]. Survivin was immobilized on the sensor chip and was shown to bind caspase-3 and caspase-7 with high affinity. In order to try to explain the discrepancies between Shin *et al.'s* caspase inhibition results and previous studies that failed to demonstrate binding, the authors noted that whereas their own experiments were carried out with all components pre-warmed to 25 °C, the survivin used in other studies was kept on ice until immediately before the assay. Using the temperature control available with the instrument, it was demonstrated that compared to experiments carried out at 25 °C, the affinity of survivin for caspase-3 was reduced 100-fold at 4 °C and may explain why the caspase : survivin interaction remained undetected in earlier assays.

4.3.3 Applications of Biacore in Clinical Research

Possibly the most attractive advantage of Biacore as a means of assessing binding partners in clinical research is the lack of a need for sample preparation or clean-up prior to running the experiment. The following examples are from reports in the literature on the analysis of clinical samples such as tissue homogenates, -plasma and urine.

During the preclinical stage of the development of potentially therapeutic antibodies, promising alternatives are tested for their affinity to the target [50,51]. In one study, chimeric and humanized murine antibodies to CD44, a marker for head and neck cancer, were affinity-ranked using Biacore and then assessed for biodistribution and efficacy in mice bearing head-and-neck squamous cell carcinoma xenografts [52]. Based on both these sets of data, the important finding was made that lower affinity antibodies to CD44 were superior to higher-affinity versions in their ability to target tumors. While tumor uptake of low affinity binders was up to 50 % higher than that of their high affinity counterparts, blood levels and uptake into non-cancerous tissue was unaffected.

Biacore has recently made its debut in the analysis of adverse events arising from the use of humanized antibodies in phase I clinical trials [53]. The monoclonal antibody, A33 recognizes a 43-kDa cell surface glycoprotein that is expressed in cancerous human colonic epithelium and which is absent from most normal tissues. A rapid and reliable means is needed to detect and quantify any host immune response to the humanized antibody which could possibly be harmful to the patient. A Biacore assay was used in which human immunoglobulin was immobilized on a sensor chip and whole patient serum was used as the analyte to detect any potentially harmful anti-human antibody response. This study showed that Biacore is sufficiently robust and consistent between sample runs to be used as an on-site method in clinical trials.

The detection and measurement of antibody markers in human plasma is a common prognostic and diagnostic means of patient evaluation, as well as in monitoring clinical changes in response to therapy. A Biacore assay has been used to detect antibodies to gangliosides found in patients with neuropathies suspected of having an autoimmune etiology [54]. By immobilizing gangliosides on a sensor chip, crude diluted patient sera could then be used to detect the presence of anti-ganglioside autoantibodies. The assay was faster than ELISA partly due to the lack of sample preparation and further, had a superior limit of detection that could be even further improved through signal amplification by using isotype-specific antibodies after serum injection (Figure 4.13).

Several further examples exist of Biacore's SPR technology in clinical evaluation including the detection of fucosylation of α1-Acid Glycoprotein (AGP), an acute

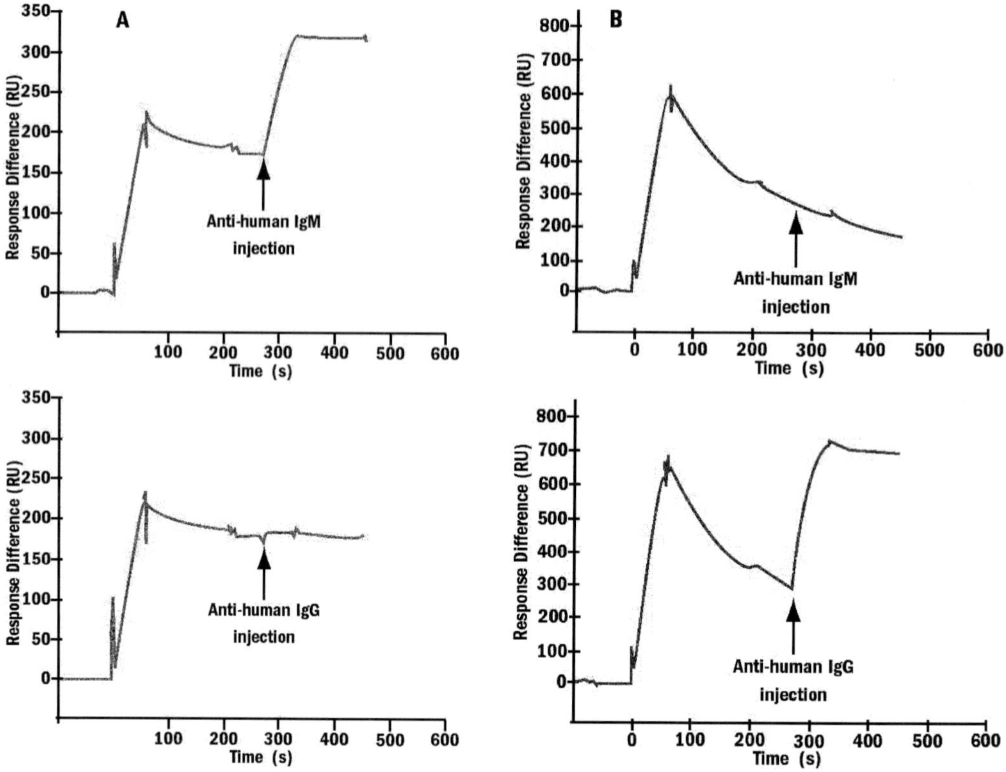

Figure 4.13 Use of secondary antibodies for identification of the anti-GM1 autoantibody isotype. (A) Analysis of a serum sample from a patient with multifocal motor neuropathy with anti-IgM and anti-IgG antibodies, indicates that the antibodies in the serum are IgM. (B) A similar analysis of a sample from a patient with Guillain–Barré syndrome indicates the presence of IgG antibodies. (Reprinted with the permission of Dr Armin Alaedini, Department of Neurology & Neuroscience, Cornell University).

phase event in inflammation. Antibodies in this case were firstly immobilized on a sensor chip to sequester AGP directly from human serum and the level of fucosylation was determined by the injection of a fucose-binding lectin [55]. The technique was used to follow the inflammatory status of severe burns patients.

Other examples of the measurement of autoantibodies directly from patient plasma with little sample preparation include the detection of anti-DNA antibodies [56] and anti-protein S antibodies [57] in systemic lupus erythematosis and anti-factor XII antibodies in patients with anti-phospholipid syndrome [58].

Biacore in combination with mass spectrometry (SPR-MS) has been shown in a proof-of-principle study to be applicable to the clinical situation [59]. Human urine or plasma were spiked with physiological levels of cystatin C, β2-microglobulin, urinary protein-1 and retinol-binding protein and injected over specific antibodies immobilized on a sensor chip. The captured analytes were then digested and identified by mass spectrometry. SPR-MS was used to detect simultaneously β2-

microglobulin and cystatin C in a single spiked plasma sample by injecting over a chip with different antibodies immobilized in two different flow cells. Additionally the method was able to detect protein complexes such as that between retinol-binding protein (RBP) and transthyretin using an anti-RBP antibody. A mass spectrum from the chip surface showed traces of both species enabling detection of *in vivo* assembled protein complexes using SPR-MS.

4.3.4 Applications of Biacore in Neuroscience

Alzheimer's disease is associated with the deposition of polymerized amyloid β-peptide (Aβ) protein in lesions in the brain parenchyma. However, Aβ protein is expressed by almost all nucleated cells and is not secreted at elevated levels in patients compared with healthy subjects. Why the protein should be deposited to form lesions exclusively in tissues of the central nervous system is unknown. The most common plasma proteins are also present in cerebrospinal fluid (CSF), but at much lower concentrations. Biacore's SPR technology was used in a screening strategy to find out whether any proteins common to both plasma and CSF could bind Aβ protein and thus potentially inhibit deposition [60]. Two forms of Aβ protein are known; one with 40 amino acids and a longer variant that contains two extra C-terminal residues (Aβ1–42). The latter polymerizes more readily into plaque-forming fibrils called β-amyloid. Two parallel immobilization strategies were followed to monitor the binding of plasma/CSF proteins to both forms of Aβ protein. Plasma and CSF proteins were tested for their capacity to inhibit Aβ protein polymerization in a microtitre assay in which Aβ protein or Aβ1–42 were immobilized as templates to which further Aβ protein was added in the presence of the test protein. This revealed that the most inhibitory substances were albumin, α1-antitrypsin, IgA and IgG, proteins also found at a lower concentration in CSF. In Biacore experiments, albumin was run over flow cells prepared as indicated above and was shown to bind only to immobilized, polymerized Aβ protein. The authors suggest that albumin, at concentrations found in plasma, may bind and prevent the deposition of polymerized Aβ protein. In inflammation, levels of albumin in CSF may be sufficiently reduced to abrogate this protective function, enabling the formation of β-amyloid.

In addition to Aβ protein, serum amyloid P component (SAP) is also found in amyloid deposits in Alzheimer's disease and, like Aβ protein, tends to self-associate in the presence of calcium. Calumenin is one of a family of endoplasmic reticulum proteins that coordinate calcium in maintaining the structural integrity of helix–loop–helix motifs in certain proteins. One study elegantly demonstrated how Biacore could be used to define the interaction and characterize the conditions necessary for it to occur [61]. An unidentified binding partner for calumenin was first detected when a crude placental extract was run over recombinant calumenin immobilized on an affinity column in the presence of calcium. After elution and PAGE, one 30-kDa band was isolated and the protein was internally sequenced by tryptic digestion and identified as SAP. Calumenin was then immobilized as a ligand on a sensor chip. The results confirm that SAP is a ligand for calumenin and that the formation and stability of the complex requires calcium. It is therefore possible that given the tendency of both SAP and calumenin to form insoluble complexes, calumenin may participate in amyloidosis, the pathological process by which amyloid is deposited in the CNS of patients with Alzheimer's disease.

Spinal muscular atrophy (SMA) is a disease that is unequivocally linked to a gene called survival of motor neurons (SMN) [62,63]. The disease is severely disabling and is characterized by the loss of motor neurons from the lower spinal cord. There is a strong correlation between the severity of SMA and levels of SMN protein. The function of SMN protein is unclear although it is thought to be involved in mRNA splicing. To determine a possible mechanism to explain why mis-sense mutations in SMN should compromise protein function and hence cause disease, Biacore was used to discover how native peptide fragments of SMN, defined by whole, discrete exons, interact with each other [64]. To identify and select the tools necessary to map these protein interactions, a panel of monoclonal antibodies was raised to full-length SMN protein. The antibodies were then tested using Biacore's SPR technology for reactivity against the peptides encoded by each exon in the cloned gene. The antibodies were immobilized on a sensor chip and screened with all the fragments. Antibodies were thereby identified which specifically recognized certain peptides within the protein and which could then be used to characterize intra-chain interactions in SMN that may determine conformational settings of functional importance in the full-length protein. By then immobilizing each peptide fragment and screening with exon-defined peptides, the authors defined the fragments of SMN which possessed the potential to interact with each other. This information, combined with epitope mapping, enabled an experiment to be constructed which gave a wealth of information about the structural intramolecular events behind the function of a protein. One monoclonal antibody that was shown to recognize the peptide product of exon "A" alone was immobilized on a sensor chip and subsequent injections of different peptides showed that the capture of peptide A enabled peptides "B1" and "B2" to bind sequentially. These data show not only that peptide A appears to form a scaffold to allow the assembly of B1 and B2, but that the functional protein product is a result of the creation of a novel epitope formed by the combination of B1 and B2. This study is recommended as a model of how Biacore's SPR technology can be used to test the molecular mechanisms underlying the expression of disease.

Alzheimer's disease is a neurodegenerative disease in which Aβ protein participates in neurotoxic aggregates which localize to sites of cell death and neuronal apoptosis. Toxicity may also be caused by the localized release of activated inflammatory cytokines in the formation of so-called senile plaques. Aβ protein has been shown to induce the release of pro-inflammatory cytokines by cultured cells and this function was shown to be inhibited by ganglioside GM1 [65]. Biacore's SPR technology was used to demonstrate direct binding of Aβ protein and liposome-immobilized GM1, suggesting the mechanism behind this neuropathologic function. Interaction was shown to depend on the presence of sialic acid. As Aβ protein is known to be neurotoxic in a β-pleated conformation, the authors suggest that one function of GM1 may be to maintain Aβ protein in an inactive state and thus prevent the initiation of the inflammatory response of neurons in Alzheimer's disease.

In a modified approach to the liposome capture strategy of Ariga and Yu [65] described above, liposomes were immobilized on a sensor chip via an antibody [16]. In addition to ganglioside GM1, lipopolysaccharide (LPA) was incorporated into the liposome construction in this study, enabling capture by an anti-LPA antibody and mild regeneration conditions.

A rigorous investigation into the immobilization of gangliosides has been carried out in an effort to develop a simple assay for anti-ganglioside antibodies [66]. A requirement of the technique is to present the glycolipid in an appropriate orientation for antibody characterization i.e. in a manner in which the sugar moiety presented in tumor cells *in vivo* is available to the analyte and which is thus a potential target for therapeutic monoclonal antibodies. The gangliosides were immobilized via the lipid tail, as using the more common aldehyde coupling procedure via the carbohydrate domain masks the epitope. The authors present a simple and rapid protocol, yielding a stable surface that can be readily reconditioned and reused. Gangliosides were injected over the unmodified surface of a sensor chip, with ganglioside GD3 covalently immobilized by photoactivation for comparison. Although overall immobilization levels were low compared with those achieved by covalent linkage, the molar level of binding (specific binding) was much higher.

Phase I clinical trials, in which chimeric anti-ganglioside antibodies were tested in children with neuroblastomas [67], have shown that a host immune response is still induced to the murine component of the antibody. For this reason, engineered humanized antibodies wele synthesizd that retain the murine specificity-determining CDR regions, but from which the flanking V-region sequences have been replaced with human counterparts. This produces a theoretically less immunogenic protein in which the murine component is reduced from 30 to 10 %. Nakamura *et al.* [67] based the characterization of the interaction between these humanized antibodies and immobilized gangliosides GD2 and GD3 on the protocol of Catimel *et al.* [66] above. Their Biacore experiments revealed that humanization changed the kinetic characteristics of the antibodies in ways that are otherwise difficult if not impossible to predict. Biacore's SPR technology thus provides information about the molecular mechanisms underlying functional changes in humanized antibodies, a critical part of the development of novel antibodies for human immunotherapy.

Nerve growth factor (NGF), a member of the neurotrophin family, is implicated in the differentiation and survival functions of neurons [68,69]. Biacore's SPR technology has been used to characterize many types of receptor : ligand interactions, including that between NGF and the Trk family of receptors containing intracellular tyrosine kinase domains. Shc is a so-called adaptor protein comprising a C-terminal SH2 domain and an N-terminal phosphotyrosine-binding (PTB) domain. One of its known functions is to link ligand-activated growth factor receptors, including the human NGF receptor TrkA, to the Ras signaling pathway. This functions through its participation in the multi-molecular signaling complex, Shc-

Grb2-Sos. The SH2 domain recognizes phosphotyrosine indirectly via adjacent C-terminal amino acids, while the PTB domain directly binds phosphotyrosine located within a specific amino acid motif.

Biacore's SPR technology was used to study the specificity and kinetics of the interaction between the PTB domain of Drosophila Shc and phosphopeptides derived from an EGF-receptor homolog, containing structural features shared by other receptors including TrkA, such as conserved hydrophobic amino acid residues [70]. This system is therefore likely to be a suitable model for other receptor : ligand interactions. Initial experiments in which a peptide containing phosphotyrosine flanked by 11 N-terminal and five C-terminal residues was immobilized on a sensor chip, showed that the association and dissociation rates of the Shc PTB domain were somewhat slower than typical, previously determined SH2 interactions. This may imply that in the interaction between Shc and an activated receptor, Shc is likely to bind rapidly to its target via its SH2 domain while its PTB domain limits the rate of dissociation. By measuring the ability of truncated or point-mutated phosphopeptides to inhibit the binding of the Shc PTB domain to native phosphopeptide, the specificity of the target sequence surrounding phosphotyrosine in the growth factor receptor was determined in a Biacore competition assay. Phosphopeptides were truncated at various positions N-terminal to phosphotyrosine, incubated in solution with Shc and passed over native phosphopeptide immobilized on a sensor chip. C-terminal deletions had little effect, but deleting N-terminal amino acids revealed that the presence of hydrophobic residues at positions -5 and -7 were critical for Shc to effectively bind the receptor. As expected, nanomolar concentrations of unmodified peptide efficiently inhibited binding of Shc while an interaction could still be detected even in the presence of high concentrations of deletion-peptides.

In a study of some relevance to Biacore assays in the receptor : ligand interaction field in general, the receptor for mammalian TrkA was characterized [71] in which the extracellular ligand-binding domain of TrkA was over-expressed in insect cells, purified and immobilized on a sensor chip. The association rate of NGF-binding to recombinant TrkA extracellular domain was close to previously reported values using the full-length receptor. These results suggest that recombinant, purified receptors may be used as a model in the analysis of NGF : TrkA interactions using Biacore's SPR technology and that the strategy may be generally applicable to a wider range of receptor : ligand interactions.

The physico-chemical composition of the microenvironment of the binding site on a receptor for its cognate ligand is critical. Ionic strength and pH, for example, affect non-covalent bonds and so by varying the buffer composition, judgments can be made on the nature of the interaction. Biacore's SPR technology was used to discover whether local variations in pH changed the capacity of Zn^{2+} and Cu^{2+} to modulate NGF binding to its receptor [72]. To do this, the cations were included in running buffers with pH varying from 5.5 to 7.4, mimicking local conditions in cerebral acidosis, a condition frequently arising after stroke or traumatic insult. This pH range also covers the pKa value of histidine, the position and electrostatic status of which are known to influence cation binding. The authors used the strategy described by Woo *et al.* [71] to prepare and immobilize recombinant TrkA. The dissociation equilibrium constant, KD, for the NGF : TrkA interaction was calculated over a range of pH values covering the pKa value of histidine. This revealed that Zn^{2+}, but not Cu^{2+}, lost its ability to bind NGF and inhibit its interaction with TrkA under acidic conditions. The physiological consequences of this depend on the cell type and context; TrkA-expressing cells in an acidic environment and needing NGF for survival may well benefit from the inactivation of Zn^{2+}, whereas the effects might be detrimental if NGF, in contrast, initiates signals leading to cell death.

NGF functions both within and outside the nervous system, for example it is released from mast cells during allergic inflammation and is implicated in the pathophysiology of asthma. Using *in vitro* studies it was shown that a novel protein called FIZZ1, isolated from the lavage fluid of inflamed murine bronchoalveoli, dampened NGF-mediated survival and gene expression in neurons [73]. Biacore was used to show that an antagonistic mechanism, which is not based on competition for a single receptor, operates to inhibit NGF function in TrkA-bearing cells. The data suggested that rather than inhibiting the interaction of NGF with its receptor, FIZZ1 functions indirectly by binding to a different and as yet unknown receptor, possibly regulating TrkA activity by "inside-to-out" signaling.

The rationale for one Biacore study on Alzheimer's disease [74] was to identify small peptides that can inhibit aggregation and therefore the toxic neurodegenerative effects of β-amyloid (Aβ) in the diseased brain. Although Aβ aggregates are present in neurotoxic lesions, it is not known if they are causative. Short peptides directed to the self-recognition motifs in Aβ were synthesized and were used to screen the affinities of the interactions. Solution-based methods for screening Aβ inhibitors consume large amounts of the target and they are limited by the tendency of Aβ to aggregate in solution and thus present multiple target forms. This problem is avoided in Biacore assays because the target is immobilized on a planar sensor chip surface at a density and in an orientation that can be carefully controlled by the user. A 26-amino acid fragment of Aβ comprising amino acids 10 to 35 with a C-terminal cysteine was immobilized at low density on a sensor chip thus presenting a uniform surface. Further, inter-assay variability was low as a single chip surface could be regenerated several times and re-used for all test interactions. A range of concentrations of each peptide, based on variations of a five-amino acid sequence corresponding to the central hydrophobic domain of Aβ known to be responsible for self-association, was injected over Aβ on the chip surface for 500 s and the response at equilibrium (R_{eq}) was plotted against time. As the binding was weak, only relative affinities were measured, with the equilibrium response determined after 90 % of the contact time at each concentration.

Of several peptides based on multiple length C-terminal extensions of a pentapeptide in the central hydrophobic domain of Aβ (which generally bound more strongly than the original pentapeptide to Aβ), a pattern emerged in which the affinity for Aβ was sensitive to site-specific positioning of positively-charged lysine residues. For example, it was previously determined that the addition of six lysine residues directly adjacent to the basic KLVFF sequence generated a more effective inhibitor of cytotoxicity in cell-based assays. In the Biacore assay, peptides in which three of these residues were separated from the basic sequence by three negatively-charged residues had a 14-fold lower affinity for Aβ, while the affinity of a peptide of similar size bearing three directly adjacent lysine residues was almost unchanged.

The peptides were then tested in cytotoxicity microtitre assays by incubating Aβ with neuroblastoma cells in the presence of peptides and compared with the Biacore affinity data (Table 4.1). High affinity binders protected against toxicity by greater than 80 % in viability assays while low affinity binders were less effective. It is hoped that these peptides will be useful as probes of the mechanisms underlying amyloid plaque formation and in the design of drugs. Biacore's SPR technology, therefore, is proposed as a reliable means to measure the affinity of small peptides for Aβ , a protein which is notoriously difficult to handle in solution-based assays. Such data may help predict the suitability of candidate peptides as inhibitors of Aβ toxicity in Alzheimer's disease.

The glutamate receptor δ2 (GluRδ2) [75,76] is exclusively expressed on the dendritic synapses of Purkinje cells [77], structurally unique cells that are among the most organizationally complex neurons in the mammalian nervous system. Although they receive an enormous amount of data from neurons of the spinal cord and elsewhere, they are nevertheless able to convert this input into inter-

Table 4.1 Peptides based on C-terminal extensions to a central KLVFF fragment derived from the central hydrophobic domain of Aβ were tested for their affinity to Aβ immobilized on Pioneer Chip B1. Affinity (K_D) is related to the ability of the same peptide to inhibit Aβ-mediated cell cytotoxicity in a cell-based assay.

*Peptide sequence**	*KD (μM)*	*Viability (%)*[+]
KLVFFRRRRRR	40	> 90
KLVFFKKKKKK	40	80–90
KLVWWKKKKKK	40	80–90
KLVFFKKKK	37	70–80
KLVFWKKKKKK	65	70–80
KLVFFKK	80	70–80
KLVFFKKKEEE	90	60–70†
KKKKLVFF	180	60–70†
KLVFFEKEKEK	300	60–70†
KKKKKK	400	60–70†
KLVFFEEEKKK	1300	60–70†

[+]Cellular viability of human neuroblastomas cells was assessed in a microtiter assay using an MTT assay in which cells were exposed to Aβ in the presence or absence of peptides.
*K, lysine; L, leucine; V, valine; F, phenylalanine; R, arginine; W, tryptophan; E, glutamic acid.
†This is a similar level of viability as seen after treatment with negative controls.

pretable information and are, in fact, the sole means of output in the cerebellar cortex. That GluRδ2 is limited to these cells possibly implies a function in this extraordinary feat of information processing. Biacore analysis showed that an intracellular protein called delphilin co-localizes with GluRδ2 at parallel fiber (input)–Purkinje cell (output) synapses [78] and that recombinant delphilin PDZ domain binds to the C terminus of GluRδ2 with moderate affinity, mostly accounted for by rapid dissociation. The authors also showed that delphilin bound profilin, an actin-binding protein, and dissociated very slowly. It is possible therefore that delphilin acts as a molecular bridge linking GluRδ2 to the actin cytoskeleton via profilin with rapid turnover of the GluRδ2 : delphilin interaction contributing to the capacity of the Purkinje cell to process a very large amount of input data.

Spinal muscular atrophy (SMA) is predominantly a childhood disease of motor neurons in the spinal cord, affecting the voluntary muscles and causing the impediment of functions such as locomotion and head and neck control. The disease is strongly linked to mutations in the *SMN1* (survival motor neuron 1) gene. It has recently been suggested that the clinical symptoms of the disease may be due to increased motor neuron death through the inability of mutated SMN protein to dimerize, and thereby sequester and inhibit the pro-apoptotic function of the tumor suppressor protein, p53 [79].

An elegant surface plasmon resonance strategy was followed in which Biacore was used initially to confirm conventional immunoprecipitation (i.p.) data using an antibody to p53 showing that SMN associated with p53. This finding was substantiated using Biacore in sequential binding experiments that demonstrated how the technology might be used as an alternative to immunoprecipitations and may represent a more convenient tool for detecting molecular interactions in whole cell extracts (Figure 4.14). Briefly, anti-p53 antibody was coupled to a sensor chip. Next, whole cell extracts from SMN-transfected cells (treated with a protease inhibitor that protects the labile p53) were passed over the chip, followed by exposure to anti-SMN antibody. Cell extracts that did not contain detectable levels of p53 were used in parallel and demonstrated the specificity of the interaction.

SMA patient-derived mutations were also tested for p53 binding. Two SMN proteins carrying C-terminal mutations found in patients with severe type I SMA and a peptide from a patient with less severe type II SMA were immobilized on a sensor chip and exposed to p53. The magnitude in reduction of SMN peptide binding strongly correlated with the clinical subset. This defect is likely to be attributable to the inability of the SMN mutant proteins to efficiently self-associate. Young et al. [79]then applied the sequential Biacore approach to show that this very inability to self-associate may prevent SMN from sequestering and inhibiting p53 function. In these experiments, p53 failed to bind to immobilized monomeric SMN and was only induced to bind when the same chip was supplemented with dimeric SMN.

The significance of these findings is supported in the same body of work by results from indirect confocal immunofluorescence experiments showing that SMN and p53 co-localize in the Cajal bodies (sites of small nuclear ribonucleoprotein particles involved in RNA splicing) of normal fibroblasts, but not in those derived from SMA patients.

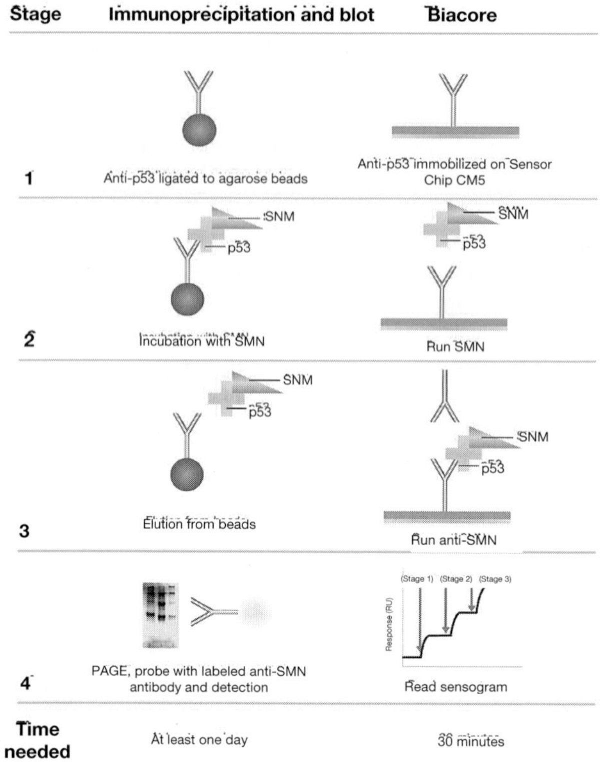

Figure 4.14 Comparison between immunoprecipitation/blotting and Biacore in the detection of molecular complexes. Both assays reveal that SMN protein derived from whole cell extracts is associated with p53 in the cell. Results achieved by immunoprecipitation are available within 1 day whereas the Biacore assay takes minutes. (see Colour Plate p. XXII).

4.3.5 Drug: Plasma Protein Interactions

The clinical potential of drug compounds is greatly affected by the nature of their interactions with circulating plasma proteins, such as human serum albumin (HSA) and α1-acid glycoprotein (AGP). The effects of plasma protein binding by drugs are varied and can influence factors such as the free concentration of drug compound in the circulation, transport and distribution around the body and the duration of drug action. Biacore offers the user the ability to run rapid, automated assays with high sensitivity and relatively low sample consumption. The sensitivity of the technology varies from system to system but association rate constants (k_a) of up to 5 x 10^6 $M^{-1}s^{-1}$ and dissociation rate constants (k_d) of between 10^{-4} and 1 s^{-1} are accurately and reproducibly attainable with the latest available systems. Lead compounds can be analyzed from sub-nanomolar to millimolar concentrations.

Biacore's latest model, Biacore® S51, is specifically designed for drug discovery applications from simple ranking, to detailed characterization of the interactions. Hit selection assays may be employed for ranking of lead compounds for binding to one or two plasma proteins, with a throughput of 96 compounds in 12 h. The data produced from single concentration analyses permit the ranking of compounds in terms of binding response to single or multiple plasma proteins. Based on this ranking assay, compounds for which a high-resolution analysis is required can then be selected. These are then taken for detailed compound characterization assays which use concentration series of compounds to generate data sets from which K_D values can be calculated using appropriate fitting models. The K_D values may then be transformed to give percentage-bound figures, which can be calculated for single or combined plasma protein binding, as appropriate. It may also be possible to maximize the level of the compound characterization to include kinetic analyses of drug–plasma protein interactions in some cases.

4.3.6 Drug : DNA Interactions

A number of drugs bind cellular DNA and inhibit targeted protein–DNA interactions. A prerequisite for the design of new types of compounds is rapid penetration of cells and interaction with a specific sequence of DNA. Such compounds, structurally predicted from the gene sequence and functionally tested using Biacore's SPR technology, could have many therapeutic and biotechnological uses. Relatively small molecules can be remarkably specific in their sequence recognition. A "prodrug" of furamidine, for example, binds strongly to the minor groove of DNA and is selective for AT-rich sequences [80]. Biacore has been used in combination with DNA footprinting methods [81], which are employed to quickly characterize compound interactions with defined sequences of various sizes to discover new DNA-binding drugs. Biacore is then used to resolve questions of stoichiometry, relative affinities and binding kinetics. The characterization of DNA sequence recognition and DNA binding of indolocarbazole antitumor drugs is described elegantly in two studies [81,82]. Thermodynamic characteristics of binding of organic cations to AT and mixed sequence DNA sites are shown by isothermal calorimetry and Biacore to be very dependent on structure, solvation and sequence of the DNA binding site (see also [83] and [84]).

Recently Biacore was shown to be a rapid and reliable means to detect DNA base pair mismatches by using a synthetic low molecular weight ligand that specifically binds G : G mismatches. This work has far-reaching implications for SPR in genome analysis [85]. The technique depends on the development of reagents that specifically recognize certain types of aberrations in native DNA. Today, single nucleotide polymorphisms (SNPs) are routinely detected using methods that rely largely on the separation of DNA fragments by gel electrophoresis or HPLC, neither of which are readily suited to high-throughput analysis.

The sheer volume of sequence information from the human genome project demands the development of a tool for rapidly scanning DNA for SNPs and Sando *et al.* [85] have used Biacore's SPR technology as an alternative means to this end. They designed and synthesized a low molecular weight ligand that specifically binds double-stranded DNA containing G : G mismatches arising from native C : G pairings and which has minimal reactivity with any of the other seven possible mismatch combinations (A : A, A : C, A : G, T : T, T : C, T : G or C : C). The ligand, naphthyridine, is modeled to recognize the characteristic open bulge that is created between DNA strands at the site of a G : G mismatch, at which no complementary base (cytosine) is present with which to form a hydrogen-bonded base pair. Naphthyridine is synthesized as a dimer and is therefore able to form simultaneous hydrogen bonds with both guanines on each DNA strand.

Experiments have shown that the melting temperature, i.e. that at which half of the double-stranded DNA in a sample dissociates into single chains and which is a measure of DNA stability, increased in the presence of dimeric, but not monomeric naphthyridine, to a much higher threshold than that of native DNA or DNA of an otherwise identical composition containing other mismatches. The synthetic naphthyridine ligand contains an aminoalkyl chain and can therefore be covalently linked by amine coupling to carboxyl groups on a sensor chip. 27-mer double-stranded DNA was passed over the prepared chips and binding was seen only if the DNA contained G : G mismatches. DNA containing other SNPs or native DNA did not bind to the ligand. The sensitivity of the assay was shown by a linear increase in response with concentration of analyte from 125 to 1000 nM, a range that was optimized by changes in salt concentration. The intensity of response was a function of both DNA concentration and length.

Other non-gel based mismatch screening methods such as melting curve analysis or MutS capture of mismatched fragments rely on detection systems such as labeling with a fluorescent dye or the use of labeled primers. Sando *et al.* [85] showed that Biacore's SPR technology is sensitive, specific, requires no primers or probes and can be adapted for high throughput analysis. This is the first report in which Biacore has been used for DNA mutation analysis and while the design and synthesis of ligands that are specific for other SNPs will be difficult, the principle of Sando *et al.'s* work is a promising development in accelerating DNA mutation analysis (see also [86] for a commentary on this exciting work). One further study shows that a second class of small molecule, pyrrole-imidazole-containing polyamides, can also function as DNA mismatch recognition agents [87].

4.3.7
Nucleic Acid Structure and Analysis

In addition to identifying prospective binders to DNA in the search for novel drugs, Biacore has also been used to investigate the effects of deliberate alterations in the structure of nucleic acids on their function. For example, it has been shown that the substitution of the sugar-phosphodiester backbone of DNA to create so-called pyrrolidine-amide oligonucleotide mimics (POM) [88] enabled the synthetic single-stranded DNA analogs to bind to complementary native strands with higher affinity than did other native strands. Additionally it was shown that POMs have a much faster association rate for binding to RNA than for binding to DNA.

Simple cations and polyamines are also shown to affect the binding affinity of ssDNA in duplex formation and the rate of DNA duplex formation and dissociation. Affinity and kinetic analyses have been performed on several different types of DNA structures including perfect linear matches, bulged loops, mismatched duplexes and hairpin duplexes. Of possibly critical importance in cancer, Biacore studies have shown that polyamines (commonly found in tumor cells) may stabilize mismatched duplexes via increasing the thermal stability of the duplex and causing a reduction in k_d [89]. Protein : DNA interactions have also been tested by immobilizing a range of different structural forms of DNA on a sensor chip surface and testing affinity and kinetic parameters with interacting proteins [90]. This conceptually interesting work shows that DNA polymerase binds quantitatively and kinetically more efficiently to primed DNA than to blunt-end dimers.

"Morpholinos" (MORFs) are a recently developed class of synthetic nucleotide oligomers that carry nucleotides in the same defined sequences as their parent nucleic acid but in which the sugar backbone is replaced with a phosphorodiamidate backbone [91]. MORFs are reported to be more nuclease-resistant, more soluble and more rigid than native DNA and thus potentially may be highly efficacious as carriers of sequence-targeted radiopharmaceuticals in cancer treatment [92].

Anti-sense DNA is a frequently investigated strategy aimed at the targeted inhibition of the translation of oncogenic protein genes *in vivo*. The efficacy of such treatment, however, is limited by the difficulty of ensuring delivery of anti-sense oligonucleotides into cells. Anti-sense oligonucleotides were conjugated with cell-penetrating peptides and Biacore was used to confirm that the binding affinities of these oligonucleotides for their targets after conjugation were similar to unconjugated oligonucleotides [93]. An extension of the anti-sense approach is a strategy in which double-stranded DNA is targeted, the so-called anti-gene approach. Here, a DNA triple helix is transiently formed between a specific dsDNA and a single-stranded oligonucleotide and which may potentially inhibit transcription of specific genes. These transient structures, however, are unstable *in vivo* and Biacore was used in one study to monitor the stability of the complexes formed between dsDNA and chemically-stabilized oligonucleotides [94].

Interactions between small deoxyribozymes and RNA are also open to Biacore on-chip analysis. In one interesting study, the role of higher order RNA structure in self-association has been investigated [95]. The authors propose that mRNA transcribed from mutated genes can form structural anomalies that are regulatory parameters in themselves i.e. that the shape of the RNA in addition to its sequence is important. The implication is that even if a mutation in a codon does not result in an amino acid change in the translated protein, the structural alteration of the transcript may nevertheless have functional consequences. Accurate mRNA splicing for example, is sensitive to the maintenance of RNA structure in order to allow interactions with other cellular components. The authors immobilized short sequences of RNA which recognized and cleaved RNA sequences containing a loop resulting from one such mutation. The cleavage function was shown to be both loop size- and sequence-dependent and further, the reaction could be monitored in real time on the chip surface.

Most real-time "on-chip" reactions have been concerned with monitoring real-time changes in molecular weight of an immobilized substrate during passage of an enzyme analyte although DNA synthesis during the passage of DNA polymerase has also been studied [96]. For selected reading on monitoring enzyme activity on Biacore chips through real-time alterations of substrate molecular weight, see also [97] (Design of helical proteins for real-time endoprotease assays); [98] (Proteolysis of the exodomain of recombinant protease-activated receptors: prediction of receptor activation or inactivation by MALDI mass spectrometry); [99] (Monitoring of the refolding process for immobilized firefly luciferase with a biosensor based on surface plasmon resonance); and [100] (Peptide chips for the quantitative evaluation of protein kinase activity).

4.3.8
Protein : RNA Interactions

The association of specific proteins with RNA molecules begins immediately as the nascent RNA chain is produced during transcription. From this point onwards, a series of different proteins are bound to and released from the RNA and are involved in co- and post-transcriptional RNA processing events such as splicing, export to the cytoplasm, regulation of RNA stability and translation. These interactions are highly dynamic and are critical in the regulation of the various steps involved in RNA processing. In addition, these individual RNA : protein interactions occur in the environment of extremely large and complex nucleic acid–protein complexes such as the transcriptosome and spliceosome. Until recently, however, studies on RNA-binding proteins have relied heavily on techniques which examine binding under equilibrium conditions (such as electrophoretic mobility shift assays). Although this type of affinity-based data is valuable in many instances, it does not address the key issue of the dynamics of the RNA : protein interactions. This requires a true kinetic analysis that can deliver individual association and dissociation rate constants (k_a and k_d).

Biacore's SPR technology can be used to provide detailed kinetic data on the interaction of RNA-binding proteins with their specific RNA targets and this information can provide invaluable insights into their functions. In one study, detailed kinetic analysis was used to elucidate a novel two-stage binding interaction between the spliceosomal protein, U1A and its RNA target [6].

Although the range of surface chemistry approaches that are compatible with Biacore's technology enables great flexibility in selecting which interaction partner should be immobilized on the sensor surface, the nucleic acid is most often immobilized in RNA : protein studies. One reason for this is the relative ease with which RNA oligonucleotides can be chemically synthesized with a terminal biotin modification which enables them to be immobilized in an oriented fashion on a commercially-available streptavidin surface (Sensor Chip SA).

In one particular study on binding of the neuronal protein, HuD to RNA recognition motifs (RRMs), individual RRMs that had remained "hidden" during previous equilibrium analysis-based studies were shown to make important kinetic contributions to the overall binding function (see [22] and [101]). Many RNA-binding proteins contain multiple RRM regions and the results from the HuD study may therefore represent a general model for how investigation into this type of protein should be approached. In the same study, the analysis of spliceosomal protein U1A binding to its target RNA revealed a hitherto unknown two-stage mechanism of binding. This data was also able to identify key amino acid and nucleotide residues involved in the individual interaction stages and provided evidence for the types of forces involved. This model may also prove to have a much wider relevance, since a number of other RNA-binding proteins are known to contain similar distributions of basic amino acids around their RNA binding sites. In both cases, crucial new information regarding the mechanism of the protein : RNA interaction was obtained. These studies relied heavily on the ability of Biacore's SPR technology to provide high-quality kinetic data. In addition to furthering our understanding of the many vital biological interactions between proteins and RNAs, kinetic approaches may also be invaluable in designing therapeutic small molecule inhibitors.

4.4. Current Developments and Future Perspectives

4.4.1 SPR-MS

The advent of the proteomics era has placed a new emphasis on techniques that enable the identification of proteins. Of these, mass spectrometry has been the most significant, with MALDI-TOF (matrix-assisted laser desorption/ionization-time of flight) and ESI (electrospray ionization) approaches dominating the field. MS analysis is commonly used in combination with traditional separation techniques, predominantly two-dimensional gel electrophoresis, in order to combine expression profiling with protein identification. In contrast to the array technology approaches used in expression profiling at the mRNA level, analogous developments within proteomics will be greatly complicated by the extensive and complex interactions among the proteins expressed. In practice, this means that simple expression data will be of extremely limited value without a detailed understanding of the expressed proteins' interaction partners. Combining SPR with MS would therefore provide a potentially powerful marriage of technologies within the proteomics field, with the possibility of providing protein identification which is selected on the basis of functional binding criteria. The painful lesson learned from previous genomic studies is that the expression of a gene in a particular biological context can often prove to be a meaningless observation in itself in the absence of direct functional data.

A number of approaches to combining SPR and MS have been taken thus far. It is possible, for example, to physically remove the sensor chip from a Biacore instrument following binding of the analyte to the ligand and use it directly in a MALDI-TOF analysis [102,103]. This approach produces minimal analyte loss, since all of the analyte molecules are bound within the area of the flow cell which is used as the MALDI target.

An alternative method is to elute captured analyte from the sensor surface, using a procedure designed to recover the sample in a small volume after SPR analysis [59,104]. In addition to being able to use the same sensor surface for repeated recoveries, another advantage of this approach is that it is also possible to enzymatically digest the bound analyte on the sensor surface prior to or following recovery [105], which greatly facilitates MS identification by allowing alignment of MS profiles against various databases. A disadvantage to this approach is the risk of contamination, which must always be guarded against.

The SPR route into MS analysis holds a great deal of promise, especially in the field of interaction proteomics, and as testified by the recently published work just described, much has already been achieved. Future developments will doubtless further improve the efficiency and automation of the technique. The small sample volume requirement and high sensitivity features of SPR technology, which are major advantages in terms of interaction analysis, mean that the amount of material that it is possible to obtain from a single SPR experiment for MS analysis remains somewhat restricted. This reflects the fact that the design of SPR instruments has hitherto been optimized very much from the viewpoint of the interaction analysis itself, rather than placing any emphasis on preparative purposes. As SPR-MS continues to be developed as a combined technology concept however, it can be anticipated that the two constituent technologies will adapt to each other to provide an optimal solution.

4.4.2
The Challenges of High-throughput Protein Analysis and Array Technologies

The range of currently available technologies for studying proteins all appear to be restricted in the throughput that can be easily achieved. This suggests that the main challenge for the next phase of development is principally one of scale. As interest in proteins shifts gradually away from considering isolated or restricted groups of molecules and more towards understanding how multiple proteins interact in complex, functionally interrelated networks, it is inevitable that increasingly higher throughput systems will be desirable. While protein microarray systems will certainly be developed over the next few years, it is worth considering some of the realities relating to the challenges involved.

It would be optimistic to imagine that in the near future proteomics will be conducted on a genomics-scale, since the complexity of working with proteins as compared to nucleic acids presents more of a challenge for developments in automation. There are no current proteomics technologies analogous in performance to the PCR, high-throughput sequencing and SNP analysis technologies which have enabled such astonishingly rapid progress in genomics-related disciplines. Protein array chips will certainly be important tools in proteomics research, but the scale with which they will be used will depend on the information they provide.

The word "array" may itself have very different meanings and interpretations. To the genomics field, familiar with chip-based cDNA arrays for expression profiling for example, "array" conjures up an expectation of anything from several hundred to many tens of thousands of individual DNA sequences on a single analysis surface. Despite the highly specific nucleic acid sequence hybridization principle on which such systems are based and the simple, semi-quantitative nature of the data output, this type of array technology experienced difficulties in terms of reliability, reproducibility and not least, data analysis. The challenges for large-scale protein arrays are likely to be considerably more difficult than for their DNA counterparts, with complications such as the difficulties in expressing and purifying large numbers of correctly folded proteins with the appropriate post-translational modifications, protein instability, much greater potential for cross-reactivity, greater difficulty in standardizing binding conditions and much more complex data handling. These fundamental challenges must be dealt with in some way by all the technologies that hope to contribute to the development of protein array approaches.

In practice, the development of protein array chips will almost certainly involve a trade-off between relatively low-information screening and more detailed analysis approaches, in terms of the total number of simultaneously analyzable protein "spots" that are practical to use. Protein arrays will doubtless be applied in both the areas of "discovery proteomics" and "interaction proteomics" (see Section 4.1.1), but are likely to be adapted in quite different ways to meet these distinct challenges. While it may be possible for example to spot many thousands of proteins for semi-quantitative expression profiling applications, it is highly unlikely that high-quality kinetic analysis of protein interactions will be carried out on the same scale in the foreseeable future (or even that there would be a demand for such an application).

The nature and size of protein microarrays may vary greatly therefore, depending on the intended application. Currently, it is possible to simultaneously analyze a maximum of four immobilized proteins using Biacore's SPR systems and this can be viewed as a very small-scale microarray system. The strengths of this technology are largely targeted towards producing high-quality, information-rich data (such as high resolution kinetics) and this may position it more towards the high-data content, rather than ultra-high throughput end of the protein array scale.

It is important to bear in mind that each data "spot" in a Biacore assay produces a real-time analysis of its interaction with the analyte over the entire course of binding and dissociation. In high-resolution kinetics analysis, this may involve SPR measurements taken at a rate of up to 10 s^{-1} over the course of many minutes. The amount of data produced by each interaction monitored is therefore very high and while this is one of the cornerstones of the success of Biacore's technology in a broad range of different applications, it is something that must be considered in the context of larger array situations. As was experienced by the pioneers of large-scale DNA arrays, it may be that data analysis will provide a more important practical limitation to the scope of protein arrays than the developments in technological hardware which are involved.

The technical feasibility of using SPR technology to detect interactions in a larger scale array format was announced to the media by Biacore several years ago however, and it is likely that future developments will lead to SPR array systems that are tailored in a format that is appropriate for useful applications within proteomics research, as well as in other applications in life science research and drug development. Precisely where the final balance between array size and data/information content will fall remains to be seen for all the technologies that may contribute to the development of protein microarrays, but this may well vary considerably among them and may in the end be determined more forcefully by application requirements rather than by what may be technically possible to achieve.

References

1. Liedberg, B., Nylander, C., and Lundström, I. Surface plasmon resonance for gas detection and biosensing. *Sensors Actuators.* 1983, **4,** 299–304.
2. Venter, J.C., et al. The sequence of the human genome. *Science* 2001, **291,** 1304–1351.
3. Brett, D., Pospisil, H., Valcarcel, J., Reich, J., and Bork, P. Alternative splicing and genome complexity. *Nature Genet.* 2002, **30,** 29–30.
4. Banks, R.E., Dunn, M.J., Hochstrasser, D.F., Sanchez, J.C., Blackstock, W., Pappin, D.J., and Selby, P.J. Proteomics: new perspectives, new biomedical opportunities. *Lancet* 2000, **356,** 1749–1756.
5. Park, S., Myszka, D.G., Yu, M., Littler, S.J., and Laird-Offringa, I.A. HuD RNA recognition motifs play distinct roles in the formation of a stable complex with AU-rich RNA. *Mol. Cell Biol.* 2000, **20,** 4765–4772.
6. Katsamba, P.S., Myszka, D.G., and Laird-Offringa, I.A. Two functionally distinct steps mediate high affinity binding of U1A protein to U1 hairpin II RNA. *J. Biol. Chem.* 2001, **276,** 21476–21481.
7. De Genst, E., Areskoug, D., Decanniere, K., Muyldermans, S., and Andersson, K. Kinetic and affinity predictions of a protein–protein interaction using multivariate experimental design. *J. Biol. Chem.* 2002, **277,** 29897–2907.
8. Choulier, L., Andersson, K., Hamalainen, M.D., Van Regenmortel, M.H., Malmqvist, M., and Altschuh, D. QSAR studies applied to the prediction of antigen–antibody interaction kinetics as measured by BIACORE. *Prot. Eng.* 2002, **15,** 373–382.
9. Kretschmann, E. Die Bestimmung optischer Konstanten von Mtallen durch Anregung von Oberflächen plasmonenschwingungen. *Zeitschrift Physik.* 1971, **241,** 313–324.
10. Löfås, S., and Johnsson, B. A novel hydrogel matrix on gold surfaces in surface plasmon resonance sensors for fast and efficient covalent immobilization of ligands. *J. Chem. Soc. Chem. Comm.* 1990, **21,** 1526–1528.

11. Johnsson, B., and Löfås, S. Immobilization of proteins to a carboxymethyldextran modified gold surface for biospecific interaction analysis in surface plasmon resonance. *Anal. Biochem.* 1991, **198,** 268–277.

12. Johnsson, B., Löfås, S., Lindquist, G., Edström, Å., Müller Hillgren, R.-M., and Hansson, A. Comparison of methods for immobilization to carboxymethyl dextran sensor surfaces by analysis of the specific activity of monoclonal antibodies. *J. Mol. Recog.* 1995, **8,** 125–131.

13. Tokarska-Schlattner, M., Wallimann, T., and Schlattner, U. Multiple interference of anthracyclines with mitochondrial creatine kinases: preferential damage of the cardiac isoenzyme and its implications for drug cardiotoxicity. *Mol. Pharmacol.* 2002, **61,** 516–523.

14. Caldwell, E.E., Andreasen, A.M., Blietz, M.A., Serrahn, J.N., Vandernoot, V., Park, Y., Yu, G., Linhardt, R.J., and Weiler, J.M. Heparin binding and augmentation of C1 inhibitor activity. *Arch. Biochem. Biophys.* 1999, **361,** 215–222.

15. Nieba, L., Nieba-Axmann, S.E., Persson, A., Hämäläinen, M., Edebratt, F., Hansson, A., Lidholm, J., Magnusson, K., Karlsson, Å.F., and Plückthun, A. BIACORE analysis of histidine-tagged proteins using a chelating NTA sensor chip. *Anal. Biochem.* 1997, **252,** 217–228.

16. Mackenzie, C.R., Hirama, T., Lee, K.K., Altman, E., and Young, N.M. Quantitative analysis of bacterial toxin affinity and specificity for glycolipid receptors by surface plasmon resonance. *J. Biol. Chem.* 1997, **272,** 5533–5538.

17. Sevin-Landais, A., Rigler, P., Tzartos, S., Hucho, F., Hovius, R., and Vogel, H. Functional immobilisation of the nicotinic acetylcholine receptor in tethered lipid membranes. *Biophys. Chem.* 2000, **85,** 141–152.

18. Karlsson, O.P., and Lofas, S. Flow-mediated on-surface reconstitution of g-protein coupled receptors for applications in surface plasmon resonance biosensors. *Anal. Biochem.* 2002, **300,** 132–138.

19. Cooper, M.A. Optical biosensors in drug discovery. *Nature Rev. Drug Discov.* 2002, **1,** 515–528.

20. Sjölander, S., and Urbaniczky, C. Integrated fluid handling system for biomolecular interaction analysis. *Anal. Chem.* 1991, **63,** 2338–2345.

21. Stenberg, E., Persson, B., Roos, H., and Urbaniczky, C. Quantitative determination of surface concentration of protein with surface plasmon resonance by using radiolabeled proteins. *Coll. Interface Sci.* 1991, **143,** 513–526.

22. Katsamba, P.S., Park, S., and Laird-Offringa, I.A. Kinetic studies of RNA–protein interactions using surface plasmon resonance. *Methods* 2002, **26,** 95–104.

23. Myszka, D.G. Improving biosensor analysis. *J. Mol. Recognit.* 1999, **12,** 1–6.

24. Karlsson, R., Kullman-Magnusson, M., Hamalainen, M.D., Remaeus, A., Andersson, K., Borg, P., Gyzander, E., and Deinum, J. Biosensor analysis of drug–target interactions: direct and competitive binding assays for investigation of interactions between thrombin and thrombin inhibitors. *Anal. Biochem.* 2000, **278,** 1–13.

25. Day, Y.S., Baird, C.L., Rich, R.L., and Myszka, D.G. Direct comparison of binding equilibrium, thermodynamic, and rate constants determined by surface- and solution-based biophysical methods. *Prot. Sci.* 2002, **11,** 1017–1025.

26. Myszka, D.G., Morton, T.A., Doyle, M.L., and Chaiken, I.M. Kinetic analysis of a protein antigen antibody interaction limited by mass transport on an optical biosensor. *Biophys. Chem.* 1997, **64,** 127–137.

27. Matsuda, H. Theory of the steady-state current potential curves of redox electrode reactions in hydrodynamic voltammetry. II Laminar pipe- and channel flow. *J. Electroanal. Chem.* 1967, **15,** 325–336.

28. Myszka, D.G., He, X., Dembo, M., Morton, T.A., and Goldstein, B. Extending the range of rate constants available from BIACORE: Interpreting mass transport-influenced binding data. *Biophys. J.* 1998, **75,** 583–594.

29. Sikavitsas, V., Nitsche, J.M., and Mountziaris, T.J. Transport and kinetic processes underlying biomolecular interactions in the BIACORE optical biosensor. *Biotechnol Prog.* 2002, **18,** 885–897.

30. Roden, L.D., and Myszka, D.G. Global analysis of a macromolecular interaction measured on BIAcore. *Biochem. Biophys. Res. Comm.* 1996, **225,** 1073–1077.
31. Park, B.W., Zhang, H.T., Wu, C., Berezov, A., Zhang, X., Dua, R., Wang, Q., Kao, G., O'rourke, D.M., Greene, M.I., and Murali, R. Rationally designed anti-HER2/neu peptide mimetic disables P185HER2/neu tyrosine kinases *in vitro* and *in vivo*. *Nature Biotechnol.* 2000, **18,** 194–198.
32. Berezov, A., Zhang, H.T., Greene, M.I., and Murali, R. Disabling erbB receptors with rationally designed exocyclic mimetics of antibodies: structure–function analysis. *J. Med. Chem.* 2001, **44,** 2565–2574.
33. Berezov, A., Chen, J., Liu, Q., Zhang, H.T., Greene, M.I., and Murali, R. Disabling receptor ensembles with rationally designed interface peptidomimetics. *J Biol Chem.* 2002, **277,** 28330-9.
34. Loukinova, E., Ranganathan, S., Kuznetsov, S., Gorlatova, N., Migliorini, M.M., Loukinov, D., Ulery, P.G., Mikhailenko, I., Lawrence, D.A., and Strickland, D.K. Platelet-derived growth factor (PDGF)-induced tyrosine phosphorylation of the low density lipoprotein receptor-related protein (LRP). Evidence for integrated co-receptor function betwenn LRP and the PDGF. *J. Biol. Chem.* 2002, **277,** 15499–15506.
35. Rooney, I.A., Butrovich, K.D., Glass, A.A., Borboroglu, S., Benedict, C.A., Whitbeck, J.C., Cohen, G.H., Eisenberg, R.J., and Ware, C.F. The lymphotoxin-beta receptor is necessary and sufficient for LIGHT-mediated apoptosis of tumor cells. *J. Biol. Chem.* 2000, **275,** 14307–14315.
36. Steplewski, Z., Lubeck, M.D., and Koprowski, H. Human macrophages armed with murine immunoglobulin G2a antibodies to tumors destroy human cancer cells. *Science* 1983, **221,** 865–867.
37. Valone, F.H., Kaufman, P.A., Guyre, P.M., Lewis, L.D., Memoli, V., Deo, Y., Graziano, R., Fisher, J.L., Meyer, L., Mrozek-Orlowski, M. et al.. Phase Ia/Ib trial of bispecific antibody MDX-210 in patients with advanced breast or ovarian cancer that overexpresses the proto-oncogene HER-2/neu. *J. Clin. Oncol.* 1995, **13,** 2281–2292.
38. Weiner, L.M., Clark, J.I., Ring, D.B., and Alpaugh, R.K. Clinical development of 2B1, a bispecific murine monoclonal antibody targeting c-erbB-2 and Fc gamma RIII. *J. Hematother.* 1995, **4,** 453–456.
39. Weiner, L.M., Alpaugh, R.K., Amoroso, A.R., Adams, G.P., Ring, D.B., and Barth, M.W. Human neutrophil interactions of a bispecific monoclonal antibody targeting tumor and human Fc gamma RIII. *Cancer Immunol. Immunother.* 1996, **42,** 141–150.
40. Mccall, A.M., Adams, G.P., Amoroso, A.R., Nielsen, U.B., Zhang, L., Horak, E., Simmons, H., Schier, R., Marks, J.D., and Weiner, L.M. Isolation and characterization of an anti-CD16 single-chain Fv fragment and construction of an anti-HER2/neu/anti-CD16 bispecific scFv that triggers CD16-dependent tumor cytolysis. *Mol. Immunol.* 1999, **36,** 433–445.
41. Yokota, T., Milenic, D.E., Whitlow, M., and Schlom, J. Rapid tumor penetration of a single-chain Fv and comparison with other immunoglobulin forms. *Cancer Res.* 1992, **52,** 3402–3408.
42. Casey, J.L., Keep, P.A., Chester, K.A., Robson, L., Hawkins, R.E., and Begent, R.H. Purification of bacterially expressed single chain Fv antibodies for clinical applications using metal chelate chromatography. *J. Immunol. Methods* 1995, **179,** 105–116.
43. Begent, R.H., Verhaar, M.J., Chester, K.A., Casey, J.L., Green, A.J., Napier, M.P., Hope-Stone, L.D., Cushen, N., Keep, P.A., Johnson, C.J., Hawkins, R.E., Hilson, A.J., and Robson, L. Clinical evidence of efficient tumor targeting based on single-chain Fv antibody selected from a combinatorial library. *Nature Med.* 1996, **2,** 979–984.
44. Neri, D., Carnemolla, B., Nissim, A., Leprini, A., Querze, G., Balza, E., Pini, A., Tarli, L., Halin, C., Neri, P., Zardi, L., and Winter, G. Targeting by affinity-matured recombinant antibody fragments of an angiogenesis associated fibronectin isoform. *Nature Biotechnol.* 1997, **15,** 1271–1275.
45. Bader, B., Kuhn, K., Owen, D.J., Waldmann, H., Wittinghofer, A., and Kuhlmann, J. Bioorganic synthesis of lipid-modified proteins for the study of signal transduction. *Nature* 2000, **403,** 223–226.

46. Huang, X., Gottstein, C., Brekken, R.A., and Thorpe, P.E. Expression of soluble VEGF receptor 2 and characterization of its binding by surface plasmon resonance. *Biochem. Biophys. Res. Commun.* 1998, **252,** 643–648.

47. Knibiehler, M., Goubin, F., Escalas, N., Jónsson, Z.O., Mazarguil, H., Hübscher, U., and Ducommun, B. Interaction studies between the p21Cip1/Waf1 cyclin-dependent kinase inhibitor and proliferating cell nuclear antigen (PCNA) by surface plasmon resonance. *FEBS Lett.* 1996, **391,** 66–70.

48. Kanaoka, Y., Kimura, S.H., Okazaki, I., Ikeda, M., and Nojima, H. GAK: a cyclin G associated kinase contains a tensin/auxilin-like domain. *FEBS Lett.* 1997, **402,** 73–80.

49. Shin, S., Sung, B.J., Cho, Y.S., Kim, H.J., Ha, N.C., Hwang, J.I., Chung, C.W., Jung, Y.K., and Oh, B.H. An anti-apoptotic protein human survivin is a direct inhibitor of caspase-3 and -7. *Biochemistry* 2001, **40,** 1117–1123.

50. Heider, K.H., Sproll, M., Susani, S., Patzelt, E., Beaumier, P., Ostermann, E., Ahorn, H., and Adolf, G.R. Characterization of a high-affinity monoclonal antibody specific for CD44v6 as candidate for immunotherapy of squamous cell carcinomas. *Cancer Immunol. Immunother.* 1996, **43,** 245–253.

51. Van Hal, N.L., Van Dongen, G.A., Ten Brink, C.B., Heider, K.H., Rech-Weichselbraun, I., Snow, G.B., and Brakenhoff, R.H. Evaluation of soluble CD44v6 as a potential serum marker for head and neck squamous cell carcinoma. *Clin. Cancer Res.* 1999, **5,** 3534–3541.

52. Verel, I., Heider, K.H., Siegmund, M., Ostermann, E., Patzelt, E., Sproll, M., Snow, G.B., Adolf, G.R., and Van Dongen, G.A. Tumor targeting properties of monoclonal antibodies with different affinity for target antigen CD44V6 in nude mice bearing head-and-neck cancer xenografts. *Int. J. Cancer* 2002, **99,** 396–402.

53. Ritter, G., Cohen, L.S., Williams, C., Jr., Richards, E.C., Old, L.J., and Welt, S. Serological analysis of human anti-human antibody responses in colon cancer patients treated with repeated doses of humanized monoclonal antibody a33. *Cancer Res.* 2001, **61,** 6851–6859.

54. Alaedini, A., and Latov, N. A surface plasmon resonance biosensor assay for measurement of anti-GM(1) antibodies in neuropathy. *Neurology* 2001, **56,** 855–860.

55. Liljeblad, M., Ryden, I., Ohlson, S., Lundblad, A., and Pahlsson, P. A lectin immunosensor technique for determination of alpha(1)-acid glycoprotein fucosylation. *Anal. Biochem.* 2001, **288,** 216–224.

56. Eivazova, E.R., Mcdonnell, J.M., Sutton, B.J., and Staines, N.A. Specificity and binding kinetics of murine lupus anti-DNA monoclonal antibodies implicate different stimuli for their production (In Process Citation). *Immunology* 2000, **101,** 371–377.

57. Guermazi, S., Regnault, V., Gorgi, Y., Ayed, K., Lecompte, T., and Dellagi, K. Further evidence for the presence of anti-protein S autoantibodies in patients with systemic lupus erythematosus (In Process Citation). *Blood Coagul. Fibrinolysis.* 2000, **11,** 491–498.

58. Jones, D.W., Nicholls, P.J., Donohoe, S., Gallimore, M.J., and Winter, M. Antibodies to factor XII are distinct from antibodies to prothrombin in patients with the anti-phospholipid syndrome. *Thromb. Haemost.* 2002, **87,** 426–430.

59. Nedelkov, D., and Nelson, R.W. Analysis of human urine protein biomarkers via biomolecular interaction analysis mass spectrometry. *Am. J Kidney Dis.* 2001, **38,** 481–487.

60. Bohrmann, B., Tjernberg, L., Kuner, P., Poli, S., Levet-Trafit, B., Naslund, J., Richards, G., Huber, W., Dobeli, H., and Nordstedt, C. Endogenous proteins controlling amyloid beta-peptide polymerization. Possible implications for beta-amyloid formation in the central nervous system and in peripheral tissues. *J. Biol. Chem.* 1999, **274,** 15990–15995.

61. Vorum, H., Jacobsen, C., and Honore, B. Calumenin interacts with serum amyloid P component. *FEBS Lett.* 2000, **465,** 129–134.

62. Lefebvre, S., Burglen, L., Reboullet, S., Clermont, O., Burlet, P., Viollet, L., Benichou, B., Cruaud, C., Millasseau, P., Zeviani, M. et al. Identification and characterization of a spinal muscular atrophy-determining gene. *Cell* 1995, **80,** 155–165.

63. Monani, U.R., Lorson, C.L., Parsons, D.W., Prior, T.W., Androphy, E.J., Burghes, A.H., and Mcpherson, J.D. A single nucleotide difference that alters splicing patterns distinguishes the SMA gene SMN1 from the copy gene SMN2. *Hum. Mol. Genet.* 1999, **8,** 1177–1183.
64. Young, P.J., Man, N., Lorson, C.L., Le, T.T., Androphy, E.J., Burghes, A.H., and Morris, G.E. The exon 2b region of the spinal muscular atrophy protein, SMN, is involved in self-association and SIP1 binding. *Hum. Mol. Genet.* 2000, **9,** 2869–2877.
65. Ariga, T., and Yu, R.K. GM1 inhibits amyloid beta-protein-induced cytokine release. *Neurochem Res.* 1999, **24,** 219–226.
66. Catimel, B., Scott, A.M., Lee, F.T., Hanai, N., Ritter, G., Welt, S., Old, L.J., Burgess, A.W., and Nice, E.C. Direct immobilization of gangliosides onto gold-carboxymethyldextran sensor surfaces by hydrophobic interaction: applications to antibody characterization. *Glycobiology* 1998, **8,** 927–938.
67. Nakamura, K., Tanaka, Y., Shitara, K., and Hanai, N. Construction of humanized anti-ganglioside monoclonal antibodies with potent immune effector functions. *Cancer Immunol. Immunother.* 2001, **50,** 275–284.
68. Snider, W.D. Functions of the neurotrophins during nervous system development: what the knockouts are teaching us. *Cell* 1994, **77,** 627–638.
69. Ip, N.Y., and Yancopoulos, G.D. The neurotrophins and CNTF: two families of collaborative neurotrophic factors. *Annu. Rev. Neurosci.* 1996, **19,** 491–515.
70. Li, S.-C., Lai, K.-M.V., Gish, G.D., Parris, W.E., Van Der Geer, P., Forman-Kay, J., and Pawson, T. Characterization of the phosphotyrosine-binding domain of the Drosophila Shc protein. *J. Biol. Chem.* 1996, **271,** 31855–31862.
71. Woo, S.B., Whalen, C., and Neet, K.E. Characterization of the recombinant extracellular domain of the neurotrophin receptor TrkA and its interaction with nerve growth factor (NGF). *Prot. Sci.* 1998, **7,** 1006–1016.
72. Ross, G.M., Shamovsky, I.L., Woo, S.B., Post, J.I., Vrkljan, P.N., Lawrance, G., Solc, M., Dostaler, S.M., Neet, K.E., and Riopelle, R.J. The binding of zinc and copper ions to nerve growth factor is differentially affected by pH: implications for cerebral acidosis. *J. Neurochem.* 2001, **78,** 515–523.
73. [73]Holcomb, I.N., et al. FIZZ1, a novel cysteine-rich secreted protein associated with pulmonary inflammation, defines a new gene family. *EMBO J.* 2000, **19,** 4046–4055.
74. Cairo, C.W., Strzelec, A., Murphy, R.M., and Kiessling, L.L. Affinity-based inhibition of beta-amyloid toxicity. *Biochemistry* 2002, **41,** 8620–8629.
75. Yamazaki, M., Mori, H., Araki, K., Mori, K.J., and Mishina, M. Cloning, expression and modulation of a mouse NMDA receptor subunit. *FEBS Lett.* 1992, **300,** 39–45.
76. Lomeli, H., Sprengel, R., Laurie, D.J., Kohr, G., Herb, A., Seeburg, P.H., and Wisden, W. The rat delta-1 and delta-2 subunits extend the excitatory amino acid receptor family. *FEBS Lett.* 1993, **315,** 318–322.
77. Araki, K., Meguro, H., Kushiya, E., Takayama, C., Inoue, Y., and Mishina, M. Selective expression of the glutamate receptor channel delta 2 subunit in cerebellar Purkinje cells. *Biochem. Biophys. Res. Commun.* 1993, **197,** 1267–1276.
78. Miyagi, Y., Yamashita, T., Fukaya, M., Sonoda, T., Okuno, T., Yamada, K., Watanabe, M., Nagashima, Y., Aoki, I., Okuda, K., Mishina, M., and Kawamoto, S. Delphilin: a novel PDZ and formin homology domain-containing protein that synaptically colocalizes and interacts with glutamate receptor delta 2 subunit. *J. Neurosci.* 2002, **22,** 803–814.
79. Young, P.J., Day, P.M., Zhou, J., Androphy, E.J., Morris, G.E. and Lorson, C.L. A direct interaction between the survival motor neuron protein and p53 and its relationship to spinal muscular atrophy. *J. Biol. Chem.* 2002, **277,** 2852–2859.
80. Laughton, C.A., Tanious, F., Nunn, C.M., Boykin, D.W., Wilson, W.D., and Neidle, S. A crystallographic and spectroscopic study of the complex between d(CGCGAATTCGCG)2 and 2,5-bis(4-guanylphenyl)furan, an analogue of berenil. Structural origins of enhanced DNA-binding affinity. *Biochemistry* 1996, **35,** 5655–5661.

81. Carrasco, C., Facompre, M., Chisholm, J.D., Van Vranken, D.L., Wilson, W.D., and Bailly, C. DNA sequence recognition by the indolocarbazole antitumor antibiotic AT2433-B1 and its diastereoisomer. *Nucleic Acids Res.* 2002, **30,** 1774–1781.

82. Carrasco, C., Vezin, H., Wilson, W.D., Ren, J., Chaires, J.B., and Bailly, C. DNA binding properties of the indolocarbazole antitumor drug NB-506. *Anticancer Drug Des.* 2001, **16,** 99–107.

83. Wang, L., Bailly, C., Kumar, A., Ding, D., Bajic, M., Boykin, D.W., and Wilson, W.D. Specific molecular recognition of mixed nucleic acid sequences: an aromatic dication that binds in the DNA minor groove as a dimer. *Proc. Natl Acad. Sci. USA* 2000, **97,** 12–16.

84. Davis, T.M., and Wilson, W.D. Determination of the refractive index increments of small molecules for correction of surface plasmon resonance data. *Anal. Biochem.* 2000, **284,** 348–353.

85. Sando, S., Saito, I., and Nakatani, K. Scanning of guanine–guanine mismatches in DNA by synthetic ligands using surface plasmon resonance. *Nature Biotechnol.* 2001, **19,** 51–55.

86. Kwok, P.Y. Reflections on a DNA mutation scanning tool. *Nature Biotechnol.* 2001, **19,** 18–19.

87. Lacy, E.R., Le, N.M., Price, C.A., Lee, M., and Wilson, W.D. Influence of a terminal formamido group on the sequence recognition of DNA by polyamides. *J. Am. Chem. Soc.* 2002, **124,** 2153–2163.

88. Hickman, D.T., King, P.M., Slater, J.M., Cooper, M.A., and Micklefield, J. Kinetically selective binding of single stranded RNA over DNA by a pyrrolidine-amide oligonucleotide mimic (POM). *Nucleosides Nucleotides Nucleic Acids* 2001, **20,** 1169–1172.

89. Hou, M.H., Lin, S.B., Yuann, J.M., Lin, W.C., Wang, A.H., and Kan Ls, L. Effects of polyamines on the thermal stability and formation kinetics of DNA duplexes with abnormal structure. *Nucleic Acids Res.* 2001, **29,** 5121–5128.

90. Tsoi, P.Y., and Yang, M. Kinetic study of various binding modes between human DNA polymerase beta and different DNA substrates by surface-plasmon-resonance biosensor. *Biochem. J.* 2002, **361,** 317–325.

91. Summerton, J., and Weller, D. Morpholino antisense oligomers: design, preparation, and properties. *Antisense Nucleic Acid Drug Dev.* 1997, **7,** 187–195.

92. Mang'era, K.O., Liu, G., Yi, W., Zhang, Y., Liu, N., Gupta, S., Rusckowski, M., and Hnatowich, D.J. Initial investigations of 99mTc-labeled morpholinos for radiopharmaceutical applications. *Eur. J. Nucl. Med.* 2001, **28,** 1682–1689.

93. Astriab-Fisher, A., Sergueev, D., Fisher, M., Shaw, B.R., and Juliano, R.L. Conjugates of antisense oligonucleotides with the Tat and antennapedia cell-penetrating peptides: effects on cellular uptake, binding to target sequences, and biologic actions. *Pharm. Res.* 2002, **19,** 744–754.

94. Bates, P., Reddoch, J., Hansakul, P., Arrow, A., Dale, R., and Miller, D. Biosensor detection of triplex formation by modified oligonucleotides. *Anal. Biochem.* 2002, **307,** 235.

95. Okumoto, Y., Ohmichi, T., and Sugimoto, N. Immobilized small deoxyribozyme to distinguish RNA secondary structures. *Biochemistry* 2002, **41,** 2769–2773.

96. Nilsson, P., Persson, B., Uhlén, M., and Nygren, P.-Å. Real-time monitoring of DNA manipulations using biosensor technology. *Anal. Biochem.* 1995, **224,** 400–408.

97. Steinrucke, P., Aldinger, U., Hill, O., Hillisch, A., Basch, R., and Diekmann, S. Design of helical proteins for real-time endoprotease assays. *Anal. Biochem.* 2000, **286,** 26–34.

98. Loew, D., Perrault, C., Morales, M., Moog, S., Ravanat, C., Schuhler, S., Arcone, R., Pietropaolo, C., Cazenave, J.P., Van Dorsselaer, A., and Lanza, F. Proteolysis of the exodomain of recombinant protease-activated receptors: prediction of receptor activation or inactivation by MALDI mass spectrometry. *Biochemistry* 2000, **39,** 10812–10822.

99. Zako, T., Harada, K., Mannen, T., Yamaguchi, S., Kitayama, A., Ueda, H., and Nagamune, T. Monitoring of the refolding process for immobilized firefly luciferase with a biosensor based on surface plasmon resonance. *J. Biochem. (Tokyo)*. 2001, **129,** 1–4.

100. Houseman, B.T., Huh, J.H., Kron, S.J., and Mrksich, M. Peptide chips for the quantitative evaluation of protein kinase activity. *Nature Biotechnol.* 2002, **20,** 270–274.

101. Katsamba, P.S., Bayramyan, M., Haworth, I.S., Myszka, D.G., and Laird-Offringa, I.A. Complex role of the beta 2–beta 3 loop in the interaction of U1A with U1 hairpin II RNA. *J. Biol. Chem.* 2002, **277,** 33267–33274.

102. Nelson, R.W., Nedelkov, D., and Tubbs, K.A. Biosensor chip mass spectrometry: a chip-based proteomics approach. *Electrophoresis* 2000, **21,** 1155–1163.

103. Krone, J.R., Nelson, R.W., Dogruel, D., Williams, P., and Granzow, R. BIA/MS: interfacing biomolecular interaction analysis with mass spectrometry. *Anal. Biochem.* 1997, **244,** 124–132.

104. Sönksen, C.P., Nordhoff, E., Jansson, Ö., Malmqvist, M., and Roepstorff, P. Combining MALDI mass spectrometry and biomolecular interaction analysis using a biomolecular interaction analysis instrument. *Anal. Chem.* 1998, **70,** 2731–2736.

105. Williams, I., and Addona, T.A. The integration of SPR biosensors with mass spectrometry: possible applications for proteome analysis (In Process Citation). *Trends Biotechnol.* 2000, **18,** 45–48.

5
Surface Plasmon Resonance Imaging Measurements of DNA, RNA, and Protein Interactions to Biomolecular Arrays

Greta J. Wegner, Hye Jin Lee and Robert M. Corn

5.1
Introduction

Multi-component arrays of immobilized biomolecules are rapidly becoming essential tools for the screening of bioaffinity interactions. Biomolecular array technology is especially amenable to proteomics research, since much of a protein's function is accomplished through specific interactions with DNA, peptides, and other proteins. Array-based assay formats have been developed both for the detection of proteins in a sample [1], and for the identification of new protein–protein interactions [2,3]. Arrays are also now regularly employed for fundamental studies of protein binding interactions. For example, peptide arrays have been used to study the binding interactions of antibodies [4,5] and to characterize the sequence-specific activity of enzymes [6,7]. Arrays of proteins have also been used for the high-throughput screening of protein–protein interactions and to identify novel biochemical activities [2,3].

The majority of array-based assays currently employ fluorescent, enzymatic or radiolabeled biomolecules. Further advancement of analytical systems that do not require labeled biological molecules to detect affinity binding interactions is needed. Surface plasmon resonance (SPR) imaging is emerging as a label-free surface sensitive optical technique that measures the adsorption of solution phase molecules to arrays of biomolecules on thin gold films by changes in the local index of refraction. Label-free detection is particularly advantageous for the assay of small molecule-binding to proteins, since labels can interfere with the biological activity of the target molecule. To date, SPR imaging has been used to study reversible protein binding to DNA, peptide, protein, and carbohydrate arrays [8–10].

This review summarizes the recent advances that we have made in the implementation of SPR imaging to study DNA, RNA, and protein interactions with DNA, peptide and protein probes immobilized in an array format. After a brief description of the SPR imaging technique in Section 5.2, two surface attachment chemistries used for the covalent immobilization of thiol-modified DNA, peptides, and carbohydrates through the modification of self-assembled monolayers of alkanethiols are discussed in Section 5.3. In Section 5.4, an array fabrication method based on chemical protection and photopatterning is described. Several examples

Protein Microarray Technology. Edited by Dev Kambhampati

ISBN: 3-527-30597-1

of SPR imaging experiments using photopatterned arrays are presented including the study of DNA–DNA hybridization and protein–DNA binding interactions. We will conclude this rchapter in Section 5.5 with a second array fabrication method based on microfluidic stenciling and address its application to peptide–protein and protein–protein binding interactions.

5.2 Surface Plasmon Resonance Imaging

Surface plasmon resonance (SPR) measurements are surface-sensitive optical methods used to characterize organic and biological layers on gold or noble metal thin films. These measurements utilize the optical field enhancement that occurs at a metal/dielectric interface when surface plasmons are created. Surface plasmons are electromagnetic waves, excited by p-polarized light, that propagate parallel to the gold surface. The optical fields decay exponentially from the surface of the metal and have a maximum decay length of about 200 nm. Within this region, the optical fields are sensitive to changes in the index of refraction caused by the adsorption or desorption of molecules onto the surface. As a consequence, SPR experiments have been amenable for the label-free study of reversible biomolecule binding interactions on gold films in real-time using two different methods: (1) scanning angle SPR and (2) SPR imaging techniques.

Angle shift measurements are the most commonly undertaken SPR technique and have been popularized by the commercial Biacore instrument. In a scanning angle experiment, the percentage of light reflected off of a thin gold film optically coupled to a prism is measured as a function of incident angle. A prism or grating coupling arrangement is required because surface plasmons cannot be excited directly at planar air/metal or water/metal interfaces because momentum-matching conditions are not satisfied. A hemispherical prism is used to tightly focus light from a laser onto a single region of a gold thin film. Figure 5.1 shows three SPR curves and their theoretical fits where the percent reflectivity of light is measured as a function of the incident angle for light with excitation wavelengths of 1152 (Δ), 814 (❑), and 633 nm (❍) Δ [11]. For the SPR curve measured at an excitation wavelength of 814 nm, increasing the incident angle from 50 to 50.8° causes nearly 100% of the light to be reflected as the critical angle is reached and total internal reflection occurs. Further increase in the angle results in a steep drop in percent reflectivity as light is incorporated into surface plasmons until a minimum is reached at 53.5°, the surface plasmon angle. The position of the SPR angle is sensitive to changes in the index of refraction and/or thickness of the adsorbed film causing the minimum to shift. Shifts in angle position can be correlated to changes in thickness or index of refraction using N-phase Fresnel calculations as described by Hansen [12]. SPR measurements can be enhanced by using near-infrared excitation wavelengths between 800 and 1100 nm. Mid-IR light results in sharper SPR curves than those obtained at an excitation wavelength of 633 nm from a HeNe source (refer to Figure 5.1). This allows a more precise determination of the SPR angle and can be used to measure thicker films.

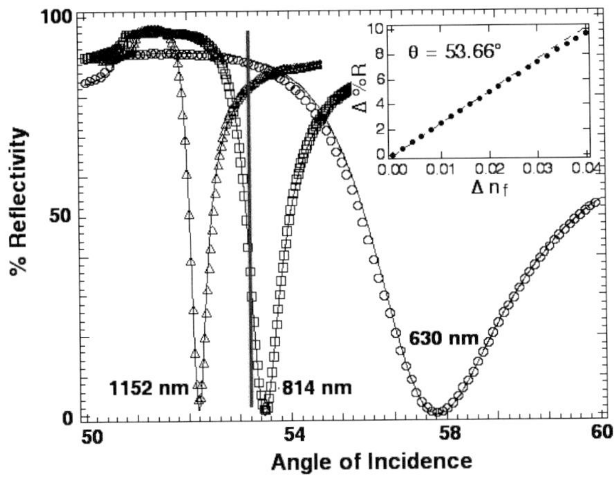

Figure 5.1 SPR reflectivity curves showing the relationship between percentage reflectivity and angle of incidence for excitation wavelengths of 1152 nm (Δ), 814 nm (❑), and 633 nm (❍) for an SF-10 glass/Au/water *(in situ)* assembly. Solid lines show the theoretical fit obtained by three-phase Fresnel calculations. Sharper SPR curves are obtained with increasing wavelength and can be applied to study thicker films with angle shift measurements and to enhance the contrast of SPR imaging experiments. The angle at which SPR imaging measurements are obtained is marked by a vertical line on the SPR reflectivity curve obtained at an excitation wavelength of 814 nm. Inset shows the relationship between the change in percentage reflectivity (Δ%R) and the change in index of refraction (Δn_f) at a fixed incident angle of 56.33°. A linear relationship is observed for Δ%R less than 10%. Reprinted with permission from *Analytical Chemistry*, 71 3928–3934, copyright 1999 American Chemical Society and *Analytical Chemistry*, 73 1–7, copyright 2001 American Chemical Society.

While scanning angle experiments are used to study a single region on a gold surface, SPR imaging is used to monitor the spatially resolved adsorption of biomolecules onto a multicomponent array. This technique is especially important for biological experiments where the simultaneous measurement of multiple interactions on a single chip is desired. SPR imaging measurements of the change in percent reflectivity are performed at a fixed angle near the SPR angle, as shown by a vertical line passing through the SPR reflectivity curve measured using an excitation wavelength of 814 nm as shown in Figure 5.1. Figure 5.2 shows a schematic representation of the SPR imaging instrument used to detect the adsorption of biopolymers in solution to surface-immobilized biomolecules such as DNA and peptides [11]. A collimated white light source is coupled with a narrow band pass filter centered at 830 nm and used in place of a laser source. By using this configuration, laser fringes which can interfere with image quality are avoided and the enhanced contrast obtained using an NIR source is maintained. Images with lateral resolutions on the order of 50 μm can be obtained using an excitation wavelength of 830 nm. The light is passed through a polarizer in a direction incident to a high index glass prism/sample assembly. Here, light impinges on the back of a sample consisting of a gold thin film chemically modified with an array of biomolecules. Analytes are delivered to the array by either a 500-μL flow cell or through a set of parallel microfluidic channels from PDMS requiring 1 μl of sample per channel. Images are collected using a CCD camera after the light reflected off the sample is passed through a narrow band pass filter. The light interacts with the biomolecules immobilized on the thin gold film to create surface plasmons, inducing attenuation in the light reflected from the surface. As biomolecules from solution adsorb onto the array, the local index of refraction changes causing an increase in the percentage of incident light reflected off each array element of the sample, at a fixed optimal angle. The lower left corner of Figure 5.2 contains a typical SPR difference image showing sequence-specific adsorption of complementary DNA onto a two-component DNA array.

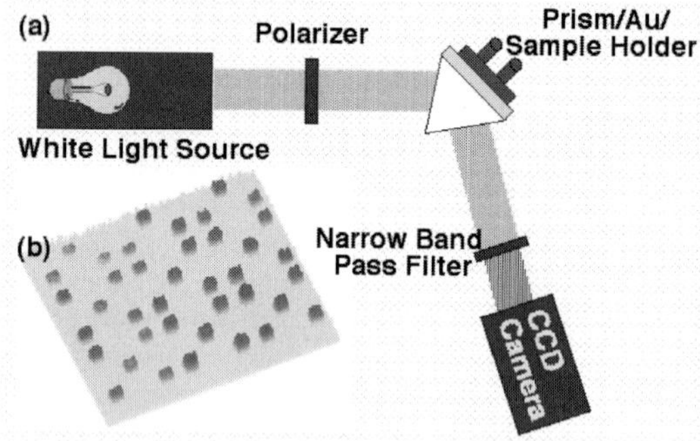

Figure 5.2 Schematic diagram of the surface plasmon imager apparatus. P-polarized white light impinges on a prism sample apparatus at a fixed optimal angle, near the surface plasmon angle. The light is then passed through a narrow band pass filter and images are collected by a CCD camera. The analyte is delivered to the array fabricated on a gold thin film using a simple 500 μL flow cell. An SPR difference image of the sequence-specific hybridization of 18-mer DNA to a two-component DNA array is shown in the bottom left corner of the figure. A maximum of 160 array elements can be studied on one SPR chip with a total surface area of 0.8 cm^2 using square elements with 500 μm widths. (see Colour Plate p. XXIII).

Like the scanning angle SPR measurements, quantitative information can be obtained from SPR imaging measurements. The inset in Figure 5.1 shows the relationship between the percentage reflectivity and the index of refraction for changes in percentage reflectivity under 10% [13]. This relationship can also be applied to determine the surface coverage of biomolecules adsorbed onto a surface. SPR imaging measurements have been used for the quantitative evaluation of the interactions between biomolecules in solution and DNA or peptide probes immobilized on the array. In the case of biomolecules that have a high molecular weight, the number of immobilized surface target sites can be decreased in order to stay within this linear Δ %R region [10]. The sensitivity of the SPR imaging approach is about 10 femtomoles (1 μL of a 10 nM solution) for 18-mer single-stranded DNA hybridized to a DNA array [14], and 1 femtomole (1 μL of a 1 nM solution) for specific antibody adsorption onto a peptide array [10]. Although SPR imaging measurements do not match the sensitivity of fluorescence or radioactive methods, it is more than adequate for many applications including the study of protein–protein interactions and has the advantage of employing direct detection of binding interactions.

5.3 Surface Attachment Chemistries

The development of well-characterized surface chemistries for the attachment of biological molecules onto gold thin films in an array format is necessary for SPR imaging measurements of biomolecular affinity interactions. Noble metal thin films are required for the propagation of surface plasmons. As a consequence, commercially available DNA arrays made on glass substrates, such as those produced by Affymetrix [15–18], or peptide arrays prepared by the SPOT synthesis technique on cellulose membranes [19–21], are not viable options for SPR imaging investigations. Instead, we have employed self-assembled monolayers of long chain alkanethiols that are ω-terminated with an amine functional group as the foundation of the array [8,9,11,22]. Chemical modification of the self-assembled monolayers is used to tether biological molecules to the surface. In this section, two methods using the bifunctional linkers SSMCC (sulfosuccinimidyl 4-(*N*-maleimidomethyl) cyclohexane-1-carboxylate) and SATP (*N*-Succinimidyl S-acetylthiopropionate) are described for the covalent attachment of thiol-containing DNA, peptide, carbohydrate and capture probe molecules.

5.3.1 SSMCC Attachment Chemistry

In the first reaction scheme, the heterobifunctional linker SSMCC, which contains both an N-hydroxysulfosuccinimide (NHSS) ester and a maleimide functionality, is used to covalently link the amine-terminated self-assembled monolayer of 11-mercapto-undecylamine (MUAM) to a thiol-modified probe molecule (Figure 5.3a) [8,22]. First, the NHSS ester moiety of SSMCC is reacted with the free amines of the alkanethiol monolayer to form amide bonds terminating with maleimide groups. Next, thiol-modified DNA or cysteine-containing peptides are introduced to the thiol-reactive maleimide monolayer and attached by the creation of a thioether bond. DNA molecules attached by this method have a surface coverage of 1.0×10^{12} molecules cm^{-2} [13].

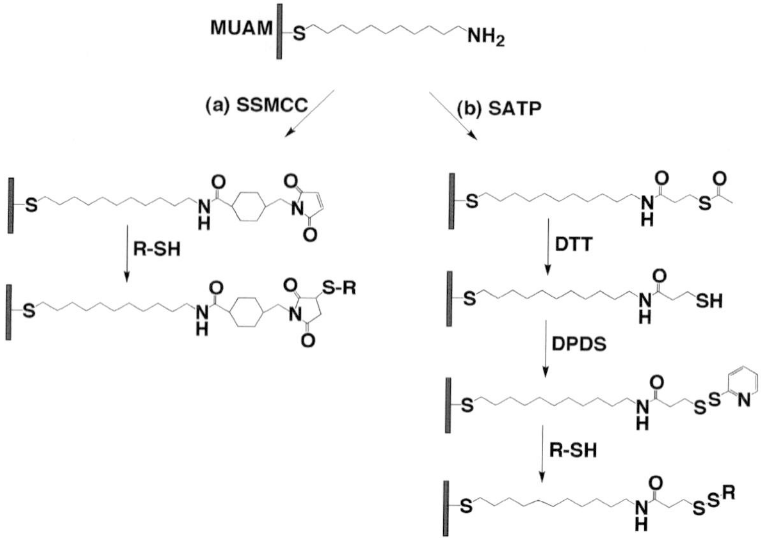

Figure 5.3 Surface attachment chemistry for the immobilization of thiol-modified DNA and cysteine-containing peptides: (a) The linker SSMCC is reacted with a well-packed self-assembled monolayer of 11-mercaptoundecylamine (MUAM) to create a maleimide-modified surface. The maleimide surface is then used to covalently attach thiol-modified DNA or cysteine-containing peptides. (b) In the second approach, SATP is reacted with MUAM to create a protected thiol surface. Upon deprotection with a basic solution containing DTT, the free sulfhydryl is reacted with dipyridyl disulfide to create a pyridyl disulfide surface. Thiol–disulfide exchange reactions are used to couple thiol-containing biomolecules to the surface.

5.3.2
SATP Attachment Chemistry

An alternative reaction scheme, more amenable for generating peptide arrays, involves the attachment of thiol-containing biomolecules onto gold surfaces through a disulfide linkage [10,23]. The reaction scheme for this surface modification process is shown in Figure 5.3b. In this approach, a self-assembled monolayer of MUAM is reacted with the molecule SATP, a bifunctional linker containing an NHS ester and a protected sulfhydryl. The NHS ester reacts with the amines present on the surface to form a stable amide bond resulting in a protected sulfhydryl surface. Exposure of the surface to an alkaline solution containing DTT removes the protecting acetyl group revealing an active sulfhydryl surface. Next, the sulfhydryl surface is reacted with 2,2'–dipyridyl disulfide to form a pyridyl disulfide surface. Thiol-disulfide exchange reactions are then performed in order to attach thiol-modified DNA or cysteine-containing peptides to the substrate, with a surface coverage of 10^{13} molecules cm^{-2}. This type of surface attachment chemistry has the advantage of being reversible, as the disulfide bond can be cleaved in the presence of DTT to regenerate the sulfhydryl-terminated surface.

5.4
SPR Imaging Experiments Using Photopatterned Arrays

Arrays used for SPR imaging experiments contain multiple individually addressable components and are prepared by a combination of self-assembly, surface attachment chemistry, and array patterning using either photopatterning or PDMS microfluidic channels. This section will explore the first array fabrication methodology developed within our laboratories for the creation of robust DNA and peptide arrays using chemical protection and deprotection and photopatterning. This multi-step fabrication procedure is outlined in Figure 5.4 [8]. The first step of the fabrication method is the creation of a temporary hydrophobic background using the bulky amine protecting group, 9-fluorenylmethoxycarbonyl (Fmoc), frequently used in solid phase peptide synthesis. The N-hydroxysuccinimide ester of Fmoc (Fmoc-NHS) is reacted with the terminal amines of a packed-self assembled alkanethiol monolayer to form a stable carbamate linkage. The surface is then exposed to UV light through a patterned quartz mask to create arrays composed of 500 by 500 μm square elements with 500 μm spacing between the elements. The gold–thiol bond is cleaved in the regions where the light shines through the mask resulting in a patterned gold surface that consists of bare gold pads surrounded by Fmoc. Next, the surface is immersed in an ethanolic solution of MUAM. The amine-terminated alkanethiols self-assemble in the bare gold pads producing a surface containing reactive hydrophilic MUAM pads surrounded by the hydrophobic Fmoc background. The lower left-hand corner of Figure 5.4 contains an image of an array showing the individually addressable hydrophilic spots surrounded by a hydrophobic background. Chemical protection allows droplets of hydrophilic DNA or peptides to be "pinned", without contaminating other sequences contained on neighboring pads on the array, and covalently attached using the linkers SSMCC or SATP. These arrays can be easily used with a mechanical array spotter to introduce multiple peptide or DNA elements onto one chip for high-throughput studies. For example, arrays composed of more than 300 elements can be produced using square elements with 250-μm widths.

The array fabrication is then completed by removing the hydrophobic Fmoc background and replacing it with polyethylene glycol (PEG) to reduce the non-specific adsorption of biopolymers. Since Fmoc is a base-labile protecting group, the original amine-terminated surface can be easily regenerated by exposure to a secondary amine. An NHS derivative of polyethylene glycol (PEG) can react with the terminal amines of the SAM to form a stable amide linkage. The formation of the temporary Fmoc surface is necessary, since PEG is too hydrophilic to prevent contamination between DNA or peptide droplets. The DNA arrays can then be analyzed by SPR imaging using a 500 μL flow cell to introduce biopolymer analytes to the array surface. DNA arrays fabricated using SSMCC attachment chemistry can be used for more than 25 hybridization cycles [13] while the SATP chemistry can be used for more than 30 cycles without degradation of the array [23].

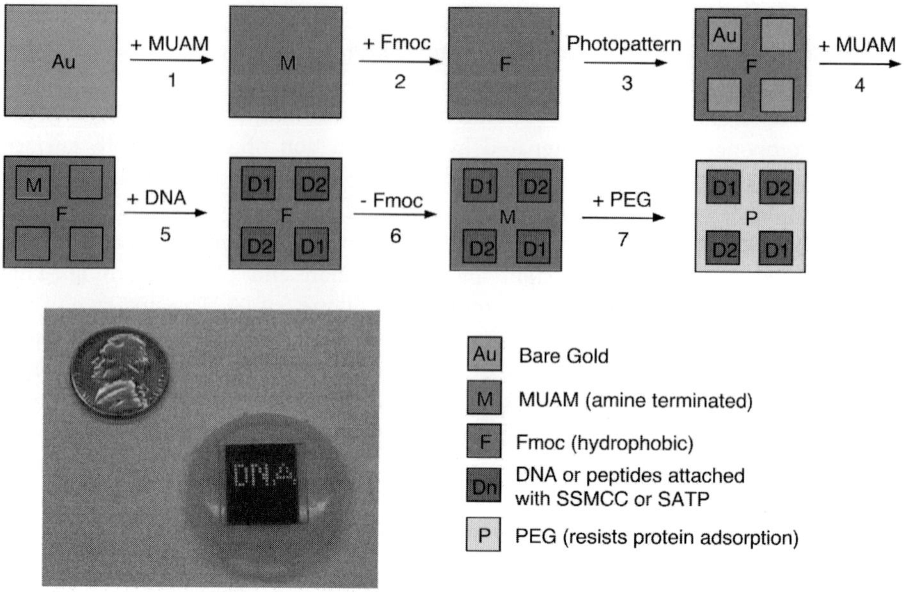

Figure 5.4 Multi-step array fabrication process using protection/deprotection chemistry and photopatterning techniques to prepare DNA and peptide arrays. The bottom left corner of the figure shows individually addressable hydrophilic drops of DNA pinned to the surface by a hydrophobic background. First, a bare gold surface is modified with a self-assembled monolayer of 11-mercapto-undecylamine (MUAM). This amine-terminated surface is reacted with the hydrophobic protecting group Fmoc. UV-light is used to break the gold–thiol bond and create bare gold pads on the surface. These pads are subsequently filled with MUAM. A bifunctional linker is used to attach DNA or peptide probes to create an array. Finally, the Fmoc is removed with base and replaced with a polyethylene derivative to prevent the non-specific adsorption of biopolymers to the background. Reprinted with permission from *Journal of the American Chemical Society*, **121**, 8044–8051. Copyright 1999 American Chemical Society. (see Colour Plate p. XXIV).

5.4.1 DNA–DNA Hybridization

The hybridization of complementary DNA to a five-component DNA array fabricated by photopatterning and SSMCC chemistry was examined by SPR imaging. All surface-immobilized DNA probes were tested by hybridization to respective perfect-match DNA complements. The probes were immobilized in a distinct geometric pattern to easily identify interacting sequences. The DNA sequences used in this experiment are listed in Table 5.1, and denoted as A–E. The surface was first exposed to a 100 nM solution of the 16-mer DNA oligonucleotide complement to probe A for 15 min (Figure 5.5a). The SPR difference image shown in this figure was produced by subtracting the images taken before and after hybridization. Regeneration of the surface was achieved by rinsing with 8 M urea. Repeated cycles of hybridization and denaturation were used to obtain Figure 5b–e. Excellent sequence-specific adsorption of the complementary DNA was observed for all probes and with little non-specific adsorption to the background. The approximately equal SPR signal under identical experimental conditions indicates that the probes are equally accessible for hybridization to target molecules in solution.

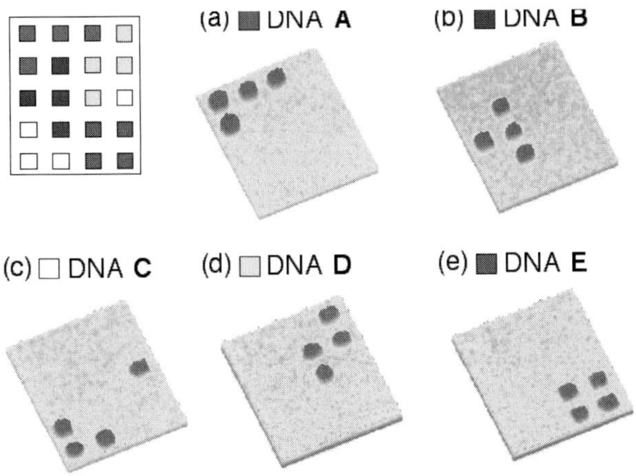

Figure 5.5 SPR difference images showing the hybridization of perfect-match DNA complements to a DNA array containing five different probes. The sequences were immobilized on 500 by 500 μm array elements in the pattern shown in the figure. (a) First, the array was exposed to a 100 nM solution of the DNA complement to probe A for 15 min. Hybridization adsorption onto the array is indicated by a change in the percentage reflectivity of incident light. (b) After rinsing with 8 M urea to regenerate the surface, the experiment was repeated with the DNA complement of probe B. Successive rounds of denaturation and hybridization of the remaining DNA complement probes resulted in images c–e. The same region of the array is shown for all images. (see Colour Plate p. XXV).

Table 5.1 DNA sequences used in Figure 5.5.

Surface-bound probe DNA	*DNA sequences (5'–3'*
Probe A	TAC TCA CCC GTC CGC C
Probe B	CTT TTA TGT TTG AAC CAT GCG
Probe C	GTG GCC GAT CAC CCT CTC
Probe D	CGT GGC TTT CTG GTT A
Probe E	CTT TGA GTT TCA GTC

Probe DNA is modified with a 5'-thiol modifier and a 15-T spacer is appended to the 5' end of the probe DNA.

5.4.2
Mismatch Binding Protein, MutS

Mismatch binding proteins, such as MutS in *E. coli* are used by cells to identify and repair mismatches, short insertions, and deletions that are caused by replication errors. The ability to identify mismatches in DNA is necessary for the technology used to detect mutations and single nucleotide polymorphism in double-stranded DNA. Such assays may be important in the future for correlating genomic information from an individual with the development of diseases [24–26]. As an example, SPR imaging was used to show the interaction of MutS with mismatch DNA sequences (A–D) immobilized onto an array [9,27]. The sequences and additional experimental information are reported elsewhere [9]. This photopatterned array was exposed to a solution containing the sequence Z. Probe A did not hybridize with sequence Z and remained single stranded. Sequence Z formed a duplex with probe B containing a two-base insertion, a perfectly matched duplex with probe C, and a duplex containing a one-base G/T mismatch with probe D. Figure 5.6 shows an SPR difference image after MutS was allowed to bind to the DNA array. Adsorption of MutS was observed to both duplex B containing the insertion and duplex D containing the G/T mismatch, which has the strongest mismatch interaction with MutS [28]. Minimal binding occurred to the perfect match or the single-stranded DNA, showing that it was possible to study the sequence-specific interactions of MutS to DNA using SPR imaging measurements of DNA arrays. The utility of mismatch binding proteins to identify single-base mismatches, as well as short insertions and deletions in double-stranded DNA will lead to the application of mismatch binding proteins in mutation and single nucleotide polymorphism detection assays.

5.4.3
Bacterial Response Regulators, OmpR and VanR

Response regulators are DNA binding proteins that are essential components of the two-component signal transduction phosphorylation relay system used by bacteria to respond to environmental stimuli and to adapt accordingly through transcriptional activation of certain genes [29]. Since few response regulators have been studied in depth, DNA arrays are a potential tool to determine the DNA sequences recognized by a particular response regulator. As an example, the sequence-specific binding of two response regulators, OmpR and VanR, to DNA probe molecules was studied [30]. OmpR is one of the most well-characterized response regulators and is important in bacterial response to environmental osmotic pressure [31], while VanR is involved in the development of antibiotic resistance [32].

A photopatterned array was prepared containing six different immobilized DNA probes known to interact with either OmpR or VanR. The DNA array was then exposed to solutions of corresponding complementary DNA to create surface-immobilized double-stranded DNA. Figure 5.7a shows an SPR difference image after the array was exposed to a 100 nM solution of OmpR. OmpR primarily

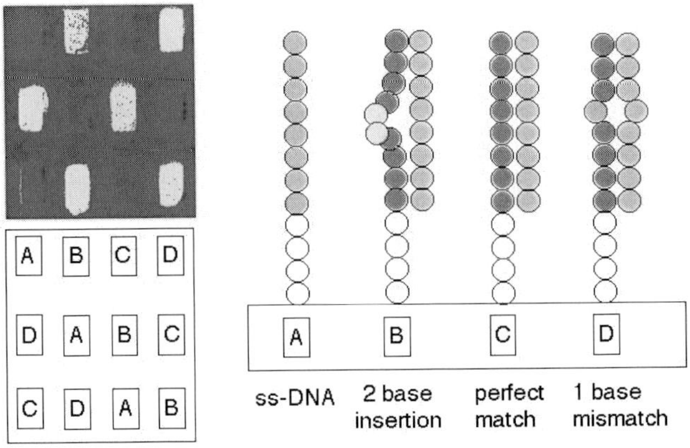

Figure 5.6 SPR imaging measurements of E. *coli* mismatch binding protein MutS adsorption onto a DNA array. This array was created with DNA probes A through D immobilized on 500 by 500 μm array elements in the pattern shown on the left of the figure. The array was then exposed to a solution containing the sequence Z. Z does not bind at all to probe A leaving it single strandedbut binds to probe B to create a duplex containing a two-base insertion, to probe C in a perfectly complementary manner, and to probe D to form a duplex containing a single-base mismatch. An SPR difference image of the binding of MutS to the array is shown on the right of the figure. The image shown is the difference between two images collected before and after exposure of the surface to MutS. Only 12 array elements from the total patterned surface area of 0.8 cm2 are presented in these images. Reprinted with permission from *Annual Review of Physical Chemistry* **51,** pp 41–63. Copyright 2000 by Annual Reviews www.annualreview.org. (see Colour Plate p. XXV).

absorbed to the DNA sequences OmpF1, F1F2, and C2 which were known to interact only with OmpR. The array was then regenerated with urea and exposed to complementary DNA before VanR (500 nM) was introduced to the array. VanR primarily bound to the DNA sequences VanH1, H2, and R1 (Figure 5.7b). Little nonspecific adsorption of either VanR or OmpR to the other sequences on the chip was observed. DNA probe sequences for VanR and OmpR response regulators can be found in the literature [30].

In addition to sequence specificity, SPR imaging measurements of DNA arrays can also be used to monitor inhibitor reactions that obstruct the binding of response regulators to DNA controlling gene expression in the bacteria. Figure 5.8 shows the effect of the inhibitor CpdA on VanR binding to the DNA sequences VanH1, H2, and R1 as measured by SPR imaging. The changes in percentage reflectivity were measured as a function of VanR concentration with and without 5 μM of CpdA. Integration of the line profile was performed to determine the average change in percentage reflectivity for the DNA sequences VanH1, H2, and R1. As can be seen in Figure 5.8, more adsorption is observed in the absence of the inhibitor than when CpdA is included with the VanR solution.

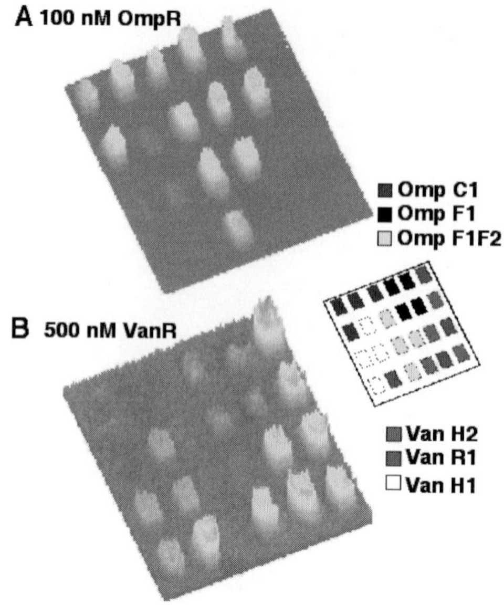

Figure 5.7 Surface plasmon imaging difference images of response regulator adsorption to double-stranded DNA immobilized on a photopatterned array composed of 500 by 500 μm array elements. (A) Specific adsorption of a 100 nM solution of OmpR to the DNA sequences OmpF1, F1F2, and C2. (B) VanR (500 nM) adsorbs to the DNA sequences VanH1, H2, and R1. OmpR is known to bind to the DNA sequences OmpF1, F1F2, and C2 and VanR is known to bind VanH1, H2, R1. There is little non-specific adsorption of the response regulators to the other sequences on the chip. DNA probe sequences for VanR and OmpR response regulators can be found in the literature. Reprinted with permission from *Langmuir*, Vol. 19, 1486–1492. Copyright 2003 American Chemical Society. (see Colour Plate p. XXVI).

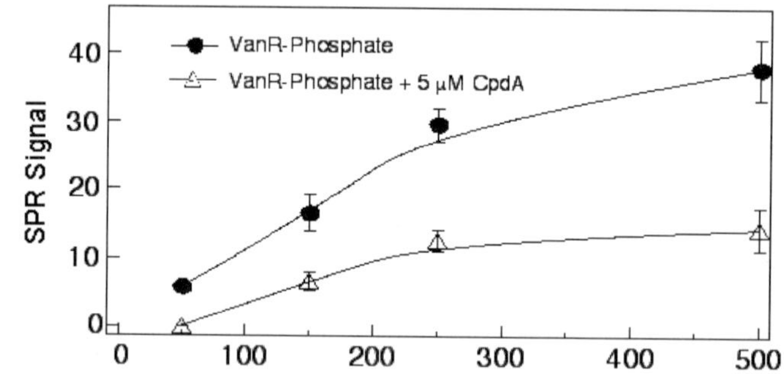

Figure 5.8 Plot showing the change in percentage reflectivity observed when VanR-phosphate (●) and VanR-phosphate plus 5 mM CpdA (○) adsorb to VanR DNA promoter sequences as a function of protein concentration. Integration of the line profile of VanR binding to DNA sequences VanH1, H2, and R1 was used to determine the average percentage reflectivity at each protein concentration. Less VanR-phosphate adsorbs to the DNA probes in the presence of the inhibitor CpdA than in its absence. Reprinted with permission from *Langmuir*, Vol. 19, 1486–1492. Copyright 2003 American Chemical Society.

5.5
SPR Imaging Experiments of Arrays Created by Microfluidic Stencils

We have recently developed a second fabrication and detection approach based on the coupling of microfluidic networks to SPR imaging measurements with the specific aim of lowering detection limits, reducing analysis time, and decreasing chemical consumption and sample volume [14]. This approach relies on the construction of microfluidic networks within the polymer polydimethylsiloxane (PDMS), which are physically sealed to chemically-modified gold surfaces. These PDMS microchannels were used in two ways: (i) to fabricate "1-D" arrays consisting of lines of immobilized peptides and DNA probes and (ii) to create "2-D" detection arrays in which a second set of PDMS microchannels were placed perpendicular to a 1-D line array in order to deliver target samples with small volume [14].

5.5.1
1-D Peptide Array for Antibody Binding Measurements

The fabrication of 1-D line arrays on gold surfaces is as follows. First, a set of parallel microchannels from PDMS were created by replication from 3-D silicon wafer masters that were created photolithographically from a 2-D chrome mask pattern (Figure 5.9) [33,34]. The microchannels were physically attached to a gold substrate modified with MUAM. A simple differential pumping system was used to introduce the chemical linkers SATP or SSMCC into the microchannels. Next, the DNA or peptide probe molecules were reacted within the microchannels. Once the immobilization was complete, the channels were removed and the alkanethiol-terminated monolayer was treated with PEG-NHS to prevent non-specific adsorption to the surface. Since the PDMS microchannels physically define where the probe molecules are immobilized, protection and deprotection chemistry is not required. By changing the spacing of the channels, up to 100 different peptide or DNA sequences can be immobilized on one chip. The array can then be coupled with a large volume flow cell (500 μL) to introduce analytes to the line array.

SPR imaging measurements were used to monitor the sequence-specific adsorption of antibodies to a two-component peptide array. The linear array contained the peptide sequences, Myc and HA denoted A and B, respectively (Figure 5.10). The introduction of a 25 nM solution of anti-Myc though a 500 μL flow cell resulted in specific adsorption to probe A. The anti-Myc was removed from the surface using pH 11.5 buffer prior to exposing the array to a solution of 100 nM anti-HA. Anti-HA primarily bound to probe B, although some non-specific adsorption probe A was observed.

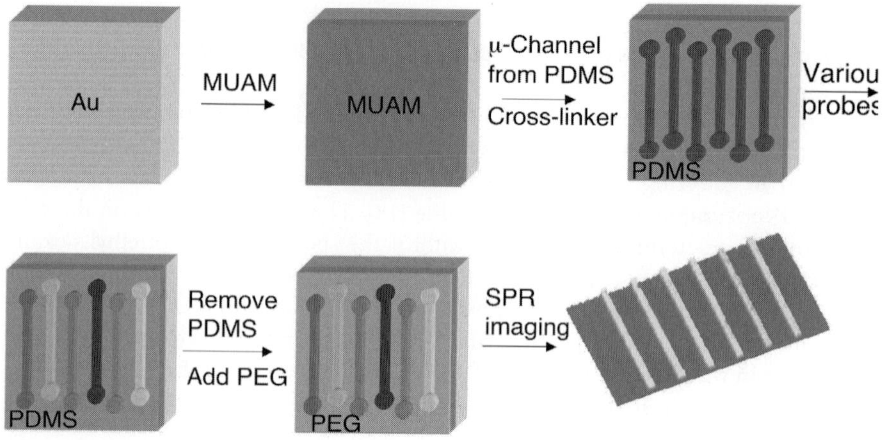

Figure 5.9 Schematic representation of the microfabrication strategy used to create DNA, peptide and protein linear arrays. A self-assembled monolayer of MUAM was formed on a clean gold surface. Polydimethyl siloxane (PDMS) microchannels were used to deliver bifunctional linkers and probe biomolecules to the surface. The microchannels were removed and the background was protected with a polyethylene glycol derivative. Adapted with permission from *Analytical Chemistry*, **73** 5525–5531. Copyright 2001 American Chemical Society. (see Colour Plate p. XXVI).

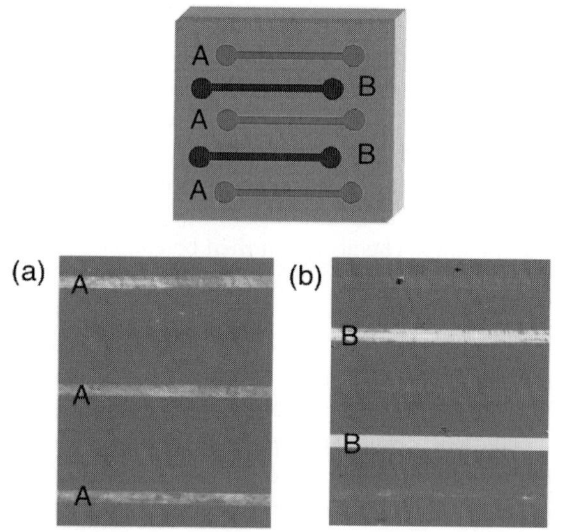

Figure 5.10 SPR difference images showing the sequence-specific binding interactions of antibodies to a peptide linear array containing two probes A and B corresponding to the peptide sequences Myc and HA. The peptides were immobilized in the pattern shown in the figure. (a) First, the array was exposed to a 25 nM solution of anti-Myc. Specific adsorption of anti-Myc to probe A is observed. (b) After rinsing with pH 11.5 buffer to regenerate the surface, the experiment was repeated 100 nM anti-HA. Anti-HA primarily bound to probe B, with little non-specific interaction with probe A. (see Colour Plate p. XXVII).

5.5.2.
2-D DNA Array for RNA Hybridization

In a second approach, microfluidic channels were used as a small volume flow cell during SPR imaging measurements to reduce the required analyte and to simultaneously introduce multiple analytes. First, a 1-D array composed of DNA elements was prepared and the background protected with PEG. Then a second set of PDMS channels was placed perpendicular to the immobilized DNA lines, forming a 2-D array, and used to introduce analytes to the array surface during SPR imaging experiments (Figure 5.11a). Figure 5.11b shows the schematic diagram of an SPR imager set-up where the large flow cell (500 µL) has been replaced with microfluidic channels constructed in PDMS reducing the required sample volume to 1 µL. An example of an SPR difference image of a three-component 2-D hybridization DNA array exposed to GUS gene ssRNA through PDMS microchannels is shown in Figure 5.12. About 20 femtomoles of RNA in a total solution volume of 1 µL was delivered through each channel. The DNA sequences and more information are reported elsewhere [14]. Varying SPR signal intensities were observed for the three different probes in Figure 5.12 and can be attributed to differences in the binding efficiency of the *GUS* gene ssRNA to the different surface-bound ssDNA probes. The ability to detect 20 femtomoles of ssRNA is sufficient for a number of biological applications, including the direct detection of messenger RNA (mRNA) from highly expressed genes and the detection of ribsomal RNA (rRNA) from complex biological samples.

5.5.3
2-D Peptide Array for Antibody Binding Measurements

SPR imaging technology integrating microfluidics was applied to monitor the sequence-specific interactions of antibodies to peptide motifs for epitope mapping applications. For example, peptide arrays were used to study the residues essential to the binding motif of anti-FLAG M2, an antibody commonly used for the purification of fusion proteins [35]. SPR imaging is currently used to study cell adhesion processes and to investigate the enzymatic modifications of peptides by sequence-specific phophatases, kinases, and proteases. Ultimately, these techniques could be used to identify the important residues in previously uncharacterized protein–protein interactions based on peptide recognition motifs.

Sequence-specific adsorption of anti-FLAG M2 to four different peptide epitopes based on the FLAG peptide tag was studied with SPR imaging [10]. Peptide line arrays were created using a microfluidic fabrication process and coupled to a second microchannel with a wraparound design (Figure 5.13a). The "worm" microchannel has a total sample volume of 5 µL and is used to deliver antibody to the peptide array. Figure 5.13b shows a difference image obtained by subtracting images of a fourcomponent peptide array before and after the introduction of a solution of 100 nM anti-FLAG M2. A line profile taken across the channels indicated that the greatest amount of antibody adsorption is observed at elements where

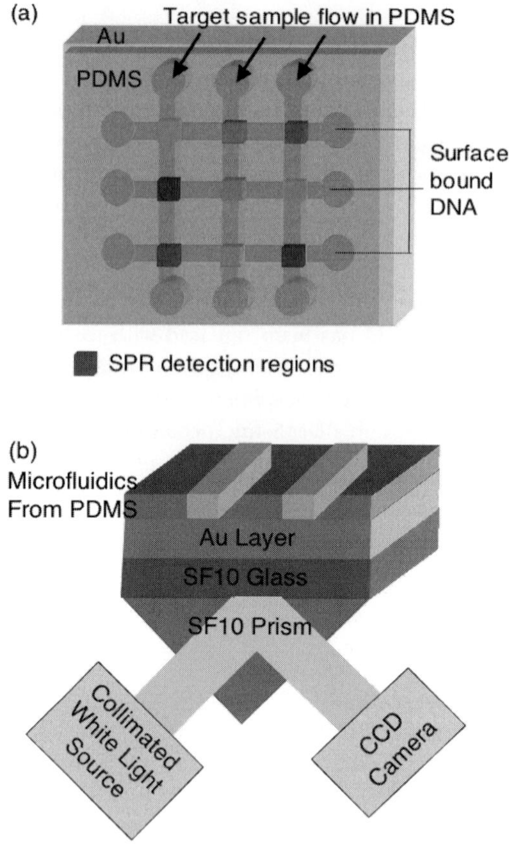

Figure 5.11 (a) Schematic diagram of parallel polydimethyl siloxane (PDMS) microchannels used to deliver small volumes of analyte (1 μl) to peptide and DNA linear arrays.
(b) Schematic representation of the SPR imaging experimental set-up incorporating PDMS microfluidics to deliver a small volume of target sample. This configuration allows different analytes to be delivered through each channel. (see Colour Plate p. XXVIII).

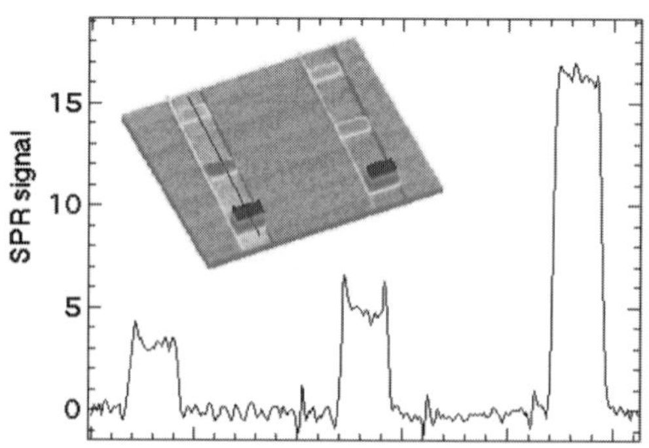

Figure 5.12 An SPR difference image showing adsorption of GUS gene ssRNA onto DNA probes. Each channel was 300 μm wide, 35 μm deep, and 1.2 μm long with 900 μm spacing between channels. Reprinted with permission from *Analytical Chemistry*, **73** 5525–5531. Copyright 2001 American Chemical Society. (see Colour Plate p. XXIX).

the original sequence, F1, was immobilized. The least amount of binding was observed to the peptide sequences F3 and F4. These peptides differed from the original sequence by the substitution of an amino acid important to the binding motif by an alanine residue. In contrast to the F3 and F4 peptides, the F2 sequence had a substantially higher signal, since this peptide contained an alanine substitution for a non-essential residue in the binding motif. These data demonstrate that SPR imaging can be used to compare the amount of protein binding to peptides differing by a single amino acid and to determine the importance of each residue for a peptide–protein interaction.

Quantitative SPR imaging measurements can also be used to simultaneously determine adsorption constants for the interactions of antibodies with multiple peptide motifs on a single chip [10]. The change in percentage reflectivity was measured for the peptide sequences F1 and F2 at increasing antibody concentrations. This data is shown in Figure 5.14 where the SPR signal resulting from antibody adsorption is plotted as a function of anti-FLAG M2 concentration, ranging from 1 to 150 nM. These data points were fitted with a Langmuir isotherm. The adsorption constants were found to be 1.5×10^8 M^{-1} for F1 and 2.8×10^7 M^{-1} for F2, showing that the original sequence had a stronger antibody peptide interaction. The detection limit for the F1 peptide with anti-FLAG M2 was 0.5 nM.

5.5.4
2-D Protein Array

With the sequencing of many genomes complete, it has become clear that high-throughput methods for studying gene products is necessary in order to elucidate the vast amount of information encoding biological function in an organism. Such studies would be enhanced by the addition of label-free analytical tools such as SPR imaging. The utility of SPR imaging for the study of protein–peptide interactions demonstrated the feasibility of this technique for measuring the adsorption of biomolecules to immobilized proteins. The first step for adapting SPR imaging measurements to the study of protein–protein interactions is the creation of a robust reproducible, biologically-active protein array. Most protein arrays are constructed by random attachment strategies, using naturally occurring lysines and terminal amines for surface attachment [36]. However, such approaches have the potential to interfere with the binding site of the protein, negatively affecting biological activity. To target such problems, we have established a general methodology to create oriented arrays of fusion proteins attached to gold thin films. In an oriented fusion protein array, the fusion protein or tag will interact with a capture agent immobilized on the gold thin film leaving the protein of interest free to interact with proteins in solution. We are currently investigating several fusion protein/capture agent pairs including the histidine peptide tag with NTA [37–39], glutathione S-transferase (GST) with glutathione or anti-GST [40,41], and maltose binding protein (MBP) with maltose [42,43]. Since these tags and capture agents are commonly used for the purification of fusion proteins, they are well-character-

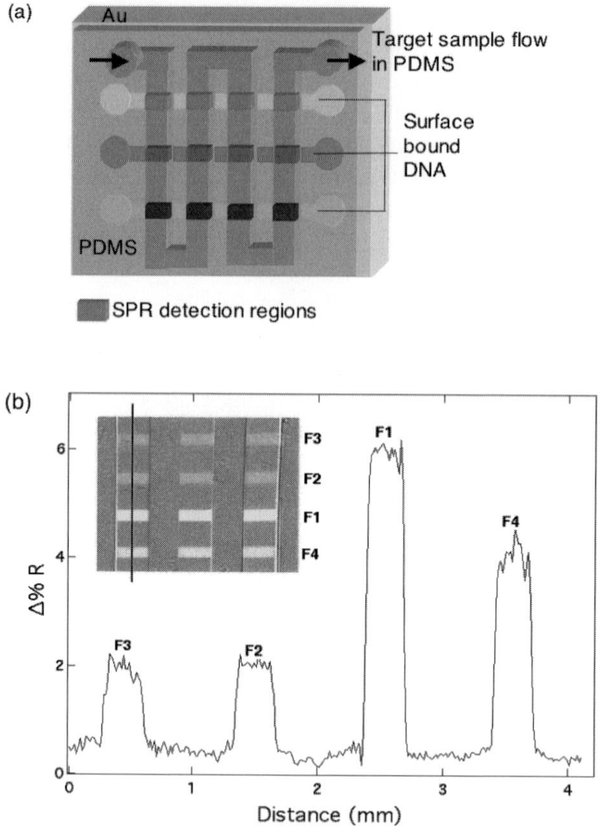

Figure 5.13 (a) A worm microchannel is used to deliver 5 μl aliquots of antibody solution to the peptide linear array. This microfluidic configuration is suitable for the delivery of a small volume of a single analyte to the array with a detection area equivalent to a configuration using a large volume flow cell. The worm channel was 500 μm wide, 35 μm deep, and 12 mm long with 500 μm spacing between channels.
(b) SPR difference image showing the adsorption of 100 nM anti-FLAG to a peptide array composed of four epitopes, differing by a single amino acid, based on the FLAG peptide sequence. The line profile reveals that the greatest adsorption occurs at elements containing the original sequence, F1. Diminished adsorption is observed at sequences, F3 and F4, containing alanine substitutions for essential residues of the binding motif. Peptide probe sequences can be found in the literature. Reprinted with permission from *Analytical Chemistry*, **74,** 5161–5168. Copyright 2002 American Chemical Society. (see Colour plate p. XXIX).

ized and the proteins can be released to follow up SPR imaging measurements with further off-chip solution phase experiments.

Protein arrays fabricated using NTA-modified gold films to capture his-tagged proteins are a promising tool for the study of protein–protein interactions using SPR imaging (see Figure 5.15a). A worm microchannel was used to pattern a self-assembled alkanethiol monolayer on a gold thin film with NTA capture agents. This microchannel was then removed and the whole surface was reacted with a polyethylene glycol derivative to prevent the non-specific adsorption of proteins to

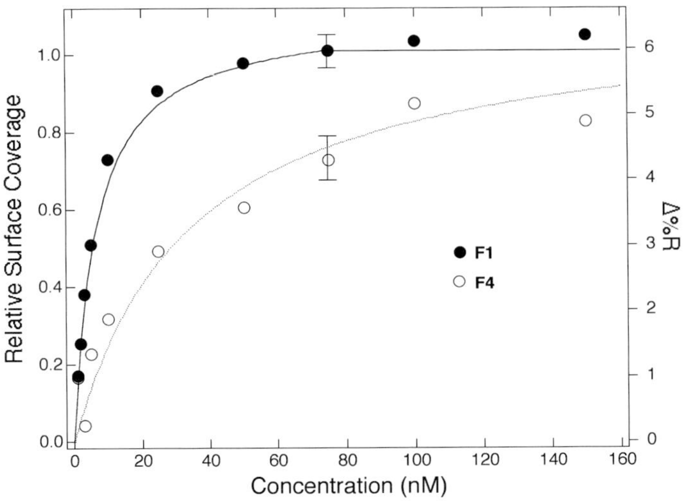

Figure 5.14 Langmuir isotherms fitted to the adsorption of anti-FLAG M2 onto an array containing the peptides F1 (●) and F2 (○). The change in percentage reflectivity was determined by integrating the line profiles for the F1 and F2 peptides at varying anti-FLAG M2 concentrations. The adsorption constants were calculated by the fit of the Langmuir isotherms, to be 1.5×10^8 M^{-1} and 2.8×10^7 M^{-1} for F1 and F4 respectively. Reprinted with permission from *Analytical Chemistry*, **74,** 5161–5168. Copyright 2002 American Chemical Society.

the background around the NTA capture agents. Next a second set of parallel microchannels oriented perpendicular to the immobilized NTA lines was used to introduce proteins to the surface (Figure 5.15b). This fabrication method provides a convenient and fast way to create arrays composed of one protein per channel. As shown in the SPR difference image in Figure 5.15c, both polyhistidine and histidine-tagged ubiquitin interact with the NTA surfaces in the presence of nickel ions, while little adsorption of these biomolecules is observed when nickel ions are not present, showing the specificity of the interaction. This protein fabrication strategy has the advantage that SPR imaging measurements can be used to evaluate protein immobilization onto the array prior to using the chip to study protein–protein interactions. Our current efforts are focused on the application of SPR imaging measurements to study the interactions of protein, DNA, and peptides to protein chips. Initial results indicate the SPR imaging techniques will become an innovative tool for proteomics research.

5.6 Conclusions

SPR imaging is emerging as a promising tool for the study of bioaffinity interactions in an array format. This review presented various recent applications of SPR imaging measurements for the study of DNA, RNA, and protein interactions with

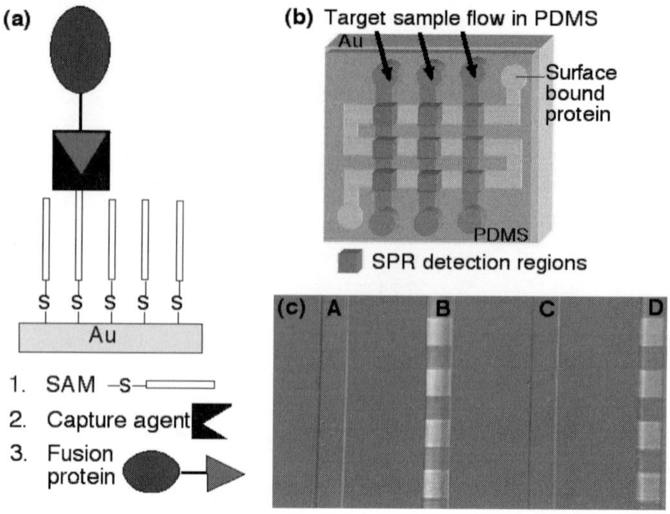

Figure 5.15 (a) Schematic representation of a general approach for the creation of an oriented fusion protein array on gold surfaces. (b) The NTA capture agent was immobilized through a worm microchannel. Next, a parallel set of PDMS channels was used to simultaneously deliver multiple histidine-tagged analytes to the NTA elements. (c) An example of an SPR difference image showing the immobilization of histidine-tagged ubiquitin **B** and polyhistidine D to NTA-modified monolayers in the presence of nickel ions. In the absence of nickel ions there is little non-specific adsorption of his-tagged ubiquitin **A** or polyhistidine **C** to the NTA surface, suggesting that a specific chelation interaction is occurring. (see Colour Plate p. XXX).

biomolecular arrays. The key components for these experiments are a well-characterized chemical modification of gold thin films to immobilize biomolecular probe molecules coupled with a robust array fabrication method based on photopatterning or microfluidics. The development of two different methods of array fabrication has increased the versatility of the SPR imaging technique; the linear arrays created from microfluidics are particularly efficient in terms of time and materials for arrays requiring only a small number (< 100) of different surface-bound biomolecules, whereas the generalized photopatterned array fabrication methodology is more applicable to systems where the number of surface-attached biomolecules is greater than 100. The successful application of DNA, peptide, and carbohydrate arrays to study protein binding interactions demonstrates the wide variety of bioaffinity interactions that can be observed with SPR imaging. The study of protein–protein interactions by SPR imaging is still in its infancy, but we predict that this technology will have a significant role to play in the high-throughput screening of protein–protein interactions, the study of synthetic chemical libraries, the binding and inhibition of cell receptors, and the elucidation of regulatory pathways in biological systems.

Acknowledgments

This research is funded by the National Institute of Health (3R01 GM59622-03S 1 and 8 ROI EB00269-02) and the National Science Foundation (CHE-0133151).

References

1. Miller, J. C., Butler, E. B., Teh, B. S., Haab, B. B. The application of protein microarrays to serum diagnostics: Prostate cancer as a test case. *Disease Markers* 2001, **17,** 225–234.
2. Zhu, H., Klemic, J. F., Chang, S., Bertone, P., Casamayor, A., Klemic, K. G., Smith, D., Gerstein, M., Reed, M. A., Snyder, M. Analysis of yeast protein kinases using protein chips. *Nature Genet.* 2000, **26,** 283–289.
3. Zhu, H., Bilgin, M., Bangham, R., Hall, D., Casamayor, A., Bertone, P., Lan, N., Jansen, R., Bidlingmaier, S., Houfek, T., Mitchell, T., Miller, P., Dean, R. A., Gerstein, M., Snyder, M. Global analysis of protein activities using proteome chips. *Science* 2001, **293,** 2101–2105.
4. Laune, D., Molina, F., Mani, J.-C., Del Rio, M., Bouanani, M., Pau, B., Granier, C. Dissection of an antibody paratope into peptides discloses the idiotope recognized by the cognate anti-idiotypic antibody. *J. Immun. Methods* 2000, **239,** 63–73.
5. Uthaipibull, C., Aufiero, B., Syed, S. E. H., Hanse, B., Guevara Patino, J. A., Angov, E., Ling, I. T., Fegerding, K., Morgan, W. D., Ockenhouse, C., Birdsall, B., Feeney, J., Lyon, J. A., Holder, A. A. Inhibitory and blocking monoclonal antibody epitopes on merozite surface protein 1 of the malaria parasiste plasmodium falciparum. *J. Mol. Biol.* 2001, **307,** 1381–1394.
6. Houseman, B. T., Huh, J. H., Kron, S., J., Mrksich, M. Peptide chips for the quantitative evaluation of protein kinase activity. *Nature Biotech.* 2001, **20,** 270–274.
7. Kilpert, K., Hansen, G., Wessner, H., Schneider-Mergener, J., Hohne, W. Characterizing and optimizing protease/peptide inhibitor interactions, a new application for spot synthesis. *J. Biochem.* 2000, **128,** 1051–1057.
8. Brockman, J. M., Frutos, A. G., Corn, R. M. A multi-step chemical modification procedure to create DNA arrays on gold surfaces for the study of protein–DNA interactions with surface plasmon resonance imaging. *J. Am. Chem. Soc.* 1999, **121,** 8044–8051.
9. Frutos, A. G., Brockman, J. M., Corn, R. M. Reversible protection and reactive patterning of amine- and hydroxyl-terminated self-assembled monolayers on gold surfaces for the fabrication of biopolymer arrays. *Langmuir* 2000, **16,** 2192–2197.
10. Wegner, G. J., Lee, H. J., Corn, R. M. Characterization and optimization of peptide arrays for the study of epitope–antibody interactions using surface plasmon resonance imaging. *Anal. Chem.* 2002, **74,** 5161–5168.
11. Nelson, B. P., Frutos, A. G., Brockman, J. M., Corn, R. M. Near-infrared surface plasmon resonance measurements of ultrathin films. 1. Angle shift and SPR imaging experiments. *Anal. Chem.* 1999, **71,** 3928–3934.
12. Hansen, W. N. Electric fields produced by the propagation of plane coherent electromagnetic radiation in a stratified medium. *J. Opt. Soc. Am.* 1968, **58,** 380–390.
13. Nelson, B. P., Grimsrud, T. E., Liles, M. R., Goodman, R. M., Corn, R. M. Surface plasmon resonance imaging measurements of DNA and RNA hybridization adsorption onto DNA microarrays. *Anal. Chem.* 2001, **73,** 1–7.
14. Lee, H. J., Goodrich, T. T., Corn, R. M. SPR imaging measurements of 1-D and 2-D DNA microarrays created from microfluidic channels on gold thin films. *Anal. Chem.* 2001, **73,** 5525–5531.
15. Pease, A. C., Solas, D., Sullivan, E. J., Cronin, M. T., Holmes, C. P., Fodor, S. P. A. Light-generated oligonucleotide arrays for rapid DNA sequence analysis. *Proc. Natl Acad. Sci. USA* 1994, **91,** 5022–5026.
16. Schena, M., Shalon, D., Davis, R. W., Brown, P. O. Quantitative monitoring of gene expression patterns with a complementary DNA microarray. *Science* 1995, **270,** 467–470.
17. Brenner, S., Johnson, M., Bridgham, J., Golda, G., Lloyd, D. H., Johnson, D., Luo, S., McCurdy, S., Foy, M., Ewan, M., Roth, R., George, D., Eletr, S., Albrecht, G., Vermaas, E., Williams, S. R., Moon, K., Burcham, T., Pallas, M., DuBridge, R. B., Kirchner, J., Fearon, K., Mao, J., Corcoran, K. Gene expression analysis by massively parallel signature sequencing (MPS) on microbead arrays. *Nature Biotech.* 2000, **18,** 630–634.

18. Winzeler, E. A., Richards, D. R., Conway, A. R., Goldstein, A. L., Kalman, S., McCullough, M. J., McCusker, J. H., Stevens, D. A., Wodicka, L., Lockhart, D. J., Davis, R. W. Direct allelic variation scanning of the yeast genome. *Science* 1998, **281,** 1194–1197.
19. Wenschuh, H., Volkmer-Engert, R., Schmidt, M., Schulz, M., Schneider-Mergener, J., Reineke, U. Coherent membrane supports for parallel microsynthesis and screening of bioactive peptides. *Biopolymers* 2000, **55,** 188–206.
20. Frank, R. Spot-synthesis: An easy technique for the positionally addressable, parallel chemical synthesis on a membrane support. *Tetrahedron* 1992, **48,** 9217–9232.
21. Reineke, U., Volkmer-Engert, R., Schneider-Mergener, J. Applications of peptide arrays prepared by the SPOT-technology. *Curr. Opin. Biotech.* 2001, **12,** 59–64.
22. Frey, B., Corn, R. M. Covalent attachment and derivitization of poly(L-lysine) monolayers on gold surfaces as characterized by polarization-modulation FT-IR spectroscopy. *Anal. Chem.* 1996, **68,** 3187–3193.
23. Smith, E., Wanat, M. J., Cheng, Y., Barreira, S. V. P., Frutos, A. G., Corn, R. M. Formation, spectroscopic characterization, and applications of sulfhydryl-terminated alkanethiol monolayers for the chemical attachment of DNA onto gold surfaces. *Langmuir* 2001, **17,** 2502–2507.
24. Hirschhorn, J. N., Sklar, P., Lindblad-Toh, K., Lim, Y. M., Ruiz-Gutierrez, M., Bolk, S., Langhorst, B., Schaffer, S. E., Winchester, E., Landler, E. S. SBE-TAGS: An array-based method for efficient single-nucleotide polymorphism genotyping. *Proc. Natl Acad. Sci. USA* 2000, **97,** 12164–12169.
25. Reich, D. E., Cargill, M., Bolk, S., Ireland, J., Sabeti, P. C., Richter, D. J., Lavery, T., Kououmjian, R., Farhadian, S. F., Ward, R., Lander, E. S. Linkage disequilibrium in the human genome. *Nature* 2001, **411,** 199–204.
26. Brookes, A. J. The essence of SNPs. *Gene* 1999, **234,** 177–186.
27. Brockman, J. M., Nelson, B. P., Corn, R. M. Surface plasmon resonance imaging measurements of ultrathin organic films. *Annu. Rev. Phys. Chem.* 2000, **51,** 41–63.
28. Su, S. S., Lahue, R. S., Au, K. G., Modrich, P. Mispair Specifity of Methy-directed DNA Mismateh Correction *in vitro*. *J. Biol. Chem.* 1987, **263,** 6829–6835.
29. Hoch, J. A. Two component and phosphorelay signal transduction. *Curr. Opin. Microbiol.* 2000, **3,** 165–170.
30. Smith, E. A., Erickson, M. G., Ulijasz, A. T., Weisblum, B., Corn, R. M. Surface plasmon resonance imaging of transcription factor proteins: interactions of bacterial response regulators with DNA arrays on gold films. *Langmuir* 2003, **19,** 1486–1492.
31. Martinez-Hackert, E., Stock, A. M. Structural relationships in the OmpR family of winged-helix transcription factors. *J. Mol. Biol.* 1997, **269,** 301–312.
32. Haldimann, A., Fisher, S. L., Daniels, L. L., Walsh, C. T., Wanner, B. L. Transcriptional regulation of the *Enterococcus faecium* BM4147 vancomycin resistance gene cluster by the VanS-VanR two-component regulatory system in *Escerichia coli* K-12. *J. Bacteriol.* 1997, **179,** 5903–5913.
33. Jackman, R. J., Duffy, D. C., Ostuni, E., Willmore, N. D., Whitesides, G. M. Fabricating large arrays of microwells with arbitrary dimensions and filling them using discontinuous de-wetting. *Anal. Chem.* 1998, **70,** 2280–2287.
34. Duffy, D. C., McDonald, J. C., Schueller, O. J. A., Whitesides, G. M. Rapid prototyping of microfluidic systems in poly(dimethylsiloxane). *Anal. Chem.* 1998, **70,** 4974–4984.
35. Slootstra, J. W., Kuperus, D., Pluckthun, A., Meloen, R. H. Identification of new tag sequences with differential and selective recognization properties for the anti-FLAG monoclonal antibodies M1, M2, and M5. *Mol. Divers.* 1996, **2,** 156–164.
36. Schreiber, S. L., MacBeath, G. Printing proteins as microarrays for high-throughput function determination. *Science* 2000, **289,** 1760–1763.
37. Bornhorst, J. A., Falke, J. J. Purification of proteins using polyhistidine affinity tags. *Methods Enzymol.* 2000, **326,** 245–254.
38. Schmitt, J., Hess, J., Stunnenberg, H. G. Affinity purification of histidine-tagged proteins. *Mol. Biol. Rep.* 1993, **18,** 223–230.

39. Ho, C. H., Limberis, L., Caldwell, K. D., Stewart, R. J. A meta-chelating pluronic for immobilization of histidine-tagged proteins at interfaces: Immoblilization of firefly luciferase on polystyrene beads. *Langmuir* 1998, **14,** 3889–3894.

40. Simons, P. C., Vander Jagt, D. L. Purification of glutathione S-transferase from human liver by glutathione-affinity chromatography. *Anal. Biochem.* 1977, **82,** 334–341.

41. Wang, C., Castro, A. F., Wilkes, D. M., Altenberg, G. A. Expression and purification of the first nucleotide-binding domain and linker region of human multidrug resistance gene product: Comparison of fusions to glutathione S-transferase, thioredoxin and maltose-binding protein. *Biochem. J.* 1999, **338,** 77–81.

42. Cattoli, F., Sarti, G. C. Separation of MBP fusion proteins through affinity membranes. *Biotechnol. Prog.* 2002, **18,** 94–100.

43. Bach, H., Mazor, Y., Shaky, S., Shoham-Lev, A., Berdichevsky, Y., Gutnick, D. L., Benhar, I. Escherichia coli maltose-binding protein as a molecular chaperone for recombinant intracellular cytoplasmic single-chain antibodies. *J. Mol. Biol.* 2001, **312,** 79–93.

6
Surface Plasmon Fluorescence Spectroscopy for Protein Binding Studies

Fang Yu, Björn Persson, Stefan Löfås and Wolfgang Knoll

Abstract

Surface plasmon fluorescence spectroscopy (SPFS) is a novel technique offering an enhanced sensitivity for biomolecular interaction analysis. The dextran matrix on the CM5 chip has been shown to be a matrix which is well suited to use in protein interaction studies carried out with SPFS because of its extraordinary sensitivity. In a sandwich assay which mimics a real application, the use of a fluorescently labeled secondary antibody increased the sensitivity by 1 ~ 2 orders of magnitude when compared with the direct label-free surface plasmon spectroscopy assay. A simplified two-component interaction model was applied to further reveal the limit of detection (LOD) which can be achieved with SPFS. Under mass transport-limited binding conditions, the linear range of the calibration curve covered more than four orders of magnitude of analyte (labeled IgG) concentration and could be extended down to 25 femtomolar solutions. This corresponds to 660 IgG molecules binding to one mm² surface every minute.

6.1
Introduction

Biological interactions at surfaces have been studied extensively employing a large variety of label-free optical biosensors [1–3]. Surface plasmon resonance spectroscopy (SPR, SPRS or SPS), which monitors refractive index changes caused by macromolecules in the vicinity of a thin noble metal film, is one of those techniques which have been well established theoretically and instrumentally. It has become one of the most important and sensitive tools in studies related to the quantification of heterogeneous biological interactions at a solid–water interface, and it is one of the configurations compatible with the fabrication of large scale multi-component read-out devices for biochips [4]. Many attempts at improving the sensitivity of SPRS further [5,6] have been reported. Improved sensitivity would extend the range of applications, especially for sensing small molecules such as oligonucleotides, synthetic drug molecules, etc. However, there seems to be little scope left for the improvement of sensitivity within the area of label-free detection. It is well known that significant increases in sensitivity can be realized by surface

Protein Microarray Technology. Edited by Dev Kambhampati

ISBN: 3-527-30597-1

electromagnetic field enhancement mechanisms. Combining this concept with the widely used, and already rather sensitive technique of fluorescence spectroscopy, we developed surface plasmon field-enhanced fluorescence spectroscopy (SPFS) in 2000 [7]. The technique uses the locally enhanced optical field of a surface-plasmon mode to excite the fluorophores at a metal–liquid interface. By this method, we have succeeded in detecting hybridization reactions between surface-attached oligonucleotide probes and single-stranded target DNA molecules binding from the bulk solution phase [8,9,]. With the capability of *in situ* and real-time monitoring of the whole interaction process which is a basic advantage inherent in SPR, we can evaluate kinetic rate constants of hybridization, k_{on} and k_{off}, in addition to the affinity constant, K_A. Single base-pair mismatches in the sequence can be easily discriminated. Moreover, with the excellent signal-to-noise ratio (SNR) of SPFS, an in-depth investigation of the more subtle details of oligonucleotide interactions, such as the influence of the mismatch position, sequence length and ionic strength, can be undertaken. Meanwhile, various novel detection modes have been introduced, such as interfacial fluorescence resonance energy transfer, by labeling both oligonucleotide strands with donor and acceptor dyes [10].

Interesting questions concerning SPFS which have not been addressed so far focus on interfacial protein interaction studies, and on how sensitive the technique can possibly be. It is not at all straightforward to extend DNA experience to protein experiments. First of all, a suitable surface matrix for protein interactions needs to be established. Generally speaking, an ideal matrix should at least meet the following requirements:

1. suppress non-specific protein adsorption
2. expose easily accessible moieties for protein immobilization
3. exhibit a robust chemistry against regeneration conditions

Among the matrices which meet these requirements, oligo-ethylene-glycol (OEG) and dextran layers are most commonly used [11,12], and are now also commercially available.

However, because of the special requirements of SPFS, the choice of possible matrices is quite limited. One of the major concerns is the significant quenching of fluorescence if fluorophores are placed too close to the metal surface (< 20 ~ 30 nm). This means that the sensitivity gained from the surface-enhancement effect would be largely compromised. Another serious problem concerns the non-linear response of the system on a planar surface matrix. For planar surface coatings with the ligands attached in a 2-D arrangement at fairly close proximity to the metal surface, the fluorescence yield from each fluorophore will depend on its distance from the metal which will be influenced by the various supramolecular architectures which in turn will result from the different packing densities of bound analyte molecules. This has been proved by recent work on a planar matrix fabricated with self-assembled monolayers (SAMs) [13].

Based on the above considerations, the most versatile chip from Biacore, the CM5 chip, was employed as our standard interaction matrix in the following studies. The carboxymethyldextran (CMD) matrix was first described by Löfås and Johnsson in 1990 [14]. Dextran is a linear polymer based on 1,6-linked glucose units and is attached to the gold surface by covalent binding to a SAM with activated end groups. Carboxylic groups are then generated along the poly-sugar backbone by treatment with bromoacetic acid. NHS/EDC chemistry can then be applied to covalently attach proteins containing accessible ε-amino groups in their lysine residues. Dextran layers provide an effective barrier between the bulk solution and the gold surface, thus minimizing the non-specific binding of ligands. One special feature of this matrix is that the CMD chains extend about 100 nm into the solution and possess numerous carboxylic acid groups along the backbone. This arrangement can effectively keep a major proportion of the bound fluorophores away from the quenching region and in this way will reduce the significant effect of distance on the fluorescence yield. On the other hand, the high protein loading capacity favors highly sensitive detection.

When using SPFS, the mass and the fluorescence signals arising from the binding of labeled analyte molecules can be recorded simultaneously. This is more significant for protein studies than for DNA experiments, because proteins are usually of sufficient size to generate a good SPR signal. Thus, it is possible to establish a correlation between the fluorescence and the surface concentration of the bound molecules. Moreover, the correlated information from SPFS allows us to extract information about the three-dimensional interfacial arrangement of biomolecules [13].

In the following, we start by briefly summarizing the theoretical basics of using surface plasmon enhanced optical fields to excite fluorescence. Subsequently, two systems which were studied with CM5 chips are presented. In the first, we demonstrate a real application of SPFS in an immuno-reaction study, by using a sandwich assay format. In the second, we simplify the interaction model, in order to assess the limit of detection (LOD) of this novel spectroscopic technique.

6.2 Fluorescence Profile at the Interface

The field enhancement mechanism of surface plasmons has been widely used in surface-enhanced Raman spectroscopy (SERS). When tuning into the resonant mode with a prism or grating coupler, the intensity of the incident light is localized at and propagates along the metal/dielectric interface. The enhancement factor (the field intensity at the interface divided by the intensity of the incident light) can be calculated using Maxwell's equations, and amounts for a factor of c. 16 for gold and c. 50 for silver (at $\lambda = 633$ nm). The surface plasmon field also extends into the dielectric medium along the normal of the surface, with the intensity decaying exponentially over several hundreds of nanometers (1/e decay is c. 150 nm). This evanescent character of the optical field guarantees that the analysis will be surface sensitive. When an appropriate fluorophore remains in the region of this field, fluorescence can be excited which directly reflects the local field intensity. However, the fluorescence yield is not only dependent of the field intensity, but is also influenced by the operation of additional de-excitation channels. For instance, it is well known that dyes at distances to the metal surface within the Förster energy transfer range will be significantly quenched. The theoretical fluorescence intensity profile is schematically depicted in Figure 6.1A, together with the exponential decay of the surface-plasmon mode. Within about two Förster radii (20 ~ 30 nm) from the metal, the fluorescence intensity decreases significantly. The best position to achieve a high fluorescence yield is right outside of the Förster region, where the enhanced optical field is just slightly lowered as compared with its maximum value at the interface. For comparison, the surface architecture applied in our studies (Figure 6.1B, C, D) is also plotted on approximately the same scale.

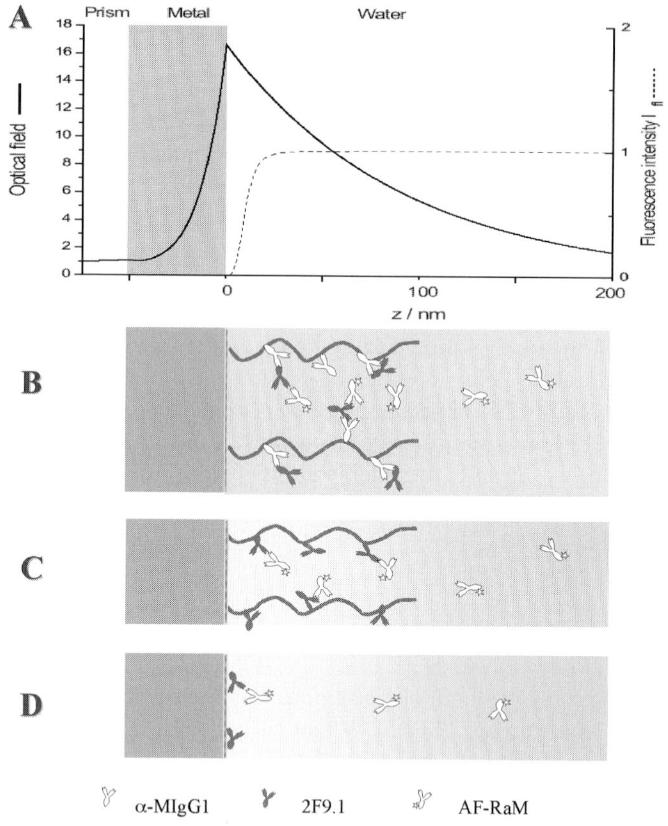

Figure 6.1 **A** Intensity profile of a surface plasmon with the evanescent field extending into the dielectric medium (water) in contact with the metal (Au) layer (solid line), and the fluorescence intensity profile of a fluorophore near a (quenching) metal surface (dashed line).
B–D Schematic representations of the interfacial architectures studied in this work.
B Sandwich detection scheme with a dextran matrix, **C** direct two-component interaction model for LOD evaluation on a dextran matrix, and
D lipoic acid-modified surface.

6.3 Instrumentation

The experimental realization of SPFS is schematically depicted in Figure 6.2A. A Kretschmann configuration has been used. The beam of a HeNe laser (Uniphase, 5 mW, 632.8 nm) is passed through a light chopper that is connected to a lock-in amplifier (EG & G) and the modulated beam then passes through two polarizers (Glan-Thompson), with which the intensity and polarization of the laser can be adjusted. A programmable shutter was installed to constantly block the laser unless data points were being recorded, thus minimizing photo-bleaching of the fluorophores. The p-polarized beam then passes through the coupling prism (Schott, LASFN9, n = 1.85 at 633 nm), and is reflected off the base of the metal-coated glass substrate which is optically attached to the prism by an index-matching fluid. A lens (f = 50 mm, Owis) focuses the reflected beam onto a photodiode detector. The fluorescence emission of the sample is collected at the back of the prism by another lens (f = 50 mm, 25 mm diameter), and then passes through an interference filter (LOT, $\lambda = 670$ nm, $\Delta\lambda = 10$ nm, LOT, 80 % transmittance) onto a photomultiplier tube (PMT, Hamamatsu). A neutral filter (attenuator) is sometimes used in order to attenuate fluorescence with too strong an intensity, in order to keep the PMT running in the linear range (< 1 ~ 2 million counts s^{-1}). The filter attenuates the intensity by a factor of 60.88 as shown with a calibration experiment. The fluorescence detection unit is mounted on the goniometer rotating together with the prism (sample) at an angle θ, while the photodiode detecting the reflected light rotates at 2θ. Data acquisition and control of the system electronics are accomplished by custom programs.

A schematic of the liquid handling system and the flow cell is shown in Figure 6.2B. The cuvette consists of a thin polydimethyl siloxane (PDMS) spacer (300 µm, with a 5 x 7 mm elliptical hole) and a quartz cover slip (Herasil glass) in which has been customized with two holes and two steel needles to serve as the inlet and outlet respectively. The flow cell is attached with Tygon tubing with an inner diameter of 0.76 mm, to a peristaltic pump (Ismatec) and the sample tube to form the circulation loop. Buffer and sample solution are exchanged manually. The sealed loop volume is around 300 µL, which results in a minimum sample consumption of about 400 ~ 600 µL in order to guarantee the desired working sample concentration. The maximum linear flow velocity of 100 mm s^{-1} at the sensing spot compares well to the flow cell in the Biacore microfluidic system.

Of special interest for this work was the use of the commercially available CM5 chips in our system. Borosilicate glass with a thickness of about 0.5 mm (n = 1.52 at 588 nm) provides the substrate of the chip. It is apparent that there is a refractive index (RI) mismatch between the prism and the substrate. However, the surface plasmon reflectivity dip and the fluorescence peak appear at the same position as usual [7] as shown in Figure 6.3. Addition of the 'extra' parallel glass layer between the prism and the metal does not change the in-plane component of the wave vector of the incident light, $k_{||}$, which has already been increased by the high refractive index prism to fulfill the momentum-matching condition for the excitation of the

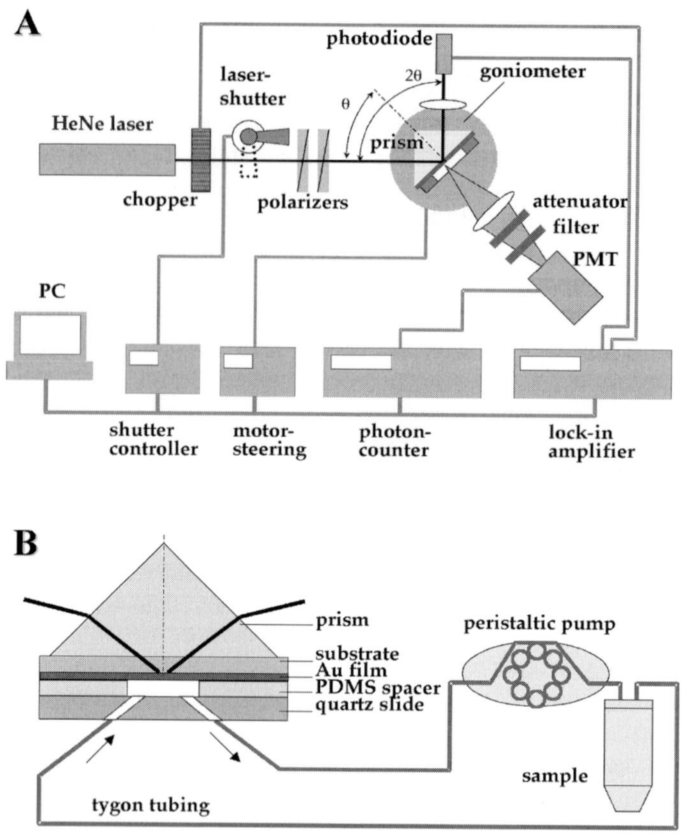

Figure 6.2 **A** Schematic experimental set-up for SPFS in a Kretschmann configuration. **B** Schematic representation of the flow cell with a thin PDMS spacer (300 μm) and a sealed circulation loop for the sample solutions.

surface-plasmon mode. On the other hand, we did observed two 'unusual' phenomena in the reflectivity signal with such a configuration. These are:

(1) An abrupt increase in reflectivity at c. 65° which is slightly higher than the angle of surface plasmon resonance minimum, which clearly indicates the commencement of another total reflection at the prism/substrate interface

(2) A weak interference pattern superimposed onto the 'smooth' reflectivity curve which presumably originates from the interference between two beams reflected from the interfaces of prism/substrate and substrate/liquid, respectively

The latter phenomenon is more problematic as it sometimes influences the precise positioning of the total reflection angle and the resonance minimum in the angular scan curve. However, neither observation compromises the fluorescence detection.

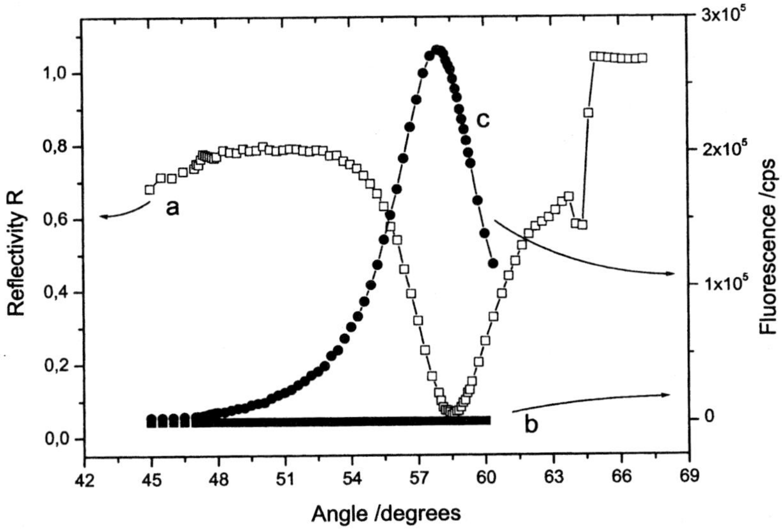

Figure 6.3 Example of an SPFS measurement with the CM5 chip. SPR reflectivity (a) and the corresponding fluorescence angular scan before (b) and after (c) the binding of Alexa Fluor 647 dye-labeled antibody.

6.4
SPR Signal Conversion

In 1991, Stenberg and his colleagues at Biacore calibrated the SPR signal, i.e. the shift of the SPR minimum with the absolute molecular surface concentration is determined using radiolabeled proteins [15]. They found the correlation to be $\Delta\theta/\sigma = 0.10 \pm 0.01°$ (ng mm^{-2})$^{-1}$, and largely independent of the size of the protein. Thus they defined 1 RU (resonance unit) as corresponding to a protein concentration of 1 pg mm^{-2}. This enables us to correlate our SPR signal to the absolute surface concentration, if we can establish a relationship between the SPR signal measured by SPFS and Biacore, respectively.

Due to the differences in the optical arrangement between Biacore and our instrument (listed in Table 6.1), the SPR responses are different for the same amount of protein binding. Using these parameters reflectivity-versus-angle curves can be simulated by Fresnel calculation for a four-layer system (Figure 6.4A). The thickness of the 'protein' layer (the third layer) was chosen at 0, 2.5, 5, 7.5 and 10 nm, respectively, assuming its refractive index n = 1.45. The corresponding angular shifts of the SPR minima have been plotted for both instruments respectively, and are shown in Figure 6.4B. Each group of data can be perfectly fitted by a straight line, which confirms the linear SPR response to such a range of film thickness variations. The difference in slope, on the other hand, indicates the difference in response. With the same thickness of film, our instrument gives a higher angular shift of the resonance minimum by a factor of 1.866 which is a direct consequence of the higher refractive index of the prism used. Thus, the surface concentration correlation for SPFS is c. $\Delta\theta/\sigma = 0.19°$ (ng mm^{-2})$^{-1}$.

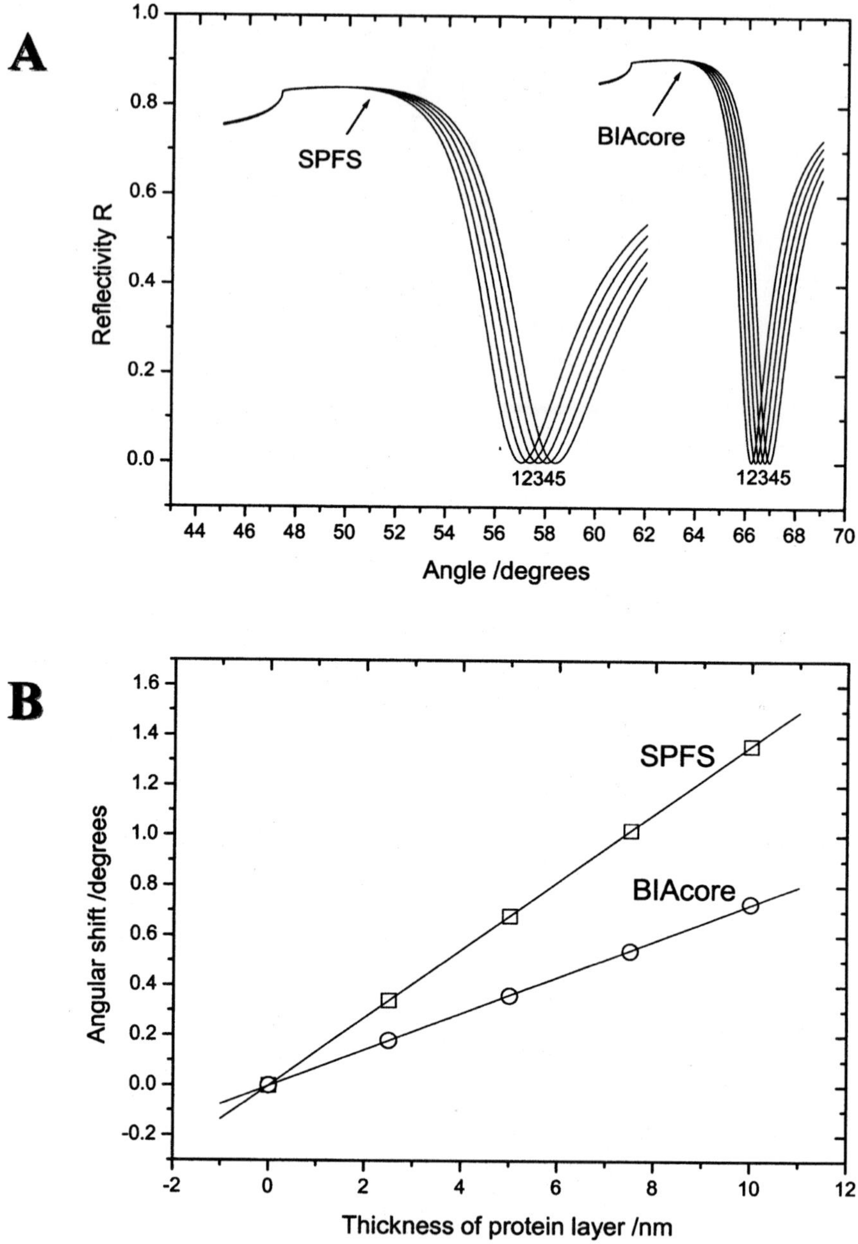

Figure 6.4 **A** Fresnel four-layer (prism, gold, protein, water) simulation curves for the Biacore and the SPFS reflectivity measurements respectively. In both cases, a 50-nm gold layer is assumed and curves 1, 2, 3, 4, and 5 correspond to protein thickness of 0, 2.5, 5, 7.5 and 10 nm respectively. **B** Simulated SPR minimum shift versus assumed thickness of the protein layer. Both solid curves are a linear fit to the data.

Table 6.1 Comparison of some of the operational parameters of the two instruments used in this study.

	SPFS	Biacore
Wavelength of laser	633 nm	760 nm
Geometry of prism	Right-angled	Half-cylindrical
Refractive index of prism	1.85 at 633 nm	1.52 at 760 nm

6.5 Sandwich Detection

The sandwich assay is illustrated schematically in Figure 6.1B. A primary antibody (rabbit anti-mouse IgG, α-MIgG1) was covalently immobilized to the dextran matrix by the usual EDC/NHS coupling procedure. Subsequently, the antigen, monoclonal mouse IgG (mouse anti-myoglobin, clone 2F9.1) was captured. And finally, after a brief rinsing step, the fluorescence signal was obtained following the injection of the Alexa Fluor 647 dye-labeled secondary antibody solution at a fixed concentration. The Alexa Fluor 647-labeled goat anti-mouse (AF-GaM, dye-to-IgG ratio: 4.5) and rabbit anti-mouse (AF-RaM, dye-to-IgG ratio: 4.8) antibodies were purchased from Molecular Probes. All other materials were obtained from Biacore unless otherwise stated. All experiments were carried out in HBS-EP buffer (Biacore) unless otherwise stated. A flow rate of 3 mL min^{-1} was used for all experiments, equivalent to Biacore's flow rate of 50 μL min^{-1}, in terms of the linear velocity at the sensing spot.

The immobilization of α-MIgG1 is straightforward by using the commercial kit and following a repeatedly tested protocol. As shown in Figure 6.5A, the dextran matrix was activated for 10 min by exposure to a fresh mixture of 1-(3-dimethylaminopropyl)-3-ethylcarbodiimide hydrochloride (EDC) and N-hydroxysuccinimide (NHS), forming the terminal NHS ester moieties ((1) in Figure 6.5A). α-MIgG1 is then injected at a concentration of 30 μg mL^{-1} prepared in sodium acetate buffer (10 mM, pH 5.5) and typically incubated with the surface for 5–10 min ((2) in Figure 6.5A). Unreacted activated groups were deactivated by injection of ethanolamine hydrochloride solution ((3) in Figure 6.5A). Finally, a pulse injection (3 min) of glycine buffer (10 mM, pH 1.7) was applied as a so-called 'chip-conditioning' step. Two angular scan curves (Figure 6.5B) were obtained (a) before and (b) after this step in order to assess the amount of protein immobilized. The angular shift was typically 1.2–1.4°, which corresponds to c. 6.3–7.3 ng mm^{-2} which is a relatively high loading.

Before recording the specific binding signal, non-specific binding (NSB) of proteins to the matrix should be quantified. AF-GaM and AF-RaM solutions made up to the highest working concentration (33 nM) were incubated with the sensor surfaces with or without the primary antibody. Figure 6.6A shows the results on the

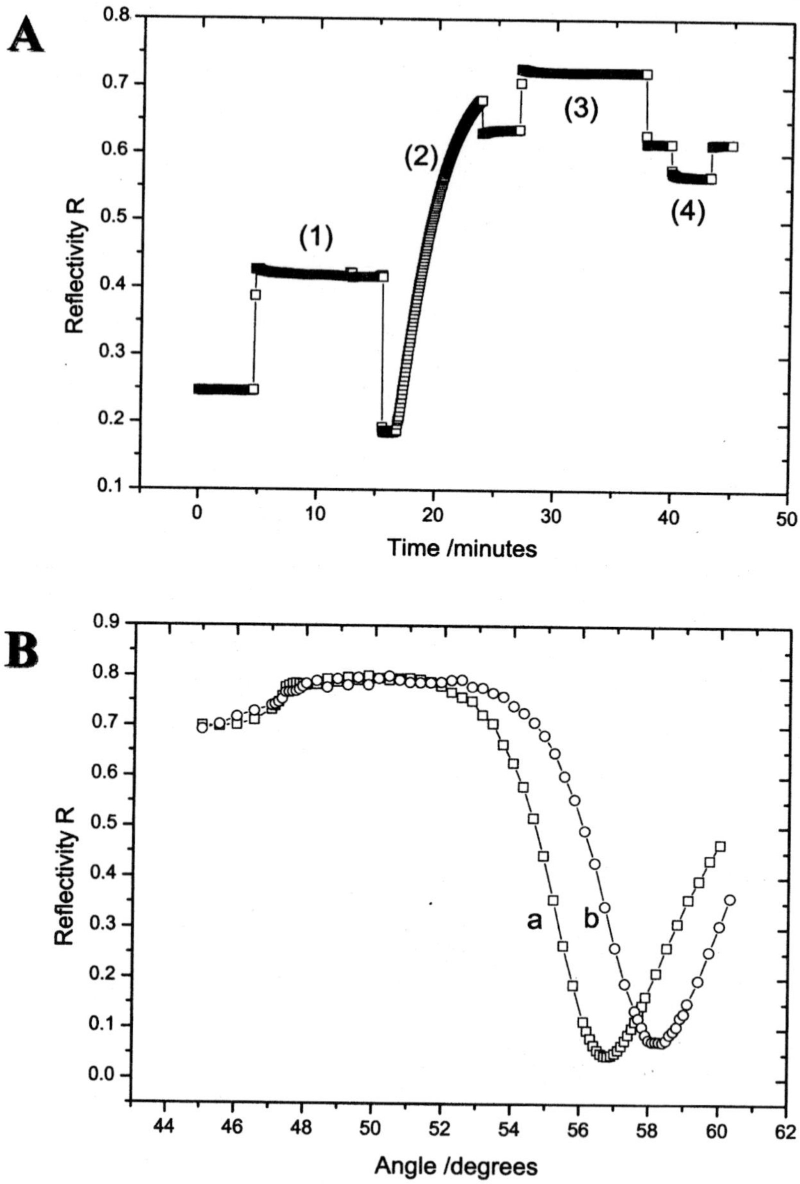

Figure 6.5 **A** Typical immobilization curve of α-MIgG1 to a dextran surface. (1) Activation of the carboxyl groups by a mixed EDC/NHS solution. The observed response is due to the differences in the bulk refractive indices between EDC/NHS solution and the running buffer HBS-EP. (2) Injection of 30 μg mL^{-1} α-MIgG1 in s odium acetate buffer. (3) Deactivation of unreacted esters and desalting of electrostatically-bound protein by injection of 1 M ethanolamine at pH 8.5. The observed response is due to the differences in the bulk refractive indices between the ethanolamine solution and the running buffer HBS-EP. (4) A pulse injection of glycine buffer as a 'chip conditioning' step.
B Angular scan curves before (a) and after (b) the immobilization of α-MIgG1.

protein-free CM5 chips. Upon the injection of the labeled protein solution, the fluorescence signal increased abruptly by a step of c. 300 counts s^{-1}econd (cps) with a 2-db attenuator in front of the PMT. However, there was no indication of any binding of either protein as the signal remained constant during the incubation period and quickly returned to the original background level after rinsing with pure HBS-EP buffer. These results show the excellent resistance of the dextran matrix to the non-specific physisorption of fluorescently-labeled IgG molecules. The abrupt increase in the signal when the solutions were exchanged originates from the small fraction of fluorophores in the bulk solution which are excited in the evanescent tail of the surface plasmon field.

The investigation of NSB to the matrix when it has been loaded with a large number of primary antibodies however, shows strong discrepancies between those two labeled polyclonal antibodies. As can be seen in Figure 6.6B (a), there was noticeable interaction between α-MIgG1 and AF-RaM since the fluorescence signal increased to c. 3E3 cps within 15 min and was relatively stable during the rinsing step. However, this level of NSB was quite satisfactory if compared with the AF-GaM test (Figure 6.6B (b)). Within the same time scale, the increase in the AF-GaM NSB signal was much faster than that of the AF-RaM signal by a factor of > 20. Thus, AF-RaM was chosen as the secondary antibody.

After loading the dextran matrix with α-MIgG1, we chose a monitoring angle of c. 57°, i.e. roughly at the 30 % reflectivity level at the low-angle side of the SPR minimum, which is a linear range for monitoring positive angular shifts in a 'kinetics' mode. The surface was then incubated with the 2F9.1 solutions of different concentrations for c. 30 min followed by a brief rinse (< 5 min). After further adjustment of the observation angle (if necessary, to keep the system in the linear zone), the 33-nM AF-RaM solution was introduced into the flow cell and allowed to flow through it for 30 min. Bound antigens and secondary antibodies were removed by a regeneration pulse of glycine buffer (10 mM, pH 1.7). A series of experiments was carried out on the same sensor chip using a decreasing concentration sequence of 2F9.1 at 333, 33, and 3.3 nM, and 333, 33 and 3.3 pM, respectively.

The binding curves obtained by using 2F9.1 for the capture process are shown in Figure 6.7A. In curve (a), the signal continued to increase after the initial rapid rise, instead of approaching an equilibrium level. This indicates that there are two binding phases in this interaction. For lower concentrations such as those shown in curve (c; 3.3 nM), the binding rate (slope) within the first 6–8 min remained constant, indicating a mass transfer-limited process. Binding curves for the lowest concentrations tested (333 pM (d); 33 pM (e); 3.3 pM (f)) are barely distinguishable. From these curves, we infer that the direct SPR detection limit of IgG is in the sub-nanomolar range even for such a large protein.

Figure 6.7B shows the fluorescence binding curves of AF-RaM to the captured antigen. The binding behavior in each curve looks similar. The binding signal for the lowest concentration of antigen (ie. 3.3 pM) can be distinguished above the signal for the negative control. From the summarized dose–response curves shown in Figure 6.7C, it can be claimed that a sensitivity increase of at least 1 ~ 2 orders of

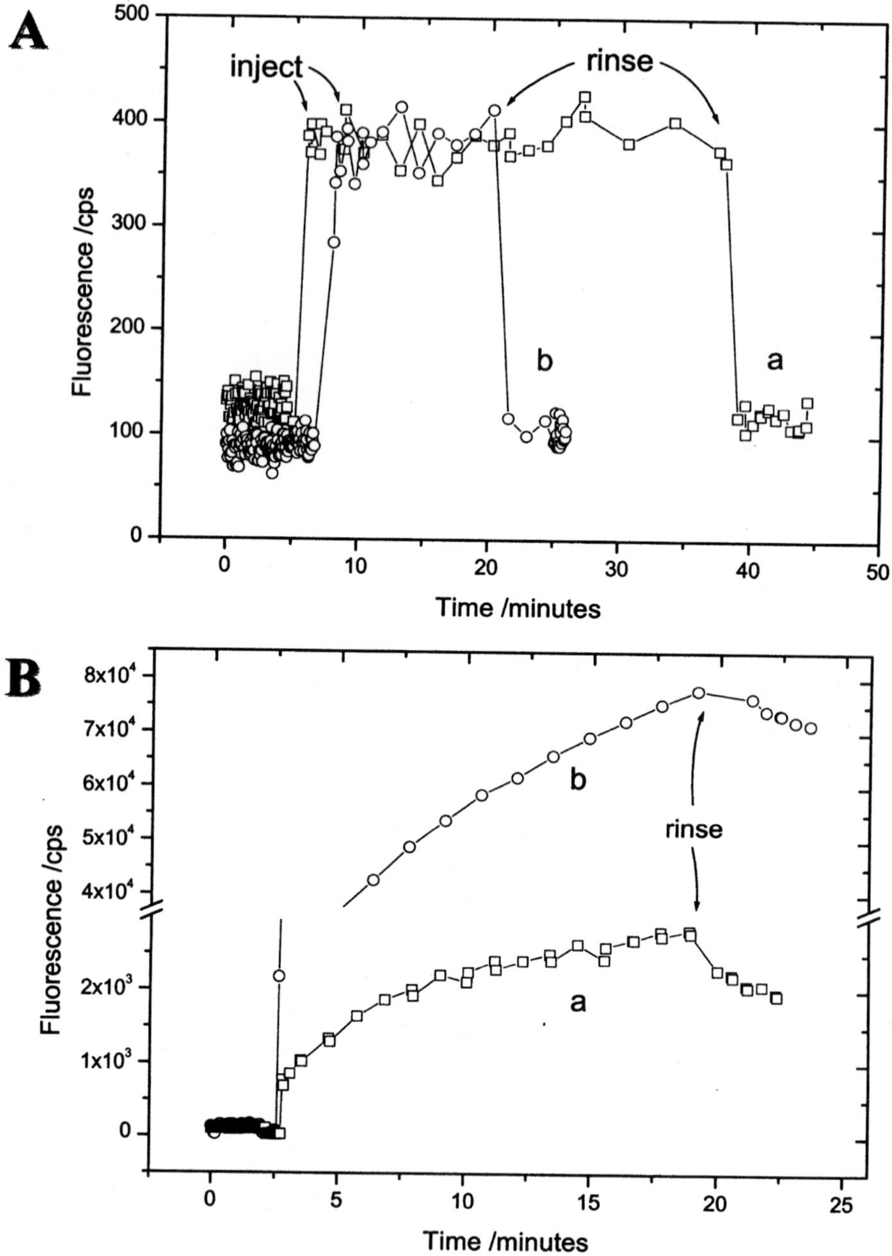

Figure 6.6 **A** Non-specific interaction of labeled antibody AF-RaM (a) and AF-GaM (b) to protein-free dextran surfaces. A 2-db attenuator was used. **B** Non-specific interaction of labeled antibody AF-RaM (a) and AF-GaM (b) with α-MIgG1-immobilized dextran surfaces. A 2-db attenuator was used.

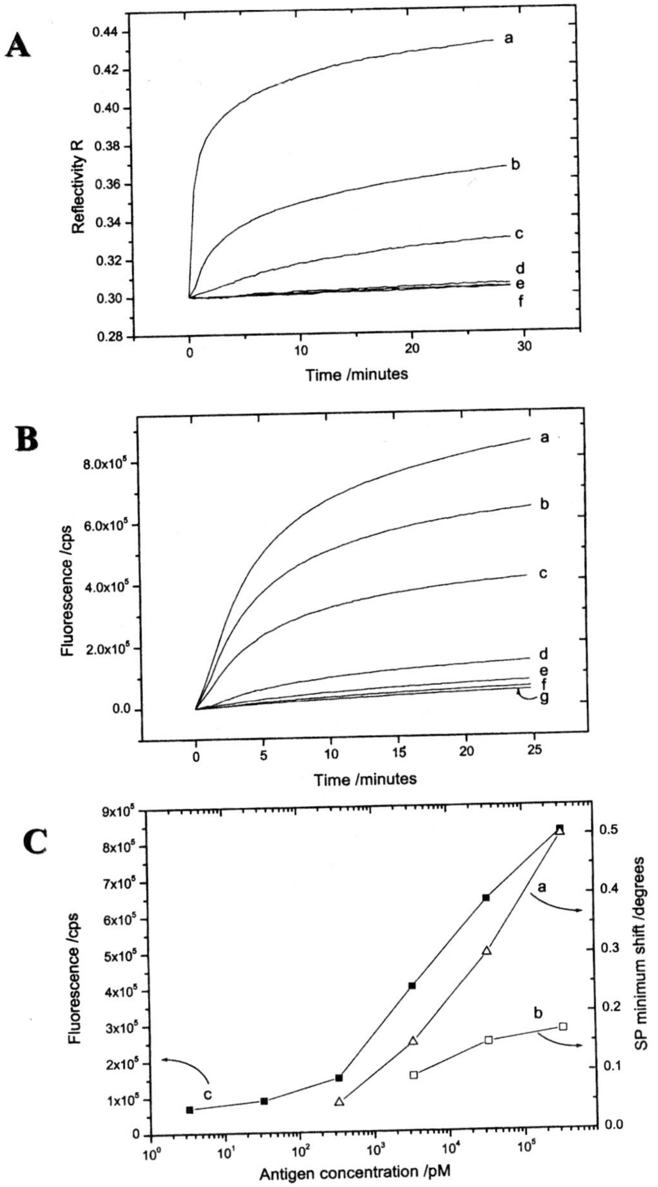

Figure 6.7 **A** Reflectivity binding curves of various concentrations of 2F9.1 solution (a) 333 nM, (b) 33 nM, (c) 3.3 nM, (d) 333 pM, (e) 33 pM, (f) 3.3 pM.
B Fluorescence binding curves of a 33-nM AF-RaM solution to a dextran surface treated with various concentrations of 2F9.1 solution (a) 333 nM, (b) 33 nM, (c) 3.3 nM, (d) 333 pM, (e) 33 pM, (f) 3.3 pM. Curve (g) is the negative control showing binding of AF-RaM in the absence of treatment with 2F9.1. A 2-db attenuator was used. All starting points of the fluorescence curves were shifted to zero for better comparison.
C The dose–response curves of (a) the SPR signal once 2F9.1 has bound, (b) the SPR response once AF-RaM has bound and (c) the fluorescence signal of the AF-RaM binding.

magnitude for antigen detection can be gained by adding the secondary fluorescently-labeled antibody, as compared to the direct SPR monitoring of antigen binding. Actually, adding the secondary antibody did not 'amplify' the response of 2F9.1 binding, on the contrary, it was substantially less by a factor of 1.5 ~ 3 (comparing curve (a) with (b) in Figure 6.7C). Comparing the SPR and the fluorescence signal both originating from the secondary antibody binding (curve (b) and (c) in Figure 6.7C), it can be seen that the lowest detectable concentration differed by 2 ~ 3 orders of magnitude.

Despite this the extra sensitivity gained is not completely satisfactory and is limited by the following two major factors. Firstly, as mentioned above, there is significant NSB between the primary and the secondary antibody. In fact, the binding of the secondary antibody shown in Figure 6.7B was the sum of specific and non-specific contributions. If the specific signal decreases, the non-specific signal becomes increasingly dominant and eventually totally obscures the former. Secondly, the regeneration strategy that we employed might be another limiting factor. Although the glycine buffer (10 mM, pH 1.7) is one of the best available buffers for the dissociation of antigen/antibody complexes used by Biacore, it was not completely satisfactory for 'cleaning up' the fluorescence signal. As Figure 6.8 shows, a series of different injections were applied for regeneration purposes, namely, glycine buffer (10 mM, pH 1.7) (1); 1 M NaCl solution (2) (for alleviating the electrostatic interactions); 100 % ethylene glycol (3) (for alleviating the hydrophobic interactions); and glycine buffer (10 mM, pH 1.5) (4), respectively. After these pulse injections, a 10 % residual fluorescence signal remained, while no meaningful signal could be seen in the SPR curve. Owing to this incomplete regeneration strategy, a significant amount of antigen/secondary antibody remained bound to the matrix which could also interfere with the specific signal in the subsequent steps. Experimentally, an upstream titration strategy provided a 'cleaner' matrix for lower concentrations and did improve the sensitivity, only by a factor of < 2 however (data not shown).

6.6
LOD Evaluation

A simplified interaction platform seemed to be necessary and essential for exploring the minimum number of target molecules per unit area that can be detected by SPFS. In the following, we focused on the interaction between mouse IgG and its labeled antibody, by covalently immobilizing the 2F9.1 on the surface and then studying its interaction with a series of concentrations of AF-RaM solutions. As shown in Figure 6.1C and D, two types of surface architecture were investigated, namely a lipoic acid functionalized surface (LA chip) and a dextran surface for demonstrating the advantage of the three-dimensional matrix in boosting the sensitivity of SPFS.

Lipoic acid is a versatile reagent used to functionalize gold surfaces with a significant number of carboxylic acid groups [16]. LASFN9 substrates attached to a fresh-

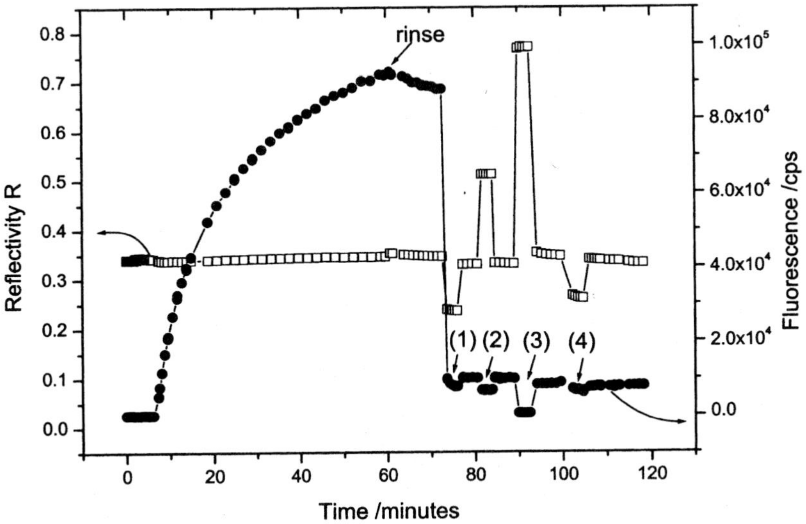

Figure 6.8 Example of the incomplete regeneration of the CM5 chip after it has been bound with a fluorescently-labeled antibody: (1) glycine buffer (10 mM, pH 1.7), (2) 1 M NaCl solution, (3) 100 % ethylene glycol and (4) glycine buffer (10 mM, pH 1.5) were applied sequentially. A 2-db attenuator was used.

ly evaporated 50-nm gold film were immersed for t > 24 h in a 2 % (w/v) alcoholic LA solution at room temperature. Before use, the substrates were rinsed with a copious amount of ethanol and dried under a stream of nitrogen.

Typical immobilization protocols for 2F9.1 are shown in Figure 6.9, following the standard procedures for α-MIgG1 at room temperature. The 2F9.1 loading level on the LA chip (angular shift of the SPR minimum, $\Delta\theta = 0.18$) was much lower than that on the CM5 chip (Figure 6.9 (a)), due to the difference in the number and accessibility of the carboxylic acid groups between these 2-D and 3-D matrices, respectively. By controlling the contact time of the 2F9.1 solution, we were easily able to manipulate the loading level for a better comparison between the matrices. The CM5 chips were loaded at four different levels, corresponding to 1.51, 1.33, 0.61 and 0.088° of the SPR minimum shift, respectively.

For analysis of concentration by SPFS, in a manner similar to the Biacore approach, we used mass transport-limited binding conditions in the association phase. Usually, under conditions of a very low concentration of analyte bulk or a very high concentration of surface ligand, the initial binding rates will be limited by mass transfer and this can be expressed as follows:

$$d[AB]/dt = k_M\,[A_0]. \qquad (1)$$

where [AB] is the concentration of the analyte/ligand complex, $[A_0]$ is the bulk concentration of the analyte, k_M is the mass transport rate constant and t is the time. The binding rate is then constant and independent of both the surface coverage of

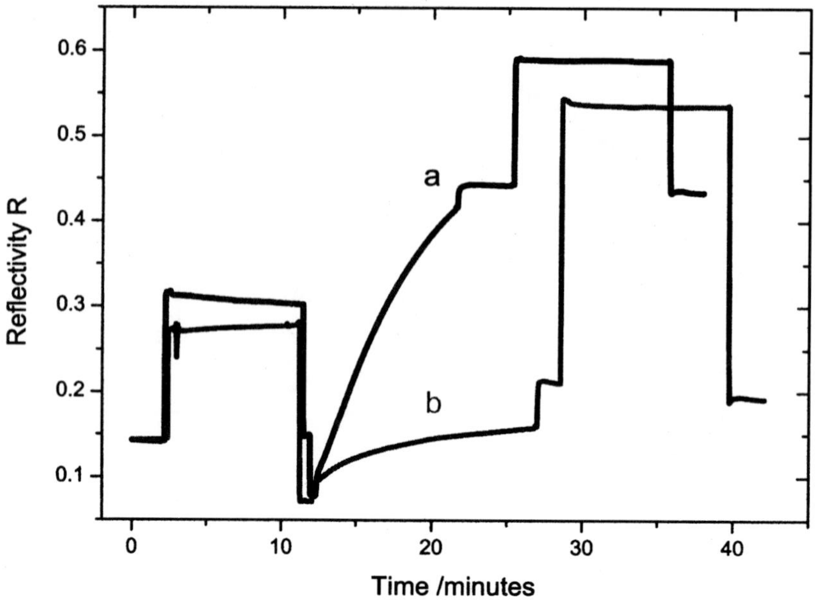

Figure 6.9 Immobilization of 2F9.1 on (a) a dextran matrix and (b) a planar lipoic acid-modified surface. The starting points of both curves have been adjusted to facilitate a better comparison.

the AB complex and the interaction affinity. k_M is obtained as the proportionality factor between the binding rate and the analyte concentration. Thus, concentration analysis under mass transport-limited conditions becomes very simple. The binding curve is a straight line, and its slope is proportional to the (active) analyte concentration. The linear zone of the calibration curve at the lower concentration end is theoretically unlimited.

Before circulating the AF-RaM solutions over the sensor surface, five replicates of HBS buffer injection as 'blanks' were applied in order to quantify any baseline deviation at room temperature. One example is shown in Figure 6.10A. Five slopes were obtained by linear fitting, and analyzed statistically. The sum of the mean and three times the standard deviation (SD) is considered to represent the baseline fluctuation which is typically 3–5 cps min^{-1}.

A typical working curve at room temperature is shown in Figure 6.10B on a CM5 chip with a considerably high load of 2F9.1 (1.33° degree). As can be seen, a constant slope was obtained in every binding curve which shows that mass transport-limited conditions were established at all the AF-RaM concentrations studied. The interaction period for each concentration was typically 30 ~ 60 min and was at least duplicated in order to produce a calibration curve. What should be noted is the observation that regeneration with glycine buffer was still incomplete after higher concentrations were applied, but visually better in the case of lower concentrations, which was probably due to the poorer SNR condition in the latter case. One draw-

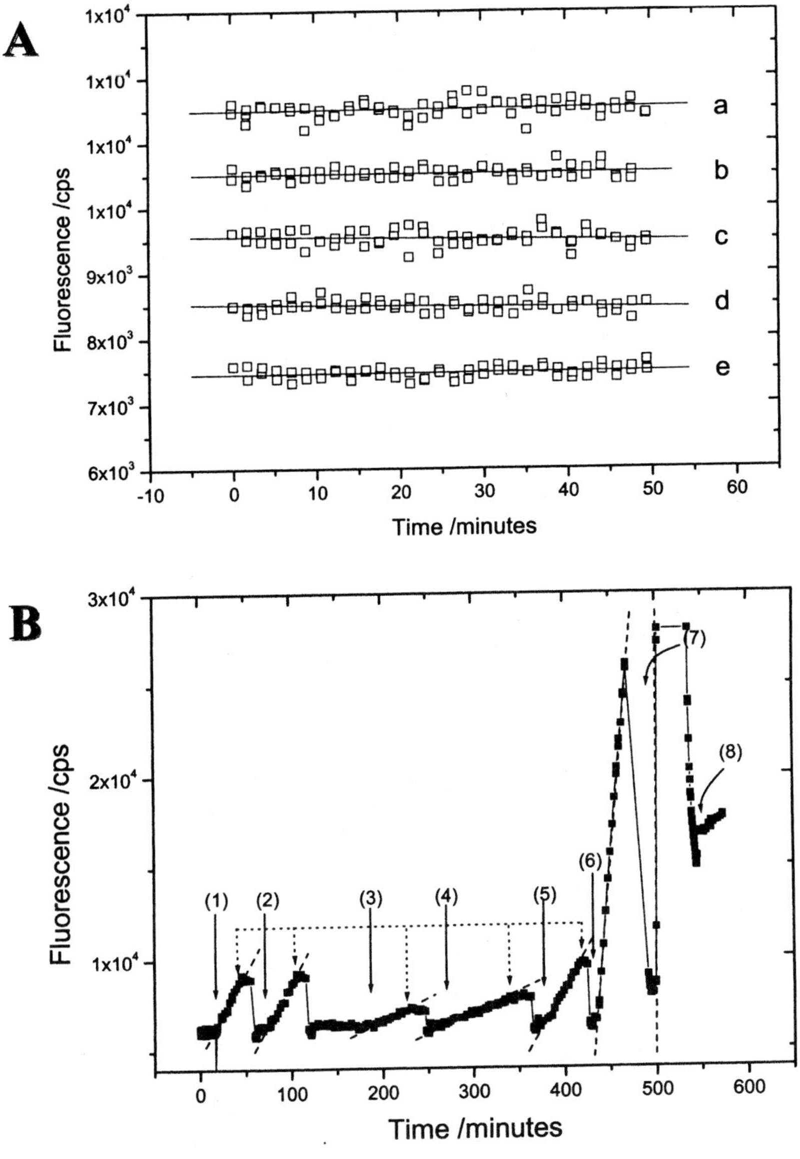

Figure 6.10 **A** Fluorescence responses after five injections of HBS-EP buffer as a blank presented as a vertical spread. All original curves represent the same fluorescence intensity. Solid lines represent the linear fit of the data which was used to calculate the different slopes, (a) 0.87, (b) 0.49, (c) – 1.25, (d) – 1.21 and (e) 0.61. The mean and SD of these values are – 0.098 and 1.043, respectively. Thus, mean + 3 x SD is 3.031 cps min^{-1}. **B** One of the working curves for LOD evaluation. The 2F9.1 loading level is 1.33°. (1) 333 fM, (2) 333 fM, (3) 67 fM, (4) 67 fM, (5) 333 fM, (6) 3.3 pM, (7) 33 pM of AF-RaM and (8) pure HBS-EP buffer were injected sequentially. The connected dotted arrows indicate the points at which a brief rinse by HBS-EP buffer was initiated. Regeneration was undertaken after every injection. Dashed lines are the linear fits to the binding curves. All fluorescence signals were monitored without an attenuator.

back of the current regeneration strategy is that the sensor chip may be 'contaminated' when the analyte solution is analyzed at a high concentration, which makes the baseline rather high and unstable and is no longer suitable for the analysis of lower concentrations (e.g. the baseline after an HBS injection at the point indicated by the arrow (8) in Figure 6.10B).

The calibration curves produced on different surfaces are presented in Figure 6.11A. Apparently the slopes of the calibration curves, representing the binding probability of AF-RaM are dependent on the level of 2F9.1 loading on the CM5 chips as shown in Figure 11B (a), (b), (c), (d). The curve which links these points deviates negatively towards a linear relationship which can be explained by the following consideration. Firstly, the epitopes are mutually blocked at high surface antigen density, leading to a decrease in the number of recognition sites. Secondly, one antibody can actually access two antigens if the distance between the antigen molecules is statistically closer when there is a greater density of molecules present. Thirdly, a greater density of antigen molecules will result in a higher probability of cross-linking with the dextran matrix which will in turn lead to an increase in the viscosity of the interaction environment. As expected, a much lower sensitivity was obtained on the LA chip (Figure 6.11B (e)). Comparing (d) with (e) in Figure 6.11B, the LA chip yielded a calibration curve with only a gentle slope even though it was loaded with a relatively greater density of 2F9.1. The lower metal-induced fluorescence quenching is an apparent advantage of the CM5 chip. Another possibility is the difference in antibody diffusion behavior between the two matrices since the dextran chain extends c. 100 nm into the bulk solution giving a nearly homogenous environment in which biomolecular interactions can take place and thus a higher probability of interaction. This can be further clarified by comparing the calibration curves obtained from the pure SPR response, with a much higher analyte concentration.

The highest sensitivity was achieved on the CM5 chip with the highest 2F9.1 loading density as shown in Figure 6.11A and re-plotted in Figure 6.12 as a double-logarithmic graph. The curve spans more than four orders of magnitudes of the analyte concentration (from 13 fM to 333 pM). The data fit a straight line which illustrates the excellent quality of the calibration curve provided that the concentration analysis is carried out under conditions of mass transfer. Extrapolating the linear curve to the baseline deviation level (c. 3 cps min^{-1}) results in a theoretical detection limit of 7.5 fM. Practically, however, the binding curves at the lowest concentration, 13 fM, were already too weak to provide any clear evidence of binding. Therefore, the LOD is given as 25 fM for an interaction period of less than 1 h, corresponding to a constant signal growth rate of c. 9 cps min^{-1}. Obviously, by extending the interaction time the signal will become more intense and the LOD will be improved.

At higher analyte concentrations, mass and fluorescence information can be obtained simultaneously by SPR and SPFS, which provides an opportunity to convert the concentration LOD into the number of bound analyte molecules (or fluorophores). The quantitative relationship between SPR and the fluorescence signal can be established by directly comparing curve (b) and curve (c) in Figure 6.7C

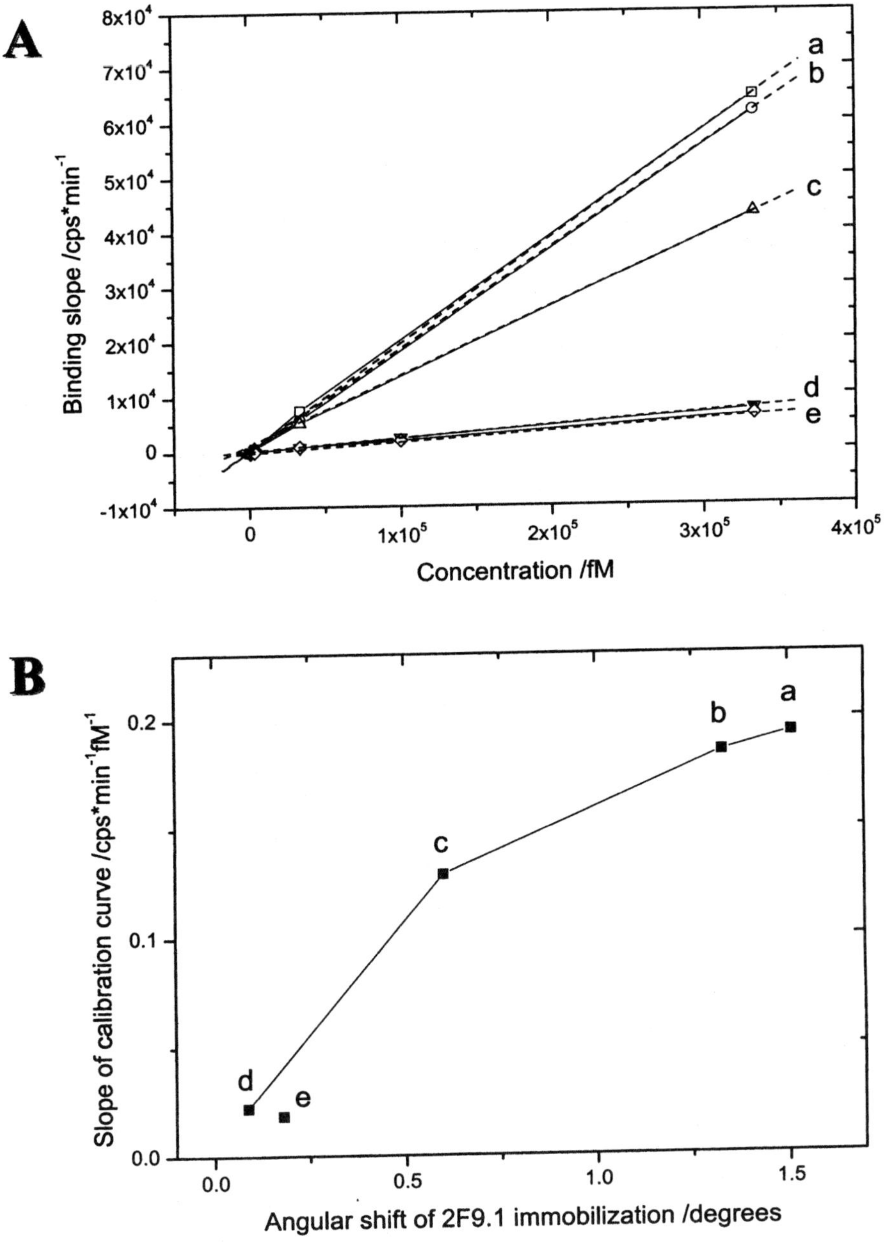

Figure 6.11 A Calibration curves of the different surfaces with various 2F9.1 loading levels. (a) CM5, 1.51°, (b) CM5, 1.33°, (c) CM5, 0.61°, (d) CM5, 0.088°, (e) LA, 0.18°. Broken lines are the linear fits. **B** Slopes of the calibration curves versus the 2F9.1 loading level. (a) CM5, 1.51°, (b) CM5, 1.33°, (c) CM5, 0.61°, (d) CM5, 0.088°, (e) LA, 0.18°. All fluorescence signals were measured without an attenuator.

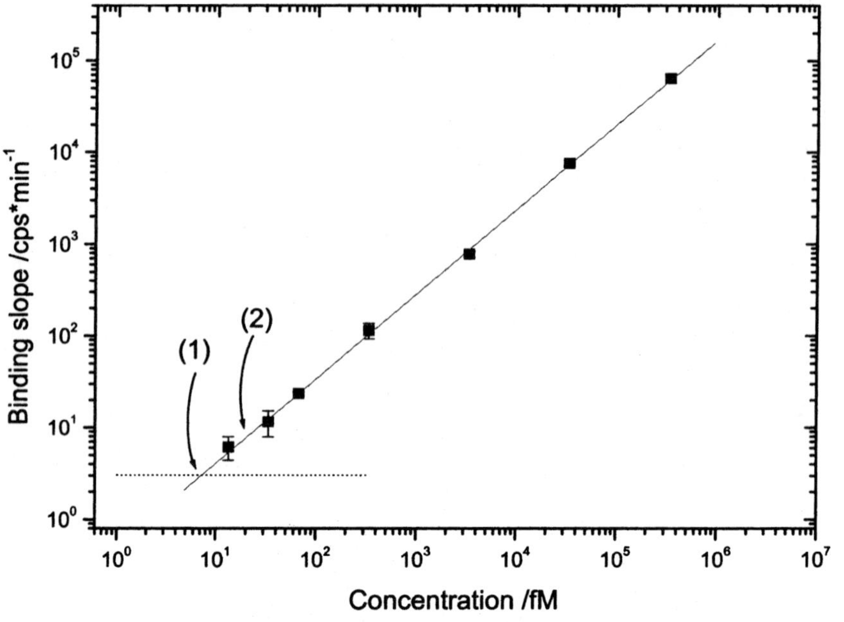

Figure 6.12 Double-logarithmic plot of the calibration curve of AF-RaM binding to the dextran surface loaded with the highest level of 2F9.1 (1.51°). Data taken from Figure 6.11A. The solid line is the linear fit. The dotted line represents the baseline fluctuation level (3 cpsmin^{-1}). Theoretical (1) and practical (2) LODs are indicated. All the fluorescence signals were measured without an attenuator.

which are largely proportional to each other. The highest angular shift $\Delta\theta$ and its corresponding fluorescence intensity ΔI are 0.17° and 5E7 cps respectively, if we consider the attenuation factor of 60.88 for the fluorescence. The minimum angular shift 0.17° corresponds to c. 900 pg mm^{-2} IgG, i.e. 3.7E9 molecules mm^{-2} (assuming the MW of IgG is 150 kDa), referring to the aforementioned surface concentration correlation. This means that the binding of c. 74 IgG molecules mm^{-2} (0.017 fg mm^{-2}) contributes an increase of 1 cps in the fluorescence signal. Thus, the signal growth rate of c. 9 cps min^{-1} (for 25 fM solution) equates to a minimum of c. 660 IgG molecules (c. 3200 AF 647 fluorophores) binding to the surface per minute, and is the lower limit that can be detected by SPFS.

6.7 Conclusions

In this paper, we have successfully implemented the Biacore sensor chip, CM5, to our SPFS instrument. The limit of detection for trace amounts of analyte has been greatly improved by this combination. The LOD of SPFS (25 fM) is more than four orders of magnitude higher than that of our SPR (about several hundred picomolar) in the IgG analysis. This also enables SPFS to record the interfacial binding process at an extremely low surface concentration (femtograms mm^{-2} for IgG) in an in situ and real-time mode of operation. However, before achieving sensitivity at the femtomolar level in a realistic detection mode (e.g. sandwich assay), a number of issues should be addressed: non-specific interaction and an un-optimized regeneration protocol are apparently two important factors which largely counter the gains made in sensitivity and need to be further improved. As for the instrument itself, it is still far from being optimized in terms of the interaction temperature, optical elements for fluorescence detection and inhibition of the background fluorescence.

References

1. Jönsson, U., Fägerstam, L., Ivarsson, B., Johnsson, B., Karlsson, R., Lundh, K., Löfås, S., Persson, B., Roos, H., Rönnberg, I., Sjölander, S., Stenberg, E., Ståhlberg, R., Urbaniczky, C., Östlin, H., Malmqvist, M. *Biotechniques* 1991, **11**, 620–627.
2. Hall, D. R., Winzor, D. J. *Anal. Biochem.* 1997, **244,** 152–160.
3. Piehler, J., Brecht, A., Gauglitz, G., Zerlin, M., Maul, C., Thiericke, R., Grabley, S. *Anal. Biochem.* 1997, **249,** 94–102.
4. Wegner, G. J., Lee, H. J., Corn, R. M. *Anal. Chem.* 2002, **74,** 5161–5168.
5. Lyon, L. A., Musick, M. D., Natan, M. J. *Anal. Chem.* 1998, **70,** 5177–5183.
6. Pei, R. J., Yang, X. R., Wang, E. K. *Analyst* 2001, **126,** 4–6.
7. Liebermann, T., Knoll, W. *Colloids. Surf. A.* 2000, **171,** 115–130.
8. Liebermann, T., Knoll, W., Sluka, P., Herrmann, R. *Colloids. Surf. A.* 2000, **169,** 337–350.
9. Kambhampati, D., Nielsen, P. E., Knoll, W. Biosens. *Bioelectron.* 2001, **16,** 1109–1118.
10. Neumann, T., Johansson, M-L., Kambhampati, D., Knoll, W. *Adv. Funct. Mater.* 2002, **12,** 575–586
11. Hodneland, C. D., Lee, Y-S., Min, D-H., Mrksich, M. *Proc. Natl. Acad. Sci. USA* 2002, **99,** 5048–5052.
12. Johnsson, B., Löfås, S., Lindquist, G. *Anal. Biochem.* 1991, **198,** 268–277.
13. Yu, F., Yao, D., Knoll, W. Analytical Chemistry 2003, **75,** 2610–2617.
14. Löfås, S., Johnsson, B. *J. Chem. Soc., Chem. Commun.* 1990, **21,** 1526–1528.
15. Stenberg, E, Persson, B., Roos, H, Urbaniczky, C. *J. Colloid Interface Sci.* 1991, **143,** 513–526
16. Berggren, C., Johansson, G. *Anal. Chem.* 1997, **69,** 3651–3657

7
The Use of Proteinchip® Arrays for Deciphering Biological Pathways

Lee O. Lomas

Abstract

The ability to understand the complexity of protein interactions will be a key component to making sense of proteomic data. Studying the complexity of such interactions using genomic-based analysis tools such as yeast-2-hybridization, or proteomic tools incorporating indirect detection methods such as fluorescence or colorimetric techniques, is generally limited to binary events and often requires subsequent validation, as in the case of yeast-2-hybridization, or the availability of high affinity, high specificity secondary reagents such as antibodies. Furthermore, it is difficult for either of these methods to evaluate interactions that involve more than two interactants at one time. Affinity-based biochips coupled directly with mass spectrometry, however, removes many limitations of the previously mentioned methods because the mass signature of each interacting protein component/subunit can be exploited when detecting multiple interactants simultaneously. Biochips based on the principles of surface enhanced laser desorption/ionization (SELDI) are now commercially available and provide a simple but sophisticated tool for detecting protein interactions. A review of the current state of such SELDI-based biochips will be provided.

7.1
Introduction

The single greatest scientific undertaking in biology in the 1990s has been to complete whole genomic sequencing for a number of vertebrate species, including human [1]. This process has provided a wealth of information related to the organization of genetic material, and the basic composition of the building blocks they encode. However, gene information alone is not sufficient to provide an understanding of why biological systems function as they do. Cellular function within a biological system is achieved only through the coordinated effect of a vast network of biomolecular interactions and understanding the function of proteins within this context will represent the proteomic 'holy grail'.

Unlike DNA where gross similarities in biochemical properties allowed for the development of generalized methodologies for massively parallel sample processing, proteins possess individually unique properties that create an enormous

Protein Microarray Technology. Edited by Dev Kambhampati

ISBN: 3-527-30597-1

hurdle in developing generic methodologies to study and characterize sets of proteins. This is further exasperated by the simple disparity in raw numbers; the 30–40 thousand estimated genes in the human genome are calculated to encode some 200 000 protein species, which can be post-transcriptionally modified to potentially millions of functionally active distinct proteins [2]. Thus, there now exists an urgent need for advancement in protein analysis tools and methodologies.

In the past decades, a primary focus has involved developing methods for global protein analysis that combine separation technologies such as one- or two-dimensional gel electrophoresis and/or chromatography with mass spectrometry (MS) techniques. This would then allow for the implementation of protein differential display approaches that compare protein profiles in control and experimental populations, analogous to strategies of differential gene display. From this the field of Proteomics has evolved, with 2D-gel electrophoresis coupled with either matrix-assisted laser desorption/ionization (MALDI) or electrospray (ESI)–MS as the primary analytical methodology. Although considerable improvements have been made in both 2D-gel electrophoresis and MS techniques, irreproducibility and insensitivity to low molecular weight proteins, particularly below 15 kDa, hydrophobic proteins and extremely acidic or basic proteins associated with electrophoretic separation still hinder this strategy.

Furthermore, confirming the identity of the individual protein species within a biological sample is only a first step. To understand even the simplest of biological processes requires understanding how proteins interact with each other and coordinate themselves to affect the specific response. The current challenge now facing researchers involves the overall complexity of the interactions at the cellular level. Simplifying the interaction environment into manageable units that are amenable for study will ultimately be necessary. For example, a protein interaction may involve simple transient associations, such as enzyme–substrate or receptor–ligand interactions, or more long-lived and relatively stable interactions such as those associated with ribosomal or structural complexes. By understanding such interactions on an individual basis, entire biochemical pathways can be recapitulated in terms of function and, in the case of disease treatment, potential therapeutic intervention.

7.2 Methods to Study Protein Interactions

A number of methods that encompass both genomic and proteomic approaches can be used to study protein interactions and it is clear that the utilization of such complementary approaches will ultimately be necessary.

7.2.1
Genomic Approaches

Genomic approaches rely on the use of cDNA or mRNA as a starting material and include gene-trapping methodologies such as yeast- or mammalian-2-hybridization techniques whereby the modular nature of transcriptional activators is exploited [3–5]. This *in vivo* assay, although easy to implement and capable of screening interactants in a high-throughput manner, has several limitations. Proteins that function as transcriptional activators or repressors cannot be studied by this method. More significantly, the investigator is unable to control either the post-translational modification state of the proteins under investigation or the temporal or special environment under which the interactions are studied, leading to false negative and false positive results believed to be as high as 30 %. Finally, there may also be a bias against larger proteins, and those that retard growth during propagation [6,7].

7.2.2
Proteomic Approaches Leading to the Development of Protein Microarrays

A more ubiquitous approach for deciphering protein interactions involves detecting interacting proteins themselves, either indirectly through the used of tags or directly by protein characterization. These affinity precipitation methods currently rely heavily on the availability of affinity tags such as antibodies and span the enzyme-linked immunoassay (ELISA), Western blotting and immunoprecipitation methods. ELISA is the classical approach for detection and quantitation of protein titer from complex biological samples. This method relies on the availability of dual affinity-based reagents, typically antibodies, to the target protein; the first to capture and retain the protein while the second is generally linked to a reporter amplification process such as horseradish peroxidase (HRP) or fluorescene. This method is generally sensitive and can easily be automated to the extent whereby thousands of samples can be analyzed in a single day in a cost-effective manner. The primary drawback, however, is that well-characterized antibodies generated against a single epitope are necessary to confer a high degree of specificity. If an antibody binds a common epitope across multiple protein species, a "composite" signal will be generated and additional assays are necessary to decipher the binding event. For example, amyloid β is a small peptide that, when inappropriately produced through the specific cleavage of its amyloid precursor protein, is linked to the development of Alzheimer's disease. To distinguish between the non-disease-associated peptide, amyloid β 1–40, and the disease-associated peptide, amyloid β 1–42, at lease three distinct antibodies are required; two antibodies specific for each of the amyloid β variants and a third antibody linked to the reporter system.

Nevertheless, methodologies that combine genomic arraying techniques with ELISA-based methodologies are now gaining considerable attention [8,9]. This approach strives to deposit nanoliter quantities of bait reagents, such as antibodies, recombinant peptides or small drug libraries on an addressable, high-density (potentially thousands of features) microarray in a process analogous to genomic microarrays. These approaches are now possible due to advances in recombinant protein technologies that allow the rapid production and purification of affinity-based reagents such as antibodies, ScFv, aptamers and peptide libraries. Although many groups have demonstrated the utility of protein microarrays, there still exist a considerable number of obstacles associated with commercialization. Attachment of the bait protein to the array surface can be accomplished through either non-specific association or through direct covalent attachment and both methods have been demonstrated [10,11]. Critical limitations include denaturing the protein upon interaction with the surface, disruption of the interaction domain through poor bait orientation or steric hindrance caused by physical closeness to the support itself (particularly important with small bait molecules), or a lack of sufficient density to allow for subsequent detection. In addition, the range of biologically significant affinities (mM to pM) that must be considered when optimizing target binding conditions to retain all relevant low-affinity interactants (to prevent false negatives) while eliminating non-specific binding (to prevent false positives) is critical when working with potentially hundreds of baits simultaneously.

7.2.3
Protein Biochips Utilizing Surface Enhanced Laser Desorption/Ionization (SELDI) ProteinChip® System Methodologies

As an alternative to the genomically spawned microarray methodologies that couple high-density bioaffinity arrays to a fluorescence or colorimetric detection methodology, SELDI-based ProteinChip® Arrays offer very unique capabilities that make them particularly suited to the role of both an assay tool and a discovery tool. Originally developed by Hutchens and Yip [12], biochips designed for the SELDI ProteinChip System incorporate a solid-phase extraction adsorbent on an electrically conductive support that can be interrogated directly by time of flight mass spectrometry (TOF-MS). The extraction adsorbents are functionalized with a number of different chemistries to yield complementary selection characteristics across a wide physicochemical range. Biochips can be provided as pre-activated surfaces for the covalent binding of biomolecules (baits), or chromatographic surfaces for the selective capture of proteins via charge, hydrophobicity or metal–chelate interactions (Figure 7.1). Irrespective of whether a pre-activated or chromatographic surface is utilized, the methodology for capturing specific classes of proteins from a crude biological sample is similar and is shown in Figure 7.2. After an affinity surface is chosen, the sample of interest is applied to the spot, typically in a volume of 1–5 μL but up 400 μL, and incubated for several minutes. During this time, proteins with a specific affinity to the chosen adsorbent are extracted from the sample droplet while proteins with no affinity remain in solution. After the binding step, the array is washed to remove all non-specific binding components, and then dried. The composition of the binding and washing buffer can be varied greatly to further enhance the selectivity of the surface. For example, when using a weak cationic exchanger as the adsorbent (eg CM10 ProteinChip Array; pKa~4), modulating the pH of the binding and/or washing buffer will protonate/deportonate the proteins within the sample, depending on their individual pIs, and thus fine tune the binding selectivity.

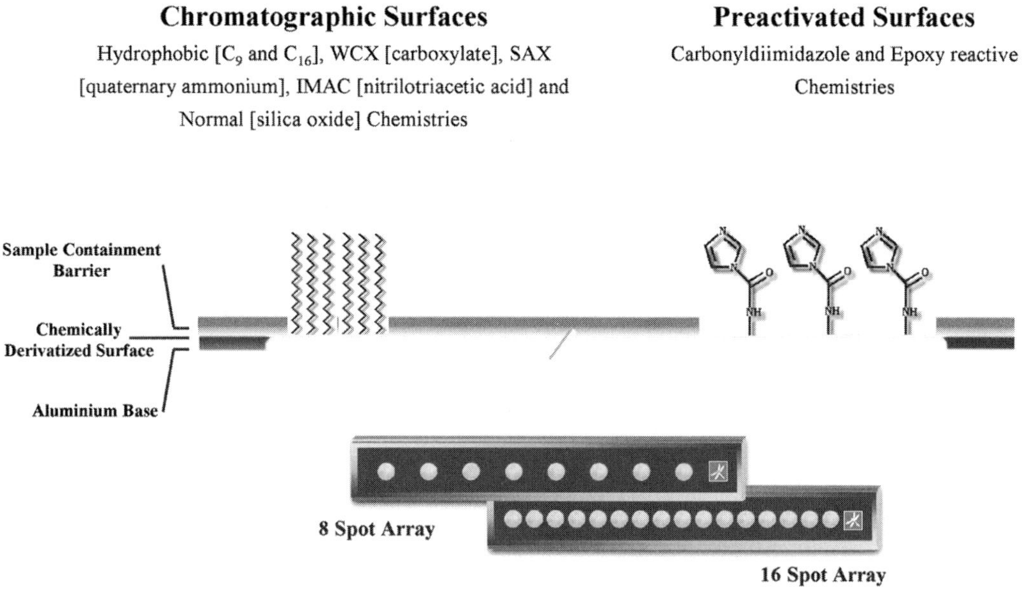

Figure 7.1 Types of ProteinChip® Array surfaces. Array surfaces fall into one of two general categories; chromatographic surfaces or pre-activated surfaces. Chromatographic surfaces utilize affinity interactions based on typical physicochemical properties such as hydrophobicity, charge or metal–chelate interactions. Pre-activated surfaces provide a covalent chemistry based on either carbonyldiimidazole or epoxy reactive groups and can be customized by adding bioaffinity molecules such as proteins (e.g. antibodies, receptors), lipids, carbohydrates or DNA.

The use of an affinity surface to selectively capture biomolecules also solves one of the biggest challenges associated with analyzing biological samples directly with laser desorption/ionization mass spectrometry. In such applications the sample must be deposited onto the MS probe in a volatile, low conductivity buffer. The presence of salts, chaotrophic salts, and detergents typically used in interaction assays to promote optimal binding often interfere with the subsequent matrix–protein co-crystallization and/or suppress sample ionization. In such instances, a second dialysis or sample clean-up step is almost always necessary, with the associated sample losses and negative impact on reproducibility. Using biochips based on SELDI, proteins selectively captured onto an array surface can finally be washed with water or volatile buffers to effectively remove any buffer components that may interfere with subsequent MS analysis. For simplicity, application types that successfully utilize SELDI-based methodologies have been divided into two types,

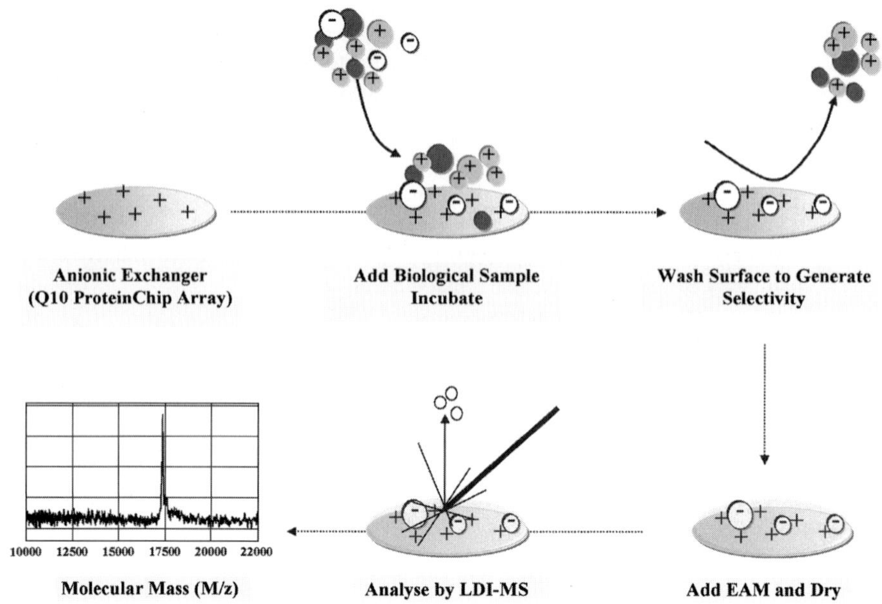

Figure 7.2 Typical methodology using the SELDI-ProteinChip System. A biological sample is incubated on an affinity surface under controlled binding conditions such as pH, salt or detergent. After incubation, the surface is washed with buffer to remove any non-specific binding. Multiple washing steps can be incorporated to fine-tune the biospecificity. After washing, the sample is dried and an energy-absorbing molecule (EAM) is added, which co-crystallizes with the retained protein. Finally, the sample is pulsed by laser light in a laser ionization source to induce sample ionization and final detection is by TOF-MS.

Expression Difference Mapping™ (EDM) and Interaction Discovery Mapping™ (IDM) applications.

7.2.3.1 Expression Difference Mapping™ Applications

Expression Difference Mapping applications are analogous to the "whole proteome approach", whereby the goal of the experiment is to detect as many proteins as possible from a biological sample. By comparing profiles between samples, e.g. control versus treatment, differential expression patterns can be generated that selectively distinguish between sample populations and can thus be used for diagnostic purposes. The speed with which these types of experiments can be completed also makes it feasible to undertake validation studies on large sample numbers (100s to 1000s) in days or weeks rather than months. This approach has now been widely demonstrated (see [13] for a general overview] and will not be further discussed here.

7.2.3.2 Interaction Discovery Mapping™ Applications

Interaction Discovery Mapping methodologies are more akin to the microarray methodologies discussed above in that the interaction generated is based on a biological affinity between a bait and target molecule. The unique difference, however, is the incorporation of mass spectrometry as a direct detection methodology, giving an additional level of information not achievable using conventional fluorescence or colometric detection schemes. As a demonstration of this, consider the amyloid β example mentioned previously where the distinction between amyloid β 1–40 and 1–42 required three antibodies raised against unique epitopes. Using an Interaction Discovery Mapping approach, a single antibody can be used (e.g. monoclonal antibody 6E10 (raised against aa 1–17) Signet Laboratories USA) to capture amyloid β 1–40, 1–42 as well as a range of additional amyloid β C-terminal cleavage products (Figure 7.3). This particular assay is now well established [14].

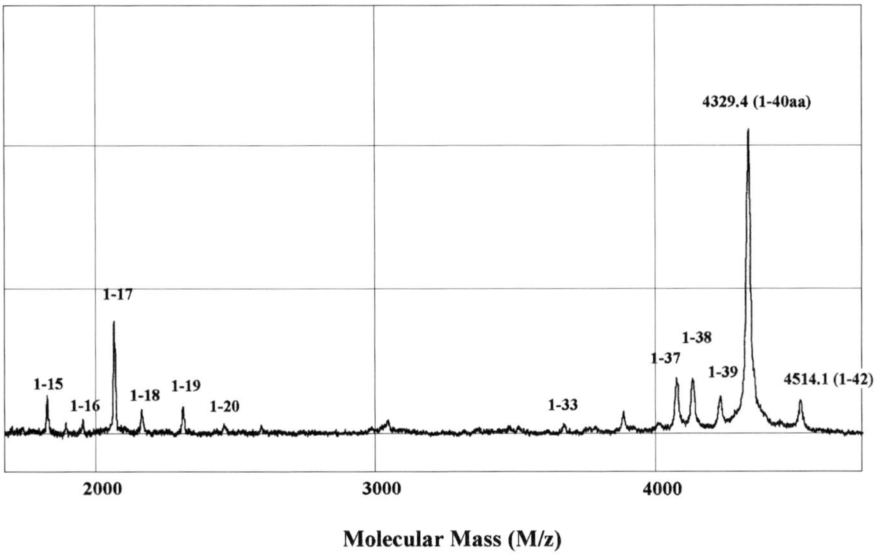

Figure 7.3 Detection of amyloid β fragments using pre-activated ProteinChip Arrays. Monoclonal 6E10 anti-amyloid _ antibody was covalently attached to the surface of an activated array, followed by incubation with cell culture media. After incubation, subsequent washing to remove non-specific binding, and analysis by LDI-MS, specific peaks were detected corresponding to amyloid β 1–40 (4329.4 Da), amyloid β 1–42 (4514.1 Da) as well as a number of additional C-terminally cleaved fragments.

Protein interactions often result in the formation of multi-component protein complexes, which provides additional difficulties in analysis when using an indirect detection methodology. Often specific affinity reagents to all members of the complex are not available and discovery relies on capture of the complex through a single subunit. Biochips utilizing mass spectrometry as a detection method can offer significant advantages in both simplicity and speed for analyzing multi-component protein complexes due to the mass discriminating power of the methodology. Using an antibody or affinity tag against one component of a multi-component complex coupled covalently to the surface of a biochip, an immunoprecipitation experiment can be carried out directly on the array surface. After the prerequisite washing step to remove non-specific binding and any salts or detergents which may interfere with the MS analysis, all components of the complex can be distinguished from one another based on their unique mass signature. Such an example is illustrated in Figure 7.4, whereby the troponin complex, consisting of troponin-C, troponin-I and troponin-T, was immunocaptured on a reactive ProteinChip Array with subsequent MS analysis. Using an anti-troponin-I antibody covalently attached to the surface of a PS20 ProteinChip Array, a crude heart tissue extract could be incubated directly on the array surface. After a stringent wash with buffer containing 1 M urea, 0.1 % CHAPS and 0.5 % NaCl to remove nonspecific binding, the surface was washed in 5-mM ammonium carbonate buffer to remove the urea, detergent and salt components that would interfere with MS analysis. Finally, upon MS analysis the troponin complex consisting of troponin-I (containing the epitope for 8I7 antibody)–troponin-C–troponin T could be distinguished based on their corresponding mass signatures.

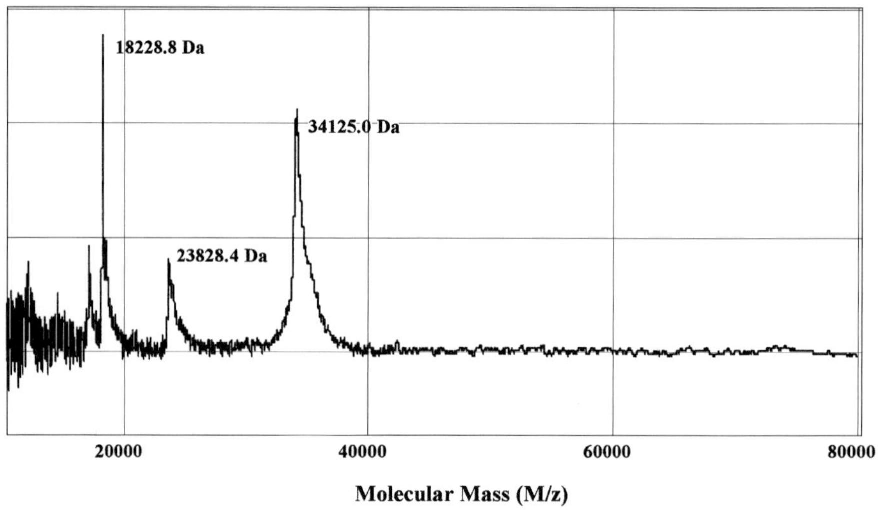

Figure 7.4 Capture of the troponin complex from heart tissue using pre-activated ProteinChip Arrays. Anti-troponin I antibody (clone 8I7) was covalently coupled overnight to a PS20 pre-activated surface. After blocking and washing to prevent non-specific binding, 5 μL of a crude heart tissue extract homogenized in 1 M urea containing 0.1 % CHAPS and 0.5 M NaCl was incubated on the array for 2 h. After subsequent washing with the homogenization buffer, the array was analyzed by mass spectrometry. Troponin I (23.8 kDa) was captured directly by the anti-troponin I antibody. Additionally, troponin C (18.2 kDa) and troponin T (34.1 kDa) could be captured and detected through complex formation with troponin I. All species were individually confirmed by MS/MS sequencing.

7.2.4
The Use of Biochips in Mechanistic Studies

The suitability of using biochips as a discovery tool for mechanistic studies is also now becoming feasible. As an example, methodologies are now being developed for the simultaneous analysis of multiple kinase activities from a single tissue sample. The methods rely on a number of synthetic kinase substrates that show high single kinase specificity. Although antibody capture methods can be used to enrich for kinases of interest if desired, the method involves incubating a number of these kinase-specific peptide substrates directly with the crude tissue extract. After an incubation period of 5–30 min, the phosphorylated products are enriched using an immobilized metal affinity capture surface (IMAC30 ProteinChip Array) preloaded with gallium. This surface shows a particularly high affinity for the phosphorylated product over the non-phosphorylated substrate. Through the use of peptide substrates with unique mass signatures, these assays can easily be multiplexed together to provide an activity profile of a variety of kinases in response to biological perturbation (Figure 7.5). An additional benefit of a mass spectrometric detection method is the applicability for characterizing post-translational modifications, such as phosphorylation. The current limitation of this methodology is in the design of highly specific peptide synthetic substrates and before such experiments are conducted, it is essential that the level of cross phosphorylation by a population of kinases is fully appreciated. However, progress continues to be made in understanding, and providing methodologies to overcome these limitations.

7.3
Conclusions and Outlook

Protein biochips will continue to generate attention as an essential tool for deciphering proteomic information and it is likely that two formats will ultimately emerge. One such format will no doubt mimic genomic arrays and consist of hundreds, if not thousands, of bait proteins immobilized onto a solid support. By coupling such arrays to fluorescence detection methodologies, the potential as an assay platform for the detection of known proteins or enzyme activity will be considerable. For this approach to be fully realized the current limitations, including generating sufficient protein content to be used as bait, immobilization methodologies that allow retention of the native state and biological affinity/activity, generation of secondary tagged detection methods and optimization of assay conditions to reduce non-specific binding must be addressed. However, this approach will never suffice as a method of discovering new interactants. Protein biochips which integrate a broad range of surface affinities with direct detection methodologies will ultimately prevail as true discovery tools. Currently SELDI-based biochips are the only commercial products that accommodate this strategy. Both chromatographic affinity surfaces and bioaffinity surfaces allow for rapid discovery and characterization of new proteins correlated with disease or cell perturbation. As

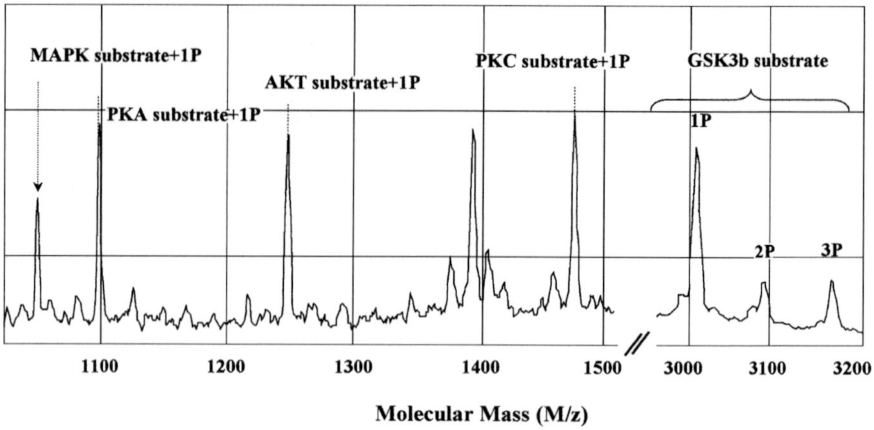

Figure 7.5 Detection of multiple kinase activities from a crude primary cortical neuronal (PCN) cell extract. MAPK, PKA, AKT, PKC and GSK3β kinase activity was detected by adding synthetic peptide substrates directly to the cell extract. After incubation, phosphorylated peptides were enriched on an IMAC-Ga ProteinChip Array and analyzed by LDI-MS. The peaks indicated correspond to the phosphorylated MAPK, PKA, AKT, PKC and GSK3β specific substrates.

methodologies develop which allow for the rapid micropurification of such biomarkers, these biomarkers can then become bait proteins themselves to generate useful information as to disease mechanisms. This in turn leads to discovery of additional proteins within a pathway in a self-perpetuating manner. As with fluorescence-based microarrays, the current focus for SELDI biochips continues to increase surface selectivity through the development of new surface affinity ligands, reduce non-specific binding, and improve immobilization efficiencies. Improved surfaces combined with advances in mass spectrometry instrumentation and methodology will be a significant tool in understanding biological function.

Acknowledgements

We thank our collaborators, Dr Steve Finkbeiner at the Gladstone Institute of Neurological Disease for generously providing us with primary cortical neuronal cell lysates from mouse and Dr Jennifer Van Eyk, Cardiomics Inc. for supplying heart tissues.

References

1. Lander, E.S. et al. Initial sequencing andanalysis of the human genome. *Nature* 2001, **409,** 860–921.

2. Kenyon, G.L. et al. Defining the mandate of proteomics in the post-genomics era. *Mol. Cell. Proteomics* 2002, **1,** 763–780.

3. Fields, S., and Song, O.K. A novel genetic sysem to detect protein–protein interactions. *Nature* 1989, **340,** 245–246.

4. Chien, C.T. et al. The two-hybrid system: a method to identify and clone genes for proteins that interact with a protein of interest. *Proc. Natl. Acad. Sci. USA* 1991, **88,** 9578–9582.

5. Fields, S., and Sterngianz, R. The two-hybrid system: an assay for protein–protein interactions. *Trends Genet.* 1994, **10,** 286–292.

6. Phizicky, E.M. et al. Biochemical genomics approach to map activities to genes. *Methods Enzymol.* 2002, **350,** 546–559.

7. Grayhack, E.J., and Phizicky, E.M. Genomic analysis of biochemical function. *Curr. Opin. Chem. Biol.* 2001, **5,** 34–39.

8. Arenkov, P. et al. Protein microchips: use for immunoassay and enzymatic reactions. *Anal. Biochem.* 2000, **278,** 123–131 ().

9. Emili, A.Q., and Cagney, G. Large-scale functional analysis using peptide or protein arrays. *Nature Biotehnol.* 2000, **18,** 393–397.

10. MacBeth, G., and Schreiber, S.L. Printing proteins as microarrays for high-throughput function determination. *Science* 2000, **289,** 1760–1763.

11. Haab, B.B. et al. Protein microarrays for highly parallel detection and quantitation of specific proteins and antibodies in complex solutions. *Genome Biol.* 2001, **2,** 1–13.

12. Hutchens, T.W., and Yip, T.T. New desorption strategies for the mass spectrometric analysis of macromolecules. *Rapid Commun. Mass Spectrometry* 1993, **7,** 576–580.

13. Issaq, H.J., Conrads, T.P., Prieto, D.A., Tirumalai, R., and Veenstra, T.D. SELDI-TOF MS for diagnostic proteomics. *Anal. Chem.* 2003, 149–155.

14. Davies, H., Lomas, L., and Austin, B. Profiling of amyloid _ peptide variants using SELDI ProteinChip® Arrays. *BioTechniques* 1999, **17,** 1258–1261.

8
Production of Protein Microarrays

Christopher J. Mann, Sarah K. Stephens and Julian F. Burke

8.1
Introduction

Microarraying is the printing of ordered arrays of biomolecules onto a solid surface in miniaturized format. The success of DNA microarrays, or biochips, in advancing the development of high-throughput screening and genome-wide expression analysis has led to a wealth of valuable information. Considerable effort is now being focussed on the transfer of this technology into proteomics. Proteins, however, are more complex and diverse than DNA, and there are many challenges to be overcome. In this chapter, we will discuss the different types of protein microarrays, and the approaches and pitfalls involved in their production.

8.2
From DNA Arrays to Protein Arrays

DNA microarrays evolved from the convergence of two very different fields: molecular genetics and computer electronics. In 1975, Ed Southern proved the concept that DNA fixed to a solid support could be used to attract complementary DNA strands in a process now known as Southern blotting [1]. In 1991, Stephen Fodor [2] at Affymax reported the fabrication of DNA microarrays on the surface of glass chips by combining the photolithographic method used to produce semiconductors, with combinatorial synthesis of oligonucleotides. In 1993, he co-founded Affymetrix, in order to develop microarrays with hundreds of thousands of different oligonucleotides. The following year, Affymetrix commenced manufacturing and selling its first DNA microarray, GeneChip®, and the DNA microarray market was born. Concurrently, Pat Brown at Stanford University developed a method for mechanically arraying and immobilizing numerous PCR-derived cDNAs using a robot to print onto simple glass slides [3]. His decision to disclose on his website how researchers could prepare their own microarrays was the key factor leading to the widespread use of DNA microarrays [http://brownlab.stanford.edu/].

Today, DNA microarrays, some incorporating tens of thousands of oligonucleotide probes per square centimetre, have become the method of choice for high-

Protein Microarray Technology. Edited by Dev Kambhampati

ISBN: 3-527-30597-1

throughput analysis of nucleic acids at their transcriptional level. Cluster analysis of such data has proven useful in linking the function of groups of genes to common metabolic pathways. However, DNA microarrays are poor at giving insight into the function of individual genes because transcriptional activity does not necessarily reflect the production or activity of the translated protein. Indeed, the correlation between mRNA and protein levels can be remarkably weak [4]. The primary reasons for this are that measurement of DNA transcripts and oligonucleotides cannot take account of post-translational modifications or protein turnover rates. So the present consensus is that analysis of protein expression and interaction should provide a superior approach to elucidating gene function. A second key argument for directly arraying proteins and studying their interactions is that almost all drugs discovered to date exert their effect through interaction with proteins.

The move towards investigating proteins on a massively parallel scale will require a paradigm shift away from conventional protein studies, which have generally focussed on a single protein at a time. Nevertheless, the most significant hurdle to overcome is how to cope with the diverse nature of the proteins themselves. Nucleic acids are made from just four nucleotide bases, and have a hydrophilic, negatively charged backbone. In contrast, proteins are made from 20 different amino acids and can be hydrophilic or hydrophobic, acidic or basic. This leads to a broad diversity of solubilities, tertiary structures and active sites, which gives each protein its individual functionality. In general, proteins are much less robust than nucleic acids and may require more stringent environmental control such as temperature and humidity than is necessary for DNA arraying. Some proteins have tolerance to external perturbation of their structure and are generally good candidates for protein arraying, for example antibodies, while other highly labile proteins are unlikely to ever retain their function when immobilized outside their native environment. Additionally, there are many more proteins than genes; latest estimates predict that the human genome has around 30 000 genes, but there may be as many as 1 million protein-based structures in the human proteome. The proteome is more complex than the genome for three important reasons: (1) many genes encode multiple variants of a protein (splice variants), (2) many proteins are post-translationally modified in a combination of different ways, and (3) the proteome is dynamic, with the proteomic phenotype evolving during onset or progression of disease, and even on a minute-by-minute basis as the needs of the organism change.

With so many protein-based structures in a proteome, it can be appreciated that the number of native protein–protein interactions is huge. An important consideration when arraying proteins for interaction studies is that while proteins have evolved the best-fit conformation for interacting with their respective protein partners, they may not have evolved a mechanism to exclude undesirable interactions with other proteins that exist in a spatially distinct environment. This is particularly important when bringing together proteins that are naturally compartmentalized.

8.3
Overview of Protein Microarray Spotting

Although the idea of highly sensitive 'microspot' arrays were conceived in the 1980s [5,6], it was not until the late 1990s that interest in developing protein microarrays gained momentum. Using a standard desktop inkjet printer, Silzel et al. printed 200-µm spots of monoclonal antibodies onto polystyrene film to form multi-analyte arrays [7]. Incubating these arrays with human myeloma proteins, they were able to demonstrate dose-dependent signals with minimal cross-reactivity. At the same time, Mirzabekov and colleagues developed 20-µm thick polyacrylamide gel pads for immobilizing functionally active proteins [8,9]. A pin device was used to manually spot the samples onto the gel surface, and microelectrophoresis was used to accelerate migration into the matrix. With the objective of making protein arraying easily accessible and compatible with standard instrumentation, MacBeath and Schreiber [10] used a high precision pin-type arrayer, designed for DNA microarrays, to array protein spots onto aldehyde-derivatized glass microscope slides. They printed over 10 000 spots of protein G onto one half of a slide (1600 spots per square cm), and within this printed a single spot of FKB12-rapamycin binding domain. It was demonstrated that this single spot could be detected using a fluorescently labeled interacting protein. These researchers also developed an activated albumin surface suitable for measuring transient enzyme and protein–small molecule interactions [10].

In 2001, a major step towards producing an entire 'proteome-on-a-chip' was achieved by Zhu et al. [11]. In this they describe the cloning, expression and arraying of proteins from 5800 yeast open reading frames, covering 80 % of yeast proteins. The proteins were expressed with a HisX6 tag and immobilized by interaction with a nickel-coated slide. Haab et al. addressed the issue of quantification of protein microarrays [12] by using a comparative fluorescence assay to measure the ligand recognition of 115 antigen–antibody pairs. A robotic arrayer was used to print separate arrays of antibodies and antigens on amine-derivatized glass slides (poly L-lysine). Complex target solutions were prepared by making different Cy5-labeled mixtures of all the cognate antigens or antibodies, and then mixing each with a common Cy3-labeled reference mixture. The study demonstrated that good specificity, sensitivity and accuracy could be obtained for this type of array. The world's first commercially available functional protein microarray was launched in 2002 by Procognia (then known as Sense Proteomic), who introduced their p53 array. These arrays, manufactured by Genetix Ltd, employed a proprietary technique for expressing biotin-tagged fusion proteins and immobilizing them to a streptavidin-coated surface, while maintaining full protein functionality. Researchers have continued to search for alternate array surfaces, including printing into the bottom of microtitre plate wells [13]. The last 2 years have seen a gradual increase in the number of reports where protein arrays have proven valuable, notably for the detection of cytokines [13–16].

8.4 Types of Protein Microarrays

Protein microarrays have evolved into different classes, which are summarized in Table 8.1. The principle of antibody arrays, and their converse counterparts, antigen or reverse arrays, is to use high affinity ligands to detect the presence of specific proteins and biomarkers in a complex mixture. Knezevic et al. [17] selected 368 antibodies specific to cancer-related proteins. Exposure of the printed arrays to six categories of histologic lysates identified 11 proteins that showed consistent change in expression or state of phosphorylation. Sreekumar et al. [18] used an array of 146 distinct antibodies to monitor alterations of protein levels in colon carcinoma cells.

Table 8.1 Types of protein microarrays.

Array type	*Description*
Antibody array	Polyclonal or monoclonal antibodies are arrayed and used to detect and quantify specific proteins in a biological sample. An antibody array is effectively a parallel series of miniature immunoassays
Antigen or reverse array	The converse of an antibody array, this chip has immobilized antigens that are used to detect and quantify antibodies in a biological sample
Functional array	Purified proteins are arrayed on the surface and used to detect and characterize protein–protein, protein–DNA or protein–small molecule interactions
Capture array	Non-protein molecules that interact with proteins are immobilized on the surface. These may be broad capture agents based on surface chemistries such as the Ciphergen Protein Chip, or may be highly specific such as molecular imprinted polymers or oligonucleotide aptamers
Solute array	The potential next generation of arrays is to have nanowells containing coded microspheres or barcoded nanoparticles in solution

For antigen or reverse arrays, the antigens are immobilized on the solid surface and the slides are probed with antibody. Paweletz et al. [19] used this technique to reproducibly quantify the status of signal proteins in 3-nL spots containing whole protein repertoires of lysed cell populations. Antigen arrays are also proving valuable for serodiagnosis. Robinson et al. [20] developed autoantigen microarrays in which 196 immobilized pure proteins, peptides and nucleic acids were incubated with patient serum to detect and characterize autoantigens in the serum. Bacarese-Hamilton and co-workers have used antigen arrays to detect serum antibodies against allergens and infectious antigens [21,22].

Where antigen–antibody arrays are essentially miniaturized, highly parallel ELISAs for quantitation of protein expression, functional arrays are concerned with elucidating novel proteins interactions, and are thus more akin to the yeast 2-

hybrid system [23,24] and co-immunoprecipitation studies [25,26]. The active proteins themselves are immobilized onto the support. Functional arrays have the potential to advance the mapping of interacting proteins on a system-wide or genome-wide scale. Major advantages of using this *in vitro* technique are that the researcher can control the conditions of the experiment, modify the state of the proteins under investigation, and study the interaction of proteins with non-proteinaceous molecules. These principles have been demonstrated by several investigators. Ge [27] arrayed 48 purified human proteins onto a nitrocellulose membrane and probed with different proteins, nucleic acids and small organic compounds. He was able to show that a double-stranded DNA probe bound more tightly to a phosphorylated form of protein PC4 than to the unmodified form. Zhu et al. [11] successfully used functional arrays to identify 150 putative phospholipids-binding proteins. Important to simplifying the mass production of functional microarrays, MacBeath and Schreiber [10] showed that the activities of purified proteins can be retained when spotted onto chemically-derivatized glass slides. Functional arrays are not limited to whole proteins. Indeed, Espejo et al. [28] used immobilized glutathione S-transferase-fused protein interaction domains to fish out proteins from total cell lysates.

A further type of protein microarray is the capture array, in which non-protein molecules immobilized onto the surface interact with proteins in the solute phase. The capture molecules may be broad capture agents based on chromatography-type surface chemistries such as ion exchange, hydrophobic and metal affinity functionality, or they may be highly specific such as molecular imprinted polymers or oligonucleotide aptamers. However, they are not the focus of this chapter, and will not be discussed here.

8.5 Protein Arrayers

The single most important investment for production of protein microarrays is the arraying robot. The reliability and reproducibility of the fabricated arrays are entirely dependent on the quality and morphology of the microspots produced by the arrayer.

All arrayers have three basic components: (1) a source plate location for placing the samples to be arrayed, usually for 384-well plates; (2) a device or head for printing the samples; and (3) a destination zone for placement of the supports to be arrayed, usually 1 x 3-inch (25 x 75-mm) slides. There are basically two types of arraying robot, those that use pin or contact printing (Figure 8.1) and those that use Piezo-electric non-contact printing. Non-contact printing is faster at multi-dispensing a single sample but slower at changing between samples, compared to contact printing. The choice of arrayer type will thus be influenced by whether a few samples are to be arrayed many times, or whether a greater number of different samples will be arrayed on each slide. The currently available Piezo instruments have a tendency to smear the spots or produce uneven intensity across the spot, and the Piezo tips are delicate and very expensive to replace. Contact printing has the major advantage that it can produce protein spots of the highest quality (Figure 8.2). Contact printing pins may either be solid tip pins or split pins. Solid pins need to be dipped into the sample between each spotting. The split pins, which hold a larger volume of sample by capillary action, dispense a defined volume each time the pin contacts a slide. This avoids the need for repeated re-loading of the pins and makes printing multiple spots of the same protein quicker. Solid pins however, have the advantage of being cheaper, more robust and not prone to blockage. A concern particular to arraying of proteins with split pins is that the inherent variability in viscosity between protein samples affects the volume dispensed for each protein. Solid pins are quite resistant to this issue, and so are the preferred pins for protein array studies at Genetix. Pins are available in a range of sizes from 75 to 300 μm in diameter.

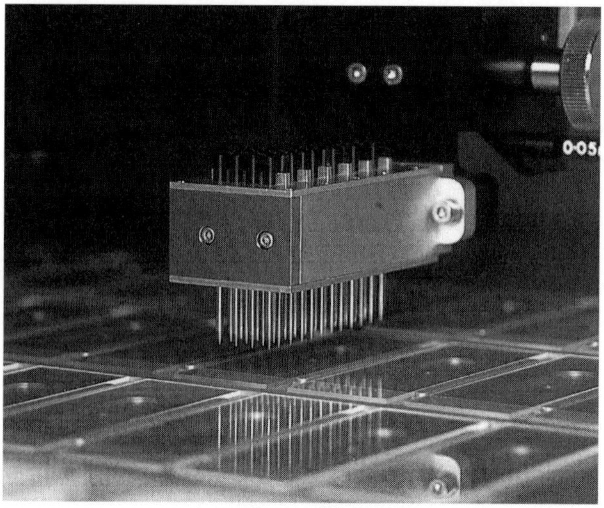

Figure 8.1 A high precision 48-pin arraying head for contact printing.

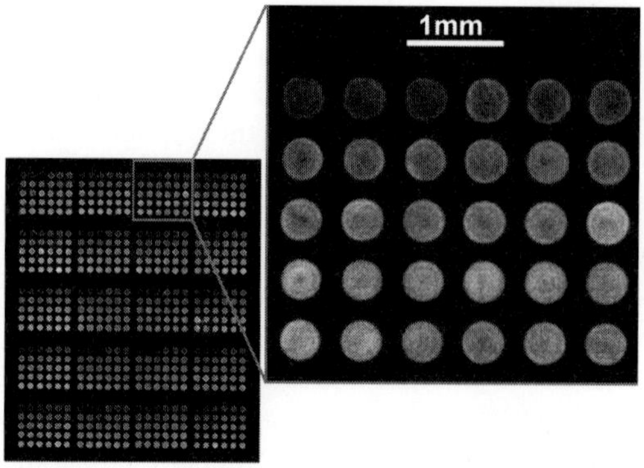

Figure 8.2 A close-up image of contact printing of proteins. A dilution series (0 to 100 μg mL^{-1} in 10 μg mL^{-1} increments) of human Cy3-labeled IgG was prepared in Printing Buffer, and then arrayed onto a Genetix Aldehyde Coated Slide using twenty 300-μm solid aQu tungsten pins. Each pin printed the dilution series in adjacent triplicate spots. After blocking and washing, the array was imaged by laser scanning (Genetix aQuire) at 10 μm resolution. The panel created by one pin has been expanded to show the spot morphology.

The pins are washed and dried in between pick-up of different samples to eliminate carryover between spots. With the Genetix QArray robots, this is normally achieved by high pressure washing with water to remove protein from the pins, and then rinsing with ethanol to promote rapid drying. Drying is accomplished with a high velocity airflow produced either via a vacuum pump or compressor. At the end of each day, the pins should be removed and washed in a solution of 1 % degreasing detergent to ensure they remain in pristine condition. If split pins are used these should be cleaned in a 60 °C sonicating water bath containing the 1 % detergent solution.

In choosing an arraying robot, the user should consider the number of arrays that are likely to be produced and the spot density of those arrays. Arraying robots vary widely in capacity for slides, source plates and unattended run times, and the price of the machines varies accordingly. Another consideration is the robustness of the robot, since an arrayer with a reported high positional accuracy, in reality will be unable to attain such accuracy if the machine is prone to vibration. Generally the most satisfactory and reproducible results will be obtained if the machine can be sited in a room with air-conditioning and which can be kept relatively clean and dust-free. Nevertheless, consideration should be given to requirements for environmental controls, such as cooling and humidification. These features are much more important for protein arraying than for nucleic acid arraying and are unlikely to be adequate on many dedicated DNA arrayers.

Another important aspect to consider when choosing a microarrayer is ease of use of the software, and tracking of data. Some arrayers require considerable software familiarization before use and are limited in their options, whilst others can print any array pattern and guide the user with step-by-step illustrations (Figure 8.3). Data tracking options can be quite minimal, or can form part of a fully-integrated Laboratory Information Management System (LIMS) allowing information from multiple experiments to be collated into one database. Data tracking can be aided by bar-coding in situations where the microarrayer is able to read and record barcodes on 384-well plates and array slides. As a minimum, an arraying robot should have a "spot tracking" function. On an experiment by experiment basis, this takes information on the content of each source plate well from a database, combines this with information on the microarray layout created, and outputs a file with spot co-ordinates plus sample identities in a format compatible with the software to be used in the analysis of the microarray image. Not using such a function can very quickly become confusing, and increases the likelihood of human error occurring at the analysis stage.

8.6 Surface Chemistry

A key step in the production of protein microarrays is the design of the support material. Ideally, the support should robustly attach all of the proteins in an appropriate orientation without altering their conformation, and must be amenable to

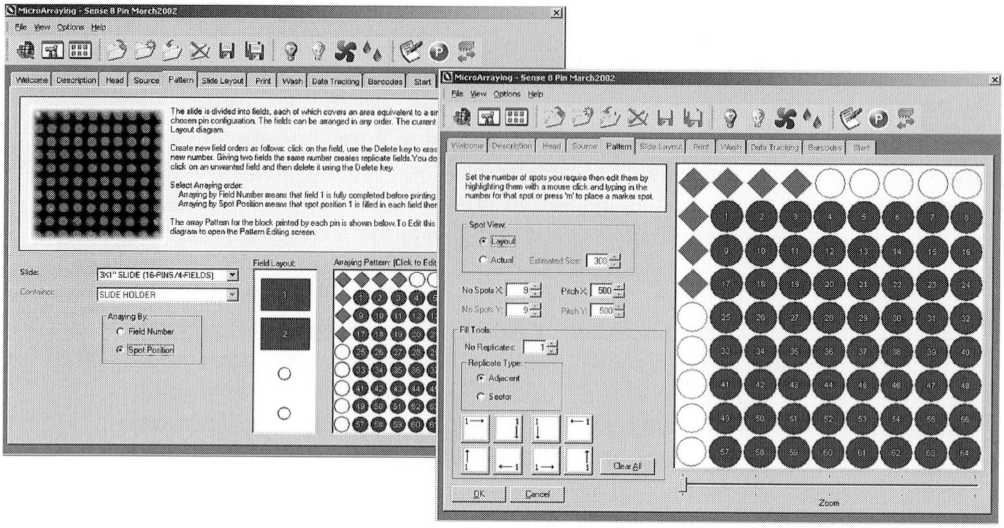

Figure 8.3 Illustrated step-by-step arraying instructions of QArray software. (see Colour Plate p. XXXI).

high-throughput manufacturing and screening procedures. The diverse nature of proteins, and even of domains within a protein, means that no solid phase support will be optimal for all proteins. Each array experiment will require testing to find the best support for the proteins under investigation, and if the experiment is driven by the need to array a defined set of proteins it may be necessary to subdivide the set onto different support materials.

Initially, protein arrays were produced on nitrocellulose and PVDF membranes or on glass slides designed for DNA arrays. Since then, a gradually increasing number of alternative surfaces have been characterized, and several groups have made comparisons [29,30]. Slides suitable for protein microarraying are summarized in Table 8.2. The merits of the different surface chemistries are discussed in the following section.

8.6.1
Derivatized Glass Slides

These surfaces are appropriate for immobilization of untagged proteins. The proteins are bound through multiple interacting groups on their surface. Attachment occurs in a variety of orientations such that different faces of the protein can interact with proteins or other molecules in solution. However, this random orientation reduces the number of protein interaction sites of any particular protein through either inappropriate orientation or inactivation due to conformational change. In our hands, the binding capacity of amine-, aldehyde- and epoxy-derivatized slides are similar (Figure 8.4).

Table 8.2 Slide surfaces suitable for protein microarraying.

Substrate	Surface	Suppliers*	Details
Planar – random	Aldehyde	Genetix Ltd	Aldehyde-coated slides
		Telechem International Inc.	SuperAldehyde
		Xenopore Corp.	Xenoslide D
	Silanated/Amine	Genetix Ltd	Silanated slides
		Telechem International Inc.	SuperAmine
		Corning	GAPS γ-aminopropylsilane for membrane proteins
		Asper Biotech	Genorama SAL-enhanced aminosilane
	Epoxysilane	Genetix Ltd	Epoxy-coated slides
		Telechem International Inc.	SuperEpoxy
		U-Vision Biotech Inc.	EasySpot Universal slides
		Xenopore Corp.	Xenoslide E
	Amine-reactive (not specified)	Quantifoil GmbH	QMT Protein slides
Planar – oriented	Streptavidin	Xenopore Corp.	Xenoslide S
		Greiner Bio-One Inc.	STRETAVIDINslides
	Nickel chelate	Xenopore Corp.	Xenoslide N
		Greiner Bio-One Inc.	NICKELslides
3-D surfaces	Nitrocellulose	Schleicher & Schuell	FAST slides
	Acrylamide gel	PerkinElmer Life Sciences	HydroGel-coated slides
		XanTec Bioanalytics	Various groups: amino, hydrazo, thiol, streptavidin
	Polymer matrix	Accelr8 Technology Corp.	OptArray Protein for amine attachment

*Not an exhaustive list.

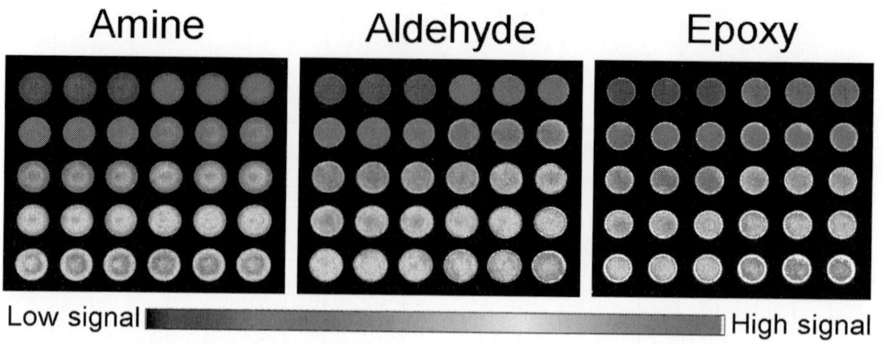

Figure 8.4 Comparison of protein immobilization capacity of three derivatized glass surfaces. Human Cy3-labeled IgG was diluted as in Figure 8.2, and then arrayed onto Genetix Amine Coated, Aldehyde Coated, and Epoxy Slides using 300-μm solid pins. After blocking and washing, the arrays were scanned at 10 μm resolution using the same laser power and PMT for all slides. The generated images have been colorized for clarity. The results show that the three surface coatings have comparable binding capacity.

8.6.1.1 Amine-coated or Poly L-Lysine-coated Slides

These are also referred to as silanated or silanized slides. Proteins are bound to the positively charged slide through electrostatic interactions to form salt bridges. They are easy to prepare manually [12] and signal uniformity and reproducibility are good, although batch-to-batch variability can be an issue.

8.6.1.2 Aldehyde-coated Slides

These are also referred to as silylated slides. Proteins are attached through the Schiff's base reaction via their side chain amino groups and the more reactive α-amino group at their NH^2 termini. The resulting covalent bonds are stronger than the salt bridge interactions of amine-coated slides. The slides are relatively easy to prepare, store and handle, and give good signal-to-noise ratios. Aldehyde slides are not suitable for printing peptides or very small proteins [10].

8.6.1.3 Epoxy-coated Slides

These slides are derivatized with epoxysilane or equivalent proprietary chemistries. Proteins are covalently attached through an epoxide ring-opening reaction primarily via their surface amino, hydroxyl and thiol groups, giving potentially higher binding affinity than amine-coated slides. The epoxy ring has susceptibility to moisture and so these slides do need to be stored in a moisture-free environment prior to use.

8.6.1.4 Bovine Serum Albumin: N-Hydroxy Succinimide (BSA-NHS) Slides

MacBeath and Schreiber developed a slide coating for peptides and very small proteins [31]. The slides are fabricated by attaching a molecular layer of BSA to the surface and then activating the BSA with *N,N'*-disuccinimidyl carbonate. The activated lysine, aspartate, and glutamate residues on the BSA react readily with surface amines on the proteins to form covalent urea or amide linkages. The slides are then quenched with glycine.

8.6.2 Oriented Surfaces for Tagged Proteins

To optimize protein interactions, expressed proteins can be tagged at their amino or carboxyl termini to enable site-specific attachment. This encourages all molecules of a protein species to be oriented in a common direction away from the support surface, and reduces structural distortion of the proteins. Peluso et al. [32] recently reported up to 10-fold better target binding to oriented capture antibodies than randomly bound antibodies.

8.6.2.1 Nickel-coated Slides

Ni-NTA-coated slides can be used to immobilize HIS-tagged proteins. Zhu et al. used this technique for production of their yeast proteome chips [11]. On the down side, the Ni-HIS interaction is neither very strong nor very stable, so dissociation with washing or with storage is an issue. Common chemicals such as EDTA and DTT also affect the reaction.

8.6.2.2 Streptavidin-coated Slides

The biotin–avidin reaction is one of strongest and most stable non-covalent interactions known, with a dissociation constant of 10^{-15} M [33]. Since avidin is toxic to cells, it is preferable to generate biotin-tagged proteins and functionalize the support with the avidin molecule. Also, due to possible non-specificity of the oligosaccharide component of avidin, non-glycosylated streptavidin is a better coating agent. The directed orientation of biotin-tagged proteins on streptavidin-coated slides does appear to be a good choice for maintaining protein functionality. Indeed, this was the principle used by Procognia in the manufacture of their p53 functional protein microarray (Figure 8.5). Similarly positive results have been reported for proteins biotinylated through the intein reaction and immobilized on avidin-functionalized slides [34].

8.6.3 Three-dimensional Surfaces

Three-dimensional slides are those that have a substantially deeper coating than the planar surfaces discussed above. This increases the surface area, and thus significantly increases the binding capacity of the slide. These are difficult to prepare manually and so commercially available slides tend to be expensive.

8.6.3.1 Polyacrylamide-coated Slides

Polyacrylamide-coated slides have a porous layer of gel matrix 20–30 μm in thickness onto which the protein is dispensed. The hydrophilic environment created by the matrix helps to reduce the rate of evaporation and thereby minimizes protein

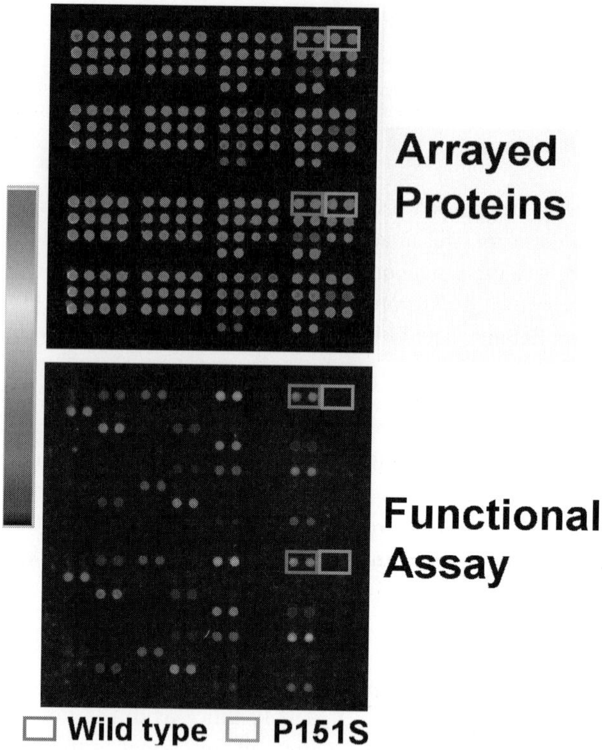

Figure 8.5 Procognia's p53 functional protein array manufactured at Genetix. Wild type and mutant p53 proteins were arrayed using 300-μm pins in adjacent duplicate spots and in two separate fields (= quadruplicate). The upper panel shows the immobilized array probed with a cy5-labeled antibody specific to a tag common to all proteins. The lower panel shows the functional assay, in this case the ability of the different mutants to bind cy3-labeled GADD45 dsDNA probe. Both assays were conducted using the same array. Images have been pseudo-colored to better visualize the intensity levels. Wild type and P151S mutant p53 are highlighted to show interaction differences. Images courtesy of Procognia Ltd., Maidenhead, UK. (see Colour Plate p. XXXI).

denaturation. Since the capacity at any spot location is higher, this also keeps the immobilized proteins better separated so they are less likely to interact together. However, the 3-D matrix makes it more difficult to change buffers. Polyacrylamide slides can be divided into the commercially manufactured slides such as those produced by PerkinElmer Life Sciences and XanTec Bioanalytics, and the manually cast slides as reported by Mirzabekov and co-workers [8,9] and Angenendt et al. [30]. The manufactured slides contain coupling reagents that fix the proteins within the matrix. PerkinElmer's HydroGel-coated slides have been shown to reach lower detection limits but with greater intra- and interfield variation [30].

8.6.3.2 **Agarose-coated Slides**

A protein microarray on agarose-film coated slides has been reported in which proteins or antibodies were covalently immobilized to activated aldehyde groups in the agarose layer [35]. These slides are simple to prepare and do permit higher sample loading but have not been sufficiently assessed for uniformity and reproducibility.

8.6.3.3 **Nitrocellulose Slides**

Nitrocellulose has been used for many years to hold proteins and seems a good choice for maintaining protein conformation. Schleicher & Schuell sell FAST slides that are coated with a 15-μm layer of nitrocellulose. Similar to the polyacrylamide slides, these permit higher sample loading. After protein printing, they do require at least 8 h of incubation to permit the proteins to immobilize in the matrix. FAST slides have been shown to retain more protein during slide processing than aldehyde-coated slides [36]. The nitrocellulose matrix tends to have a high intrinsic background level due to light scattering, but this is compensated for by the higher sample loading.

The search for optimal protein immobilization surfaces to protect proteins from denaturation will continue over the next few years. One potentially interesting development is evolving from biosensor technology, specifically the use of gold surfaces that are modified with self-assembling monolayers. These monolayers are made with derivatives of polyethylene glycol, and show good resistance to non-specific interactions [37].

8.7 The Arraying Process

The single most important aspect of protein microarraying is planning. Often, the time taken by the robot to complete the printing of the arrays is short compared to that required for preparation of a detailed work-plan. This is of particular relevance for the production of protein array batches over a period of time, for which standard operating procedures (SOPs) have to be established. SOPs not only ensure protocol consistency, their conciseness facilitates the arraying process on the day and they provide a record of each run. An example SOP is shown in Figure 8.6.

The theoretic considerations of protein microarray assays were first described by Ekins and co-workers [5,6], and have been reviewed recently by Templin et al. [38]. Ideally, the number of capture molecules for any one target protein must be sufficiently small so that the capture process does not alter the concentration of the target molecule in the sample. Thus, for a constant number of capture molecules, a smaller spot size will give a higher density and stronger signal intensity at any point within the spot. However, beyond the capacity of the arraying surface, progressively smaller-sized spots will have lower total intensity, and sensitivity will be lost. In practice, the optimal concentrations of capture molecule and target need to be determined empirically, and the need for larger rather than smaller spots is driven by the inherently low detectability of protein microarrays.

GENETIX STANDARD OPERATING PROCEDURE No. PM12-03			
Title:	**Production of IgG Microarrays on Amine Coated Slides**		
Current Date:		**SOP Date:**	6 Nov 2002
SOP Version:	03	**SOP filename:**	SOP-CM-PM12-03
Author:	Chris Mann	**Page No:**	1 of 1

AIM OF SOP

The covalent immobilization of IgG protein samples on amine-coated glass slides

EQUIPMENT

☐ 1. QArray^{Mini} or QArray
☐ 2. aQuire scanner (if using labelled protein)

REAGENTS AND CONSUMABLES

☐ 3. Genetix Amine Coated Slides (Cat. No. K2620)
☐ 4. Genetix 384-well microarray plate (Cat. No. X7020)
☐ 5. Printing Buffer (see Reagent Method 142)
☐ 6. Clean-Up Buffer (see Reagent Method 143)
☐ 7. Blocking Buffer (see Reagent Method 144). Bring to room temp.
☐ 8. Washing Buffer (see Reagent Method 145)

RELATED DOCUMENTS

☐ 9. MTP layout sheet: MTP-CM-PM12
☐ 10. QC test sheet: QC-CM-PM12

METHOD

☐ 1. Prepare 40μl dilutions of IgG samples in Printing Buffer: 1, 2, 5, 10, 20, 50, 100μg/ml
☐ 2. Pipette dilutions (6μl per well) into 384 microarray plate using plate layout in MTP-CM-PM12
☐ 3. Set up arrayer with 16 x 300μm solid pins and 75% humidity
☐ 4. Place one amine-coated slide in arrayer
☐ 5. Run test program to verify correct procedure
☐ 6. Load required number of amine-coated slides into arrayer
☐ 7. Run arrayer using program CM-26, ensuring slide number is correct
☐ 8. After completion of run, leave printed slides for 20 minutes in arrayer
☐ 9. Remove slides and immediately invert in Clean-Up Buffer for 1 minute
☐ 10. Remove slides from Clean-Up Buffer and place printed-side-up in Blocking Buffer for 30 minutes, covered and rocking gently
☐ 11. Wash slides twice for 2 minutes in Washing Buffer, then briefly in PBS
☐ 12. Air dry for 5 minutes
☐ 13. Store dry, sealed in microarray slide box at 4°C
☐ 14. If proteins are fluorescence-labelled, scan first and last slide in batch. Report results on sheet QC-CM-PM12, and attach to this sheet

Figure 8.6 A standard operating procedure for arraying proteins.

When initiating a microarray experiment, a good starting point is to print a serial dilution of the protein/antibody sample to be immobilized, in the range of 1 to 500 $\mu g\ mL^{-1}$ if the concentration is known. In this test, the protein should ideally be labeled to allow direct visualization and measurement. As a rough guide to the amount of sample printed, 150-μm spots will use 0.5–1 nL, 250-μm spots will use 2–5 nL, and 400-μm spots will use 10–20 nL. At the lower end of the dilution series, spots will contain protein in the low picogram range. However, the amount spotted will depend significantly on surface porosity and buffer composition. To enhance the accuracy of the results, all spots should be repeated at least in quadruplicate, and preferably repeated several times across the slide surface to eliminate local variations. With the QArray, a sample volume of 6–10 μL is normally required per well of a 384-well plate, although it is possible to start with as little as 3–4 μL if the volume available is limiting. This is still sufficient for replicate spotting on dozens of slides. The material with which the 384-well plate is made must also be considered. We manufacture and use polypropylene 384-well microarray plates as these adsorb less protein than plates made from either polystyrene or polycarbonate. If the proteins immobilized on the array are labeled with a fluorescent dye, such as one of the CyDyes™ (Amesham Biosciences), the slides can be visualized immediately without processing, although it must be appreciated that post-printing processing almost always alters the appearance and intensity of the spots. In most cases of simple detection, the aim here will be to establish the minimum concentration of capture agent required to saturate the slide surface.

Conditions for arraying proteins must be optimized for three important reasons: (1) to maintain a non-denaturing environment, particularly for functional microarrays; (2) to minimize non-specific interactions; and (3) to give good spot morphology, homogeneity, and spot-to-spot reproducibility. The greatest tendency for proteins to denature and lose their activity is at the liquid–air interface. This is a major problem for miniaturized assays since nanoliter spots have high surface-to-volume ratios and will therefore evaporate quickly. To reduce evaporation, protein microarrays should be printed in an environment of high humidity (70–80 %). It should be noted that the Schleicher & Schuell nitrocellulose FAST slides actually require a dry environment for optimal protein binding (C. J. Mann, unpublished observations). The sample to be arrayed can be supplemented with an agent that will reduce evaporation, such as 40 % (v/v) glycerol. This is an excellent agent to use in functional arrays as the raised viscosity reduces the kinetic energy of bombarding solvent molecules, thus protecting the proteins from denaturation. However, the higher viscosity will slow the rate of protein immobilization by reducing diffusion within the spot, and high viscosity samples are particularly difficult to array with split pins.

The printing buffer should be kept as simple as possible, unless there is a compelling reason to use different conditions. Thus, for many protein microarrays, the use of a low or physiological salt buffer at neutral pH will probably suffice. In general, a hydrophilic surface will produce larger spots while a hydrophobic surface will give smaller spots. The presence of a low concentration of mild detergent such as 0.01 % (v/v) polyoxyethylenesorbitan monolaurate (Tween 20), will help to reduce non-specific interactions, and can significantly improve spot morphology by decreasing surface tension between the buffer and the slide surface. However, higher concentrations may cause spots to spread on 2-D surfaces (Figure 8.7). Hydrophobic solvents such as ethanol will similarly lead to fatter spots, and will also encourage protein denaturation. Large molecular weight substances such as gum acacia or polyvinylpyrrolidone can be added to the printing solutions of antibody–antigen arrays to reduce electrostatic repulsive forces. Protein samples that require dilution prior to arraying should only be diluted immediately prior to use as this can cause protein precipitation which will manifest itself as poor spot homogeneity. For protein microarraying, the use of contact printing with solid pins of diameter 300 μm is recommended. Excellent spot morphology is obtained when light contact is used, in other words, setting the arrayer to print at a depth of 30–100 μm below the slide surface. Heavier contact can push the sample to the spot perimeter giving a 'do-nut' effect.

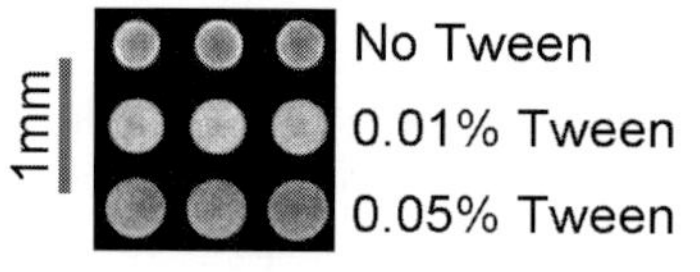

Figure 8.7 The effect of detergent on protein array spot morphology. Cy3-labeled IgG (30 μg mL^{-1}) was prepared in PBS, pH 7.4 without Tween 20, and with 0.01 % (v/v) and 0.05 % (v/v) Tween 20. The protein samples were arrayed in triplicate onto Genetix Amine Coated Slides using a single 300-μm solid pin. After blocking and washing, the array was scanned (Genetix aQuire) at 10 μm. The results demonstrate that detergents such as Tween 20 affect spot morphology. The best result was obtained with 0.01 % Tween 20.

The time required to fix the probe protein to the slide depends on the immobilization principle. In our experience, the covalent reactions such as aldehyde and streptavidin-mediated attachment to 2-D surfaces are rapid, and need only be left to incubate for 30 min or less, even in the presence of 40 % glycerol. This can be done in the humid environment of the arrayer. By contrast, the FAST slides of Schleicher & Schuell require overnight incubation at room temperature to permit fixation into the nitrocellulose matrix.

Washing of slides prior to incubation with the target protein is dependent on the immobilization surface being used. Generally, the arrays are briefly rinsed to remove unbound protein. Aldehyde slides should be treated with 0.1 mg mL^{-1} borohydride in bicarbonate buffer, pH 9.0 for 5 min to quench the remaining active groups, although a neutral buffer containing BSA is also a good for quenching purposes, in addition to reducing non-specific interactions. The arrays are then incubated in a blocking solution, such as PBS-Tween containing either non-fat milk or BSA, for 30 min at room temperature or overnight at 4 °C. If compatible with the experiment, proteolysis inhibitors or sodium azide (0.02 % w/v) can be used at this stage. Longer times should be allowed for processing of polyacrylamide 3-D slides due to the depth of the matrix.

After washing, the slides are incubated with the target sample. A typical volume for one slide is 25–500 μL, with concentration in the pg mL^{-1} to μg mL^{-1} range depending on the application. If a large volume is available, a hydrophobic pen can be used to delineate the area to be covered with target. When target volume is limiting, a hybridization chamber can be placed on each slide and the sample pipetted into it. Alternatively, the target could be pipetted under a cover slip. Target incubation time and conditions will need to be determined empirically. Longer incubations should be at high humidity to prevent the target solution from drying, which will alter the concentration of the proteins in solution. This can be done with little expense by creating a sealed chamber with an under-layer of the same buffer. After incubation, the arrays need to be washed to remove unbound target, which is usually achieved with several cycles of PBS-Tween and then with PBS. At this stage, there is always a compromise between removing the last traces of unbound target protein or non-specifically bound proteins, and the risk of dislodging specifically bound protein. His-tagged proteins on nickel-coated slides are particularly susceptible to the latter. There are a number of automated slide processing units on the market to facilitate protein array processing of multiple slides simultaneously.

8.8
Detection Issues

A major consideration to be made at the outset of protein microarraying is how to monitor the binding process. There is no protocol analogous to the fluorescent labeling of synthetic cDNAs for labeling of proteins, although fluorophores can be added chemically. If the protein sample to be used is a complex mixture such as a typical clinical sample, the proteins will be present at very different levels of abundance. In this case, the low copy number proteins may not be labeled sufficiently for detection. Similarly for functional protein microarrays, a major issue is that the affinities of native protein–protein interactions vary over several orders of magnitude. It is therefore essential to ensure that the detection system to be used is appropriate for the array experiment.

A number of different optical instruments have been used to visualize and quantify the detection signal. The most common is the confocal laser scanner, since these are the benchmark instrument for DNA array work, and are thus already established in many laboratories. Other devices that have been used are CCD cameras [39,40], which are faster but tend to have lower resolution, and even a standard flatbed scanner [17]. All microarray scanners on the market are configured to detect the most commonly used fluorescent dyes for microarray analysis, which are Cy3 and Cy5. However many are also equipped with additional lasers and filters to enable the use of a wider range of dyes. Some scanners have multiple slide loading options, which alleviate the need to manually feed each slide into the scanner.

For visualization, proteins can be labeled chemically, radioisotopically, or intrinsically by fusion with a fluorescent protein. Other label-free techniques exist which circumvent the need to modify the proteins. These methods are discussed in the following section.

8.8.1
Labeled Proteins

8.8.1.1 Chemical Labeling

Because many institutes already possess laser scanners for DNA arraying, fluorescent labeling and detection of protein arrays seems an obvious choice. The fluorescent dyes are available as NHS esters for conjugation to aliphatic amino groups, or as maleimides for reaction with free sulfhydryl groups. The latter are suited to antibody labeling as the fluorophore molecules are introduced away from the antibody binding site. Mono-reactive dyes should be used to prevent cross-linking between protein molecules. To attain sufficient sensitivity, it is necessary to aim for a dye : protein molar ratio of 5 : 1. However, incorporation of label is more difficult to control in the case of proteins than in that of nucleic acids due to the chemical heterogeneity of the former. Consequently, when labeling a complex sample it should be expected that some protein species will fail to be labeled while others will be excessively labeled. When analysing the resulting data, consideration should be given to the fact that chemical labeling could perturb the conformation of the protein-binding site either directly or allosterically. Due to the cost and labor involved in fluorescently labeling proteins, it is not feasible to use this technique to label individual probes immobilized on an array.

We have successfully used CyDye protein labeling to detect upper-picogram quantities of protein per spot using the Genetix aQuire confocal laser scanner. Because the signal obtained for each spot is dependent on the relative abundance of the probe and target proteins, and on their binding affinity, it is preferable to create a calibration curve for each probe, rather than rely on single-point data. Alternatively, Haab et al. have shown that two-color labeling systems can be applied to measure the relative intensity of a protein in an experimental sample compared against a reference sample [12]. Although the two-color approach is hampered by an inherent lack of sensitivity, the principle of differential protein microarrays is appealing to many researchers.

8.8.1.2 Radiolabeling

Some applications may be amenable to radiolabeling of the protein probe. Ge [27] used recombinant human p52 protein labeled with [γ-^{32}P]ATP in the presence of heart muscle kinase. The arrays can then be visualized with high sensitivity using a phosphorimaging device. For best results, such arrays should be created using large diameter spots (> 300 μm).

8.8.1.3 Fluorescent Fusion Proteins

To avoid the need for additional labeling of probe proteins, Kukar et al. reported the employment of recombinant proteins fused with green and red fluorescent proteins (GFP and RFP, respectively) [36]. This group demonstrated that as little as 15 pg of RFP could be detected. This technique is particularly elegant for functional arrays because GFP can be used to monitor the folding of proteins of interest.

Well-folded GFP-fused soluble proteins have been shown to be brighter than equivalent poorly folded insoluble proteins [41].

8.8.2
Sandwich Assays

Sandwich detection protocols have been used for many years in immunoassays such as ELISAs. Essentially, an immobilized antibody probe captures the target antigen, and then a second detecting antibody is used to report the presence of the antigen on the array. The value of sandwich assays in antibody microarrays is driven by the fact that there is no requirement for the target antigen to be labeled, thus giving confidence that any result obtained is true. Furthermore, a signal is only registered if two independent binding events occur, i.e. immobilized antibody–antigen, and then detecting antibody–antigen. The down side to this is that two non-competing ligands must be available for each protein, which requires initial screening tests. The early adoption of sandwich assays does mean that there is already considerable knowledge about which antibodies have good specificity and affinity for particular proteins. However, some antibodies are more suited for use on microarrays than others. For example, monoclonal anti-D4-GDI and anti-vimentin antibodies have been reported to work well in ELISA assays but not on protein microarrays [30]. Sandwich protocols may prove to be particularly useful when investigating post-translational modifications. Indeed, protein-specific antibodies could be used as capture agents on an array while a modification-specific anti-phosphotyrosine antibody could be employed to detect only those proteins that are phosphorylated.

While the secondary detecting antibody of sandwich assays can be fluorescently labeled, it is useful to take advantage of the fact that antibodies can be conjugated with enzymes, such as horseradish peroxidase, to give better signal intensity. Currently available chemiluminescence agents can be used to detect down to the low- to middle-picogram range. This is particularly important for detection of low-level proteins such as hormones and cytokines. Planar waveguide/evanescent field technology has been reported to increase the sensitivity of sandwich assays by several orders of magnitude compared to standard confocal excitation. The reader is referred to a review by Joos et al. for further information [42].

8.8.3
Direct Measurement

To avoid making any modifications to proteins, there are two systems with enough sensitivity to detect binding of proteins to immobilized ligands without the need for labeling, namely mass spectrometry and biosensor technology.

8.8.3.1 Mass Spectrometry

Mass spectrometry is proving itself to be an indispensable asset to proteomics through its ability to fingerprint proteins and to derive sequence from digested peptides. One could therefore envisage creating a functional protein array on a MALDI-compatible surface, and then using mass spectrometry to identify unknown proteins which interact with the probe proteins on the array. This system may prove to be excellent for identifying components of multi-protein complexes on a large scale. Koopmann and Blackburn [43] have recently reported a high affinity capture surface for MALDI-compatible protein microarrays. This uses tetrameric neutravidin to immobilize biotin-tagged proteins to a biotin polymer surface. The weaker aspect of mass spectrometry is that it is not useful for protein quantitation.

8.8.3.2 Biosensor technology

An alternative technique for direct detection of protein interactions is surface plasmon resonance (SPR). This principle has been used for more than a decade to measure protein interaction kinetics, and is applicable to protein microarrays. In brief, binding of a target molecule to the immobilized probe causes a change in the underlying sensing layer, which produces an electrical signal proportional to the amount of target bound. Because the measurement is based on mass, SPR cannot differentiate between different target proteins with affinity for a common probe, and similarly cannot accurately quantify the mass of a protein which is part of a multi-protein complex. The potential strength of SPR in protein microarray detection will likely be to simultaneously and quantitatively screen the binding characteristics of a single ligand in solution against thousands of immobilized proteins or antibodies. Such a system is currently being developed by HTS Biosystems (Hopkinton, MA, USA) to interrogate up to 400 antibodies in a single experiment. The reproducible manufacture of these SPR array chips on an arrayer such as the QArrayMini, would permit many different ligands to be exposed to the same panel of proteins.

8.8.4 Immunoassay Amplification

As a means to enhance the sensitivity of antibody-based microarrays, Schweitzer et al. adapted the rolling circle amplification (RCA) reporter system for the detection of antibodies [44]. By covalently attaching an oligonucleotide primer to a detecting antibody, they were able to use complementary circular DNA to initiate a polymerase reaction that replicated hundreds of copies of DNA all attached to the detection molecule. The use of fluorescently-labeled nucleotides during the reaction led to such a bright signal that femtomolar levels of immobilized antigen could be scored by counting the discrete signals arising from individual interactions. This technology has been used on a 51-feature cytokine array to measure induced secretion of cytokines from human dendritic cells [14].

8.9 Validation of Results

No protein microarray experiments should be conducted without giving due consideration to the issue of assay validity. This is important in a research context, but is critical where arrays are to be reproduced such as for diagnostic testing. Validation is an enormous subject and very much dependent on the particular application, and therefore will not be discussed in detail here. Rather, we will cite a few examples that we hope will precipitate thought as to how readers can best ensure validation of their results.

The absolute intensity of a spot is not a valid result without a calibration curve using well-defined standards. The autoantigen arrays of Robinson et al. [20] were validated using three tests: (1) antigens and antibodies were cross-compared to show that each antibody bound only its cognate ligand; (2) specific antibodies were diluted in buffer to demonstrate that detection was linear over nearly three orders of magnitude, and (3) the antigen arrays were compared directly with ELISA. The latter showed that specificity was consistent in both assays, and that the arrays were 4–8-fold more sensitive than ELISA. The researchers noted that due to differences in affinity between antibodies, there was only a relative correlation between the signal on the array and the antibody concentration in the sample.

Other groups have compared array results to ELISA. Mezzasoma et al. [21] observed greater than 90 % concordance between the techniques for some of their infectious disease assays, while one other group showed very poor comparison. Huang et al. [45] showed that comparison of multiplexed protein microarrays with ELISA had correlation coefficients that were either excellent or mediocre depending on the analyte used. These cases probably relate to the point raised earlier that not all antibodies that work well for ELISA will work for protein microarrays.

Multi-color arrays, which observe ratiometric differences between samples, overcome the need to validate absolute measurements. However, the user still needs to be confident that the results have quantitative value. Haab et al. [12] plotted the log 10 red-to-green ratio of 12 antibodies and antigens, and compared them with their true ratios. Some antibody interactions closely paralleled the ideal, while others only had semi-quantitative value. These researchers indicated the need for accurate pipetting and the use of internal control proteins in samples to compensate for variability in labeling efficiency. Knezevic et al. [17] used replicate experiments to strengthen his results, accepting only data changes that occurred in the same direction in four out of four experiments. Again, the importance of the need to have clearly defined SOPs to work from is stressed.

8.10
Stability of Protein Microarrays

The stability of arrayed proteins is an issue that everyone needs to consider, and is a major issue for manufacturers of ready-made arrays and researchers who intend to work over a period of time from a batch of arrays. By their nature, antibodies and antibody fragments are generally robust. At Genetix, we find that IgGs printed onto derivatized glass slides retain their activity when stored dry for several weeks at 4 °C. Similarly, Angenendt et al. [30] reported that storage of immobilized antibodies on non-gel coated surfaces at 4 °C showed no significant loss of signal intensity over an 8-week test period, whether the slides were stored dry or in blocking solution (3 % (w/v) fat-free milk powder in TBS-Tween 20). In contrast, these investigators found that storage of gel-coated slides dry at 4 °C gave a stronger response than storage in blocking solution [30]. Belov et al. reported that antibody arrays on nitrocellulose slides can be stored over desiccant at 4 °C for longer than 6 months without significant loss of binding activity [46]. Indeed, they observed that the stability of some antibodies within the array exceeded that of the same antibodies maintained in aqueous solution [46].

Whereas antibodies are quite stable molecules, a functional array will only be as good as its most labile protein. It is therefore best to approach the arraying of functional proteins with the assumption that the proteins will lose their activity in the instant that it takes the microspot to dry. While this does sound dramatic, it does help to keep your mind focussed on maintaining the proteins in optimal conditions at each stage of the process. This process starts with the purification of the proteins which should be conducted in appropriate buffer at 4 °C. It is not usual for all purified proteins to be available for arraying at the same time and therefore these should be snap-frozen if necessary, and stored at –80 °C until required. If a number of consecutive batches of arrays are anticipated, it would be prudent at this stage to store the proteins in appropriate-sized aliquots. With each step of the arraying procedure planned, the proteins can then be allowed to thaw on ice. Ideally, the arrayer will have a means to keep the sample microtitre plates cool, as is the case with the QArrayMini. The composition of the printing buffer for functional arrays will depend on whether a planar or 3-D surface is to be used. For planar arraying surfaces, the protein samples should be in appropriate buffer containing a high viscosity agent such as 40 % glycerol, to reduce drying. Together with a high humidity setting of 70–80 % inside the arrayer, spot evaporation can be virtually eliminated. After arraying, the slides need to be washed with glycerol-free buffer and then transferred to storage conditions. Appropriate conditions will need to be determined empirically for each array type, although a good starting point is to submerge the entire slide in the 40 % glycerol printing solution and store at 4 °C.

8.11 Future Perspectives and Challenges

The two areas where protein microarrays are anticipated to have most impact are in high-throughput analyses and medical diagnostics. The high-throughput applications will probably be in drug discovery and target validation where thousands of proteins or protein interactions could be screened simultaneously to rapidly identify the most promising drug candidates. The medical applications will predominantly be single-use protein array tests that will permit near-instant serodiagnosis of disease in hospitals and doctors' surgeries. An early example of the latter is the Triage Cardiac Panel from BioSite Diagnostics Inc., which quantitatively measures three marker proteins for acute myocardial infarction in whole blood or plasma. A hand-held meter generates the results in about 15 min. In the same way that DNA microarrays have been used to characterize expression profiles of particular disease states, it is conceivable that diagnostic antibody arrays against specific biomarkers could be used to produce distinct personal biosignatures [47]. In the more distant future, such personal medicine may be used to tailor therapy more specifically to the individual.

Assuming ways can be found to reliably immobilize proteins without altering their structure, the development of functional arrays will be simpler than protein-detecting arrays. For the latter, the greatest challenge is generation of suitable capture molecules. Indeed, availability of the natural choice, antibodies, is surprisingly limiting with the huge majority being targeted at relatively few common proteins. Finding suitable antibodies to intracellular proteins has been reported to be particularly difficult [48]. Of those antibodies that do target novel or interesting proteins, most simply do not have the appropriate characteristics. In addition to demonstrating monospecificity for the protein of interest, a suitable antibody will need to have high affinity and a low dissociation constant (K_D). One reason for this requirement is that there may well be other native binders in the target sample that compete with the ligand of interest. Classical antibodies have dissociation constants in the range 10^{-9} to 10^{-12} M, and so this is the range that should be aimed for. It is expensive and time consuming to generate antibodies by classical methods, and so there is a need to develop high-throughput methods for isolation of large numbers of protein-binding compounds that will act as capture molecules. Antibody libraries provide a good potential source of monospecific molecules, although dissociation constants tend to be low (~ 10^{-7} M) unless very large libraries are interrogated. Other alternative sources are peptide libraries and scaffolded peptides, but these also tend to produce low affinity interactants. One group of molecules that have demonstrated high affinity protein binding are nucleic acids. Since nucleic acids are easy to synthesize and store in immobilized format, they could provide a source of protein capture agents. It may also be possible to synthesize capture proteins directly. For example, Weng et al. prepared mRNA–protein fusions by in vitro translation, where each polypeptide was covalently linked to its corresponding mRNA [49]. The protein microarrays were then allowed to self-assemble via their mRNA linker to complementary surface-bound DNA probes, leading to an array of proteins with uniform orientation. The anchoring process was reported to be robust [49]. Importantly, the library of mRNA–protein molecules available for this type of array is huge because their production is not limited by transformation efficiency. Consequently, these "antibody mimics" have been shown to have dissociation constants in the range 10^{-9} to 10^{-12} M [50]. The approach taken to generate capture molecules will very much depend on the protein arraying application, and there are a substantial number of companies vying for superiority in the protein microarray market (Table 8.3).

A niche for protein microarrays is likely to be that of characterizing post-translational modifications, such as phosphorylation, glycosylation and acetylation, which cannot be addressed using DNA arrays. The greater diversity of the proteome compared to the genome is largely due to protein isoforms and transitional states. These altered protein forms are a significant means of regulating protein activity, and there remains much to be learned about how diseased states are affected by post-translational modifications. Advances in mass spectrometry as a means to detect protein modifications makes it a potential future partner for protein microarrays. There is also likely to be a diversification of array technology, exemplified by microarrays for membrane proteins [51] and protein–whole cell interactions [46].

Table 8.3 Companies developing and offering proteins in microarray format.

Adaptive Screening http://www.adaptive-screening.com/	Cambridge, UK	Deploying recombinant protein array technology, the Surrogate Proteome on a chip, to analyze protein-binding patterns of hit compounds
Affibody http://www.affibody.com/	Bromma, Sweden	Strategic alliance with Gyros of Uppsala, Sweden to jointly develop a microfluidics-based protein microarray in compact disk format
Aspira Biosystems http://www.aspirabiosystems.com/	San Francisco, CA	ProteinPrint technology uses molecular imprinting to produce capture agents. Proprietary monomers are allowed to self-assemble around proteins of interest
BD Biosciences Clontech http://www.clontech.com/	Palo Alto, CA	Currently offers the Ab Microarray 500 with over 500 monoclonal antibodies immobilized onto a glass surface. Works with two fluorescently-labeled samples
Biacore/ http://www.biacore.com/	Uppsala, Sweden	Collaborating with Millennium Pharmaceuticals to develop 40- to 50-spot protein chip for protein-protein, protein-drug interactions. Chip to debut in 2004

BioArray Solutions http://www.bioarrays.com/	Warren, NJ	Optically programmable bead array technologies for protein analysis in microfluidic chips
BioForce Nanosciences http://www.bioforcenano.com/	Aames, IA	Creating ultra-miniaturized nanoarray technologies for protein–protein interaction profiling. NanoArray chips for protein expression profiling and immunodiagnostics are slated for near-term release
BioSite Diagnostics http://www.biosite.com/	San Diego, CA	Triage® System point-of-care platform with two diagnostic devices: the Cardiac Panel of three infarct markers, and the BNP Test for congestive heart failure
Cambridge Antibody Technology http://www.cambridgeantibody.com/	Cambridge, UK	Collaborations with: Zyomyx (antibodies for arrays), Randox (chip-based system with chemiluminescent labels); Zeptosens (Planar WaveGuide technology); and Luminex (antibodies on beads)
Ciphergen http://www.ciphergen.com/	Fremont, CA	ProteinChip selectively binds proteins in a sample. Uses SELDI, a process similar to MALDI mass spectrometry, to identify protein profiles
Combimatrix http://www.combimatrix.com/	Mukilteo, WA	Developing a metal oxide semiconductor-based platform for fixing proteins to electrodes on the surface of a credit card-sized cassette with 1024 test sites
Discerna http://www.discerna.co.uk/	Cambridge, UK	Sells DiscernArray protein arrays, using the company's "protein in situ array" (PISA) technology to display functional proteins produced by cell-free synthesis
Domantis http://www.domantis.com/	Cambridge, UK	Uses grids of intersecting lines to bring different protein molecules together on the surface of an array in a multidimensional fashion
Genetix http://www.genetix.com/	New Milton, UK	Manufacturer of custom protein microarrays. Produced Procognia's p53 array, the world's first commercial functional array in 2002
HTS Biosystems http://www.htsbiosystems.com/	Hopkinton, MA	Developing protein and antibody arrays using phage display and baculovirus expression technology. Detection methods include SPR

Hypromatrix http://www.hypromatrix.com/	Worcester, MA	Sells Anitbody Array™ products and custom antibody array services
Jerini Array Technologies http://www.jerini.com/	Berlin, Germany	Sells peptide, protein domain, small molecule arrays: kinase and protease substrate chips
Large Scale Biology http://www.lsbc.com/	Vacaville, CA	Collaboration with BioSite Diagnostics to manufacture protein chips. Plans to have antibody arrays on market in 2003
LumiCyte http://www.lumicyte.com/	Fremont, CA	Uses the same SELDI technique as Ciphergen, but says it has enhanced its chips for high-throughput protein profiling service business
Luminex http://www.luminexcorp.com/	Austin, TX	xMap Technology microspheres used by Upstate as Beadlyte products. System can do up to 100 protein interaction assays per drop of fluid
MetriGenix http://www.metrigenix.com/	Gaithersburg, MD	Co-developing proteomic chips for cardiovascular disease with Temple University. Uses company's patented microchannel Flow-Thru Chips
Molecular Staging http://www.molecularstaging.com/	New Haven, CT	Developing antibody arrays using its rolling circle amplification technology (RCAT). Has defined sets of cytokines involved in autism and cerebral palsy
Nanotype http://www.nanotype.de/	Gräfelfing, Germany	Developing two-chip C-FIT (Congruent Force Intermolecular Test) sandwich system that measures "unbinding" forces between proteins
NextGen Sciences http://www.nextgensciences.com/	HuntingtonUK	Preparing breast cancer protein antibody chip for commercialization. Chips use biotin tag licensed from Avidity and substrate from Nunc
Pepscan http://www.pepscan.nl	Lelystad, The Netherlands	Sells PEP-CHIP arrays of custom peptides anchored to glass slides at N- or C- terminus
Phylos http://www.phylos.com/	Lexington, MA	In 2002, received Phase II SBIR Grant from NIAID to develop multiplex cytokine microarrays using TRINECTIN™ binding proteins
Procognia http://www.procognia.com/	Maidenhead, UK	Acquired Sense Proteomic in January 2003. With Genetix, launched world's first commercial functional array, the p53 array. Developing protein family arrays and custom SNP variant protein arrays

ProteinOne http://www.proteinone.com/	College Park, MD	Plans to generate a transcription protein array with probes from 80 transcription factors, transcriptional coactivators, and gene-specific activators
Protiveris http://www.protiveris.com/	Rockville, MD	Developing arrays of microcantilevers that serve as biosensors for multiplex protein analysis
Protometrix http://www.protometrix.com/	Branford, CT	Yeast ProtoArray with almost 5000 Saccharomyces *cerevisiae* proteins available at the end of 2003. Also developing ProtoWell nanoarrays to study protein binding in solution. Based on Mike Snyder's technology at Yale
Schleicher & Schuell http://www.schleicher-schuell.com	Keene, NH	Launched ProVision Human Cytokine Array in August 2002. Used to screen biological samples for 16 different cytokines simultaneously
SmartBead Technologies http://www.smartbead.com/	Cambridge, UK	Developing arrays of proteins or other capture molecules attached to 100 000 unique bar-coded beads, which are attached to a 3-inch diameter substrate
SomaLogic http://www.somalogic.com/	Boulder, CO	Developing arrays of photoaptamers to measure tens of thousands of proteins simultaneously
VBC-Genomics http://www.vbc-genomics.com/	Vienna, Austria	Launched Immuno Solid-Phase Allergen Chip Type 1 diagnostic chip in early 2003. Nex generation allergen chip expected and 2003.
Zeptosens http://www.zeptosens.com/	Witterswil, Switzerland	ZeptoMARK protein microarrays consist of six ready-to-use, pre-spotted microarrays, each containing up to 1000 individual recognition elements
Zyomyx http://www.zyomyx.com/	Hayward, CA	Launched their first protein array, the Human Cytokine Biochip in February 2003. The chip is ready-to-use and contains 30 biologically relevant cytokines

Adapted with kind permission from ProteoMonitor (New York), December 2, 2002.
http://www.proteomonitor.com/

8.12 Conclusion

In conclusion, there are many challenges to be overcome in the development of robust protein microarrays. Each researcher must start out with very clear objectives on their end goal. If the goal is to develop reproducible clinical arrays for quantitative detection of a small number of proteins, emphasis will need to be placed on documenting the characteristics of those proteins and on tightly defining protein production procedures. If the aim is to array hundreds or thousands of proteins in a screening application it will be more important to focus on appropriate immobilized controls or inclusion of a differentially-labeled target sample serving as an internal standard. Most importantly, the apparently daunting issues of immobilizing proteins while maintaining their native conformation, should not deter researchers from entering the laboratory and trying their own arrays, no matter how simple and basic. This is, of course, the only way that we will understand the technical aspects of handling each individual protein.

References

1. Southern, E. Detection of specific sequences among DNA fragments separated by gel electrophoresis. *J. Mol. Biol.* 1975, **98,** 503–517.
2. Fodor, S.P., Read, J.L, Pirrung, M.C. et al. Light-directed, spatially addressable parallel chemical synthesis. *Science* 1991, **251,** 767–773.
3. Shena, M., Shalon, D., Davis, R.W., and Brown, P.O. Quantitative monitoring of gene expression patterns with a complementary DNA microarray. *Science* 1995, **270,** 467–470.
4. Gygi, S.P., Rochon, Y., Franza, B.R., and Aebersold, R. Correlation between protein and mRNA abundance in yeast. *Mol. Cell. Biol.* 1999, **19,** 1720–1730.
5. Ekins, R.P. Multi-analyte immunoassay. *J. Pharm. Biomed. Anal.* 1989, **7,** 155–168.
6. Ekins, R., Chu, F., and Biggart, E. Multispot, multianalyte, immunoassay. *Ann. Biol. Clin. (Paris)* 1990, **48,** 655–666.
7. Silzel. J.W., Cercek, B., Dodson, C. et al. Mass-sensing multianalyte microarray immunoassay with imaging detection. *Clin. Chem.* 1998, **44,** 2036–2043.
8. Guschin, D., Yershov, G., Zaslavsky, A., et al. Manual manufacturing of oligonucleotide, DNA, and protein microchips. *Anal. Biochem.* 1997, **250,** 203–211.
9. Arenkov, P., Kukhtin, A., Gemmell, A., et al. Protein microchips: use for immunoassay and enzymatic reactions. *Anal. Biochem.* 2000, **278,** 123–131.
10. MacBeath, G., and Schreiber, S.L. Printing proteins as microarrays for high throughput function determination. *Science* 2000, **289,** 1760–1763.
11. Zhu, H., Bilgin, M., Bangham, R. et al. Global analysis of protein activities using proteome chips. *Science* 2001, **293,** 2101–2105.
12. Haab, B.B., Dunham, M.J., and Brown, P.O. Protein microarrays for highly parallel detection and quantitation of specific proteins and antibodies in complex solutions. *Genome Biol.* 2001, **2,** 4.1–4.13.
13. Moody, M.D., van Arsdell, S.W., Murphy, K.P. et al. Array-based ELISAs for high-throughput analysis of human cytokines. *Biotechniques* 2001, **31,** 192–194.
14. Schweitzer, B., Roberts, S., Grimwade, B. et al. Multiplexed protein profiling on microarrays by rolling circle amplification. *Nature Biotechnol.* 2002, **20,** 359–265.
15. Huang, R.P. Detection of multiple proteins in an antibody-based protein microarray system. *J. Immunol. Methods* 2001, **255,** 1–13.
16. Lin, Y., Huang, R., Santanam, N. et al. Profiling of human cytokines in healthy individuals with vitamin E supplementation by antibody array. *Cancer Lett.* 2002, **187,** 17–24.
17. Knezevic, V., Leethanakul, C., Bichsel, V.E., et al. Proteomic profiling of the cancer microenvironment by antibody arrays. *Proteomics* 2001, **1,** 1271–1278.
18. Sreekumar, A., Nyati, M.K., Varambally, S. et al. Profiling of cancer cells using protein microarrays: discovery of novel radiation-regulated proteins. *Cancer Res.* 2001, **61,** 7585–7593.
19. Paweletz, C.P., Charboneau, L., Bichsel, V.E. et al. Reverse phase protein microarrays which capture disease progression show activation of pro-survival pathways at the cancer invasion front. *Oncogene,* 2001, **20,** 1981–1989.
20. Robinson, W.H., DiGennaro, C., Hueber, W., et al. Autoantigen microarrays for multiplex characterization of autoantibody responses. *Nature Med.* 2002, **8,** 295–301.
21. Mezzasoma, L., Bacarese-Hamilton, T., Di Cristina, M., et al. Antigen microarrays for serodiagnosis of infectious diseases. *Clin. Chem.* **48,** 121–130.
22. Bacarese-Hamilton, T., Mezzasoma, L., Ingham, C., et al. Detection of allergen-specific IgE on microarrays by use of signal amplification techniques. *Clin. Chem.* **48,** 1367–1370.
23. Uetz, P., Giot, L., Cagney, G., et al. A comprehensive analysis of protein–protein interactions in Saccharomyces cerevisiae. *Nature* 2000, **403,** 623–627.
24. Ito, T., Chiba, T., Ozawa, R., et al. A comprehensive two-hybrid analysis to explore the yeast protein interactome. *Proc. Natl. Acad. Sci. USA* 2001, **98,** 4569–4574.
25. Gavin, A.C., Bosche, M., Krause, R., et al.; Functional organization of the yeast proteome by systematic analysis of protein complexes. *Nature* 2002, **415,** 141–147.

26. Ho, Y., Gruhler, A., Heilbut, A., et al. Systematic identification of protein complexes in *Saccharomyces cerevisiae* by mass spectrometry. *Nature* 2002, **415,** 180–183.

27. Ge, H. UPA, a universal protein array system for quantitative detection of protein–protein, protein–DNA, protein–RNA and protein–ligand interactions. *Nucleic Acids Res.* 2000, **28,** e3.

28. Espejo, A., Cote, J. Bednarek, A. et al. A protein-domain microarray identifies novel protein-protein interactions. *Biochem. J.* 2002, **367,** 697–702.

29. Seong, S-Y. Microimmunoassay using a protein chip: optimizing conditions for protein immobilization. *Clin. Diagn. Lab. Immunol.* 2002, **9,** 927–930.

30. Angenendt, P., Glokler, J., Murphy, D. et al. Toward optimised microarrays: a comparison of current microarray support materials. *Anal. Biochem.* 2002, **309,** 253–260.

31. MacBeath, G., and Schreiber, S.L. Printing proteins as microarrays for high throughput function determination – supplementary material. http://www.sciencemag.org/feature/data/1053284.shl [2000].

32. Peluso, P., Wilson, D.S., Do, D. et al. Optimizing antibody immobilization strategies for the construction of protein microarrays. *Anal. Biochem.* 2003, **312,** 113–124.

33. Green, N.M., and Toms, E.J. The properties of subunits of avidin coupled to sepharose. *Biochem J.* 1973, **133,** 687–700.

34. Lesaicherre, M-L., Lue, R.Y.P., Chen, G.Y.J. et al. Intein-mediated biotinylation of proteins and its application in a protein microarray. *J. Am. Chem. Soc.* 2002, **124,** 8768–8769.

35. Afanassiev, V., Hanemann, V., and Wolfl, S. Preparation of DNA and protein micro arrays on glass slides coated with an agarose film. *Nucleic Acids Res.* 2000, **28,** E66.

36. Kukar, T., Eckenrode, S., Gu, Y. et al. Protein microarrays to detect protein–protein interactions using red and green fluorescent proteins. *Anal. Biochem.* 2002, **306,** 50–54.

37. Schaeferling, M., Schiller, S., Paul, H. et al. Application of self-assembly techniques in the design of biocompatible protein microarray surfaces. *Electrophoresis* 2002, **23,** 3097–3105.

38. Templin, M.F., Stoll, D., Schrenk, M., et al. Protein microarray technology. *Trends Biotech.* 2002, **20,** 160–166.

39. Joos, T.O., Schrenk, M., Hopfl, P. et al. A microarray enzyme-linked immunosorbent assay for autoimmune diagnostics. *Electrophoresis* 2000, **21,** 2641–2650.

40. Pawlak, M., Schick, E., Bopp, M.A. et al. Zeptosens' protein microarrays: a novel high performance microarray platform for low abundance protein analysis. *Proteomics* 2002, **2,** 383–393.

41. Waldo, G.S. Standish, B.M., Berendzen, J., and Terwilliger, T.C. Rapid protein-folding assay using green fluorescent protein. *Nature Biotechnol.* 1999, **17,** 691–695.

42. Joos, T.O., Stoll, D., Templin, M.F., et al. Miniature multiplexed immunoassays. *Curr. Op. Chem. Biol.* 2001, **6,** 76–80.

43. Koopmann, J.O., and Blackburn, J. High affinity capture surface for matrix-assisted laser desorption/ionisation compatible protein microarrays. *Rapid Commun. Mass Spectrom.* 2003, **17,** 455–462.

44. Schweitzer, B., Wiltshire, S., Lambert, J. et al. Immunoassays with rolling circle DNA amplification: a versatile platform for ultrasensitive antigen detection. *Proc. Natl. Acad. Sci. USA* 2000, **97,** 10113–10119.

45. Huang, J.X., Mehrens, D., Weise, R. et al. High-throughput genomic and proteomic analysis using microarray technology. *Clin. Chem.* 2001, **47,** 1912–1916.

46. Belov, L., de la Vega, O., dos Remedios, C.G., et al. Immunophenotyping of leukemias using a cluster of differentiation antibody microarray. *Cancer Res.* 2001, **61,** 4483–4489.

47 Kodadek, T. Development of protein-detecting microarrays and related devices. *Trends Biochem. Sci.* 2002, **6,** 295–300.

48. MacBeath, G. Protein microarrays and proteomics. *Nature Genet.* 2002, **32** (Suppl.), 526–532.

49. Weng, S., Gu, K., Hammond, P.W., et al. Generating addressable protein microarrays with PROfusion covalent mRNA–protein fusion technology. *Proteomics,* 2002, **2,** 48–57.

50. Xu, L., Aha, P., Gu, K. et al. Directed evolution of high affinity antibody mimics using mRNA display. *Chem. Biol.* 2002, **9,** 933–942.

51. Fang, Y., Frutos, A.G., and Lahiri, J. Membrane protein microarrays. *J. Am. Chem. Soc.* 2002, **124,** 2394–2395.

9
Nanomechanical Cantilever Sensors for Microarrays

Marko K. Baller and Jürgen Fritz

9.1
Introduction

Microarrays allow the detection of tens to thousands of different nucleic acid sequences or proteins in parallel. By measuring fluorescence from labeled target molecules at different positions on the microarray one can identify molecules and determine their relative abundance in a sample. Fluorescence is still the most powerful detection technique to read out microarrays. This is mainly due to the high sensitivity and large dynamic range of the optical read-out and to its spatial resolution which enables the use of high density microarrays with spot sizes smaller than 10 μm.

Nonetheless, read-out techniques based on labeling systems also have their drawbacks [1]. The time required to derivatize samples with label or reporter molecule can sometimes hinder experiments where rapid detection is essential, e.g. for characterizing infectious agents, monitoring specific biomarkers for intensive care, or measuring mRNA levels during expression profiling. The overall costs for labels, labeling processes, and detection systems can be high. Inconsistencies in the yield of labeling reactions, the separation of labeled molecules from free labels, and the stability of a label are a potential source of errors. Finally, whereas fluorescent labeling of nucleic acids is now more or less a standard procedure, protein labeling techniques are not yet as well established and their complex structure renders proteins much more sensitive to chemical modifications [2]. As one possible solution, researchers are looking for alternative detection methods that do not require any label or reporter molecule to signal the presence of a molecule on a biosensor surface. Label-free methods offer the specific advantage that unmodified samples can be used. This further implies the possibility of a real-time read-out. Examples of label-free detection techniques include surface plasmon resonance [3], quartz crystal microbalances [4], and nanomechanical cantilever sensors [5]. Miniaturizing and parallelizing label-free detection schemes promises their future use as portable or point-of-care sensors. While label-free methods are still less sensitive than the established fluorescence-based methods, label-free methods could provide a faster and simpler read-out for microarrays.

Protein Microarray Technology. Edited by Dev Kambhampati

ISBN: 3-527-30597-1

Here we review a label-free detection principle, where molecular interactions are translated into a measurable mechanical motion. The nanomechanical cantilever sensor is a label-free biosensor that can monitor the binding of nucleic acids or proteins to its surface in real-time. Upon binding the cantilever bends several nanometers and thus generates an easily detectable mechanical read-out. The sensors are microfabricated, can be parallelized in arrays, and are easy to use. Detection of oligonucleotide hybridization as well as receptor–ligand, and antigen–antibody interactions by this method has already been demonstrated [5,6]. Companies are working on commercializing the technique while researchers continuously seek to improve its detection limits, miniaturize the set-up and try to understand its transduction mechanism in more detail. This contribution will focus on the use of microcantilevers for biosensing applications via surface stress detection. Recent reviews report more generally on the use of microcantilevers as chemical, thermal, or resonant mass sensors [7–9].

In Section 9.2 we take a closer look at the technology and instrumentation developed for nanomechanical cantilever biosensors including a discussion on what causes a cantilever to bend. Section 9.3 focuses on processes for sensor functionalization and Section 9.4 reviews experiments reported to date. We conclude with a brief outlook on future developments.

9.2 Basic Technology and Instrumentation

The nanomechanical sensor has its roots in atomic force microscopy (AFM) technology where a bending of microfabricated cantilevers is monitored with subnanometer resolution [10]. Cantilevers are tiny microfabricated beams with one end fixed to a support and the other end free to move up and down (see Figure 9.1). Whereas AFM uses cantilevers with an integrated sharp tip at their terminus to scan and image a surface, a nanomechanical cantilever sensor consists of freestanding cantilevers without an integrated tip and mounted in a closed detection cell (see Figure 9.2).

Nanomechanical cantilever sensors can be operated in three basic modes to detect molecular interactions. First, a mass change on a cantilever, e.g. by adsorption of molecules, can be detected by a shift in its resonance frequency [11] which has been demonstrated for gaseous analytes [12] as well as liquids [13]. In this mode detection of mass changes of about 10^{-17} g have been reported [14]. Unfortunately, the cantilever oscillation is damped in a liquid environment resulting in a significantly reduced mass sensitivity. In a second mode the bimetallic effect is used for thermal detection. Cantilevers consisting of two different materials with different thermal expansion coefficients (such as a silicon cantilever with a thin gold film on top) bend when heated up. Changes in temperature or heat on the cantilever or in its environment can be detected down to the femtojoule and μK level [15]. Again, the detection of molecular reactions on a cantilever in a liquid

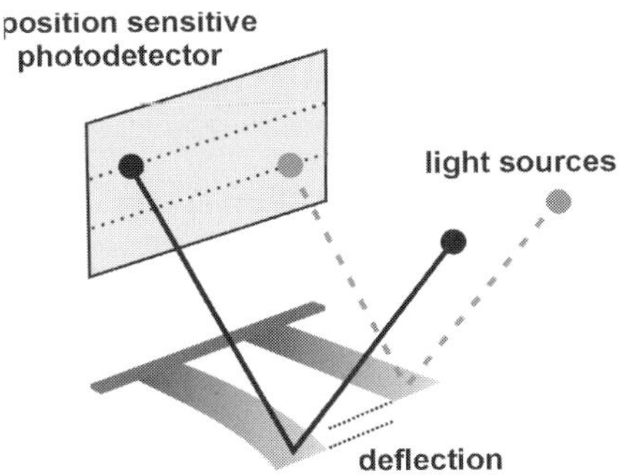

Figure 9.1 Atomic force microscopy detection principle for nanomechanical cantilever sensors. Laser light is focused at the terminus of microfabricated cantilevers and reflected off onto a position-sensitive photodetector. When cantilevers bend in response to analytes the reflected laser spot changes its position on the photodetector. The position can be converted into a deflection of the cantilevers. For cantilever arrays the light sources are multiplexed.

Figure 9.2 Electron microscopy image of a cantilever array of eight cantilevers. The silicon cantilevers are 750 μm long, 100 μm wide and 0.9 μm thick.

environment is hindered because of the almost complete and rapid dissipation of the reaction beat into the liquid.

The third mode of operation, the detection of surface stress, has been most successfully applied to biosensing and will be discussed in detail. When molecules adsorb or bind preferentially to one side of a cantilever, a ssurface stress can be created and that side will expand or contract more compared to the other one [16,17]. A difference in surface stress between the top and bottom side of a cantilever leads then to an overall bending of the cantilever structure. One can think of it as a "biochemical" bimetallic effect. Many processes and molecular interactions contribute to cantilever bending and we discuss these in more detail in Section 9.4.

Single cantilevers can be used for all three modes of operation but using at least two cantilevers in parallel can significantly improve the reliability of a measurement [9]. Cantilever arrays enable differential measurements between sensing and reference cantilevers as well as averaging over several cantilevers. This allows suppression of unspecific signals caused by changes in buffer or temperature and improves the signal-to-noise ratio. The most advanced set-up so far has eight cantilevers in parallel. Cantilevers can easily be functionalized with metals, organic monolayers, biomolecules, or polymers depending on the specific application or which analytes are to be detected.

9.2.1 Cantilever Design and Properties

At the heart of a nanomechanical sensor are cantilevers that bend in response to molecular adsorption or binding. These are fabricated by standard silicon micromachining technology and hundreds of cantilevers can be produced in parallel reducing the cost of a single cantilever to about 10 dollars [18]. Cantilevers are made of silicon or silicon nitride and are typically several hundreds of micrometers long, about 100 μm wide and 0.5 to 1 μm thick. Single cantilevers for AFM which have an integrated tip at their end to scan a surface. Cantilevers designed for stress sensors do not need an integrated tip and can be fabricated in dense arrays (see e.g. eight cantilevers close together, Figure 9.2).

For operation in the stress mode, it is necessary that the cantilever top and bottom surfaces are chemically different. Only then different molecular interactions on the two surfaces will lead to an overall bending of the cantilever (for example if the top surface expands more then the bottom surface). Most cantilevers have already two different surfaces: one is a native silicon oxide or silicon nitride surface and the other one is covered with a this layer of gold to facilitate the bending read out. The gold surface also opens up a convenient pathway for chemical derivatizations as we describe in Section 9.3.

The mechanical properties of cantilevers are described by their spring constant and resonance frequency, which are determined by the cantilever material and geometry. The spring constant can be calculated by [19]:

$$k_{spring} = \frac{E \cdot w \cdot t^3}{4l^3} \qquad (1)$$

where E is elasticity or Young's modulus, w is width, t is thickness, and l is length. This formula holds for a rectangular cantilever with homogeneous cross-section. A typical value for E is 1.7×10^{11} N m^{-2} for silicon and 2.1×10^{11} N m^{-2} or more for silicon nitride, since these values can vary with preparation, composition, and crystalline axis. The force acting on a cantilever can be calculated from the measured deflection using Hooke's law: force = k_{spring} x deflection.

The resonance frequency of a cantilever provides an easy-to-measure mechanical parameter for comparing different cantilevers and calibrating sensors (see Section 9.3). The dependence of resonance frequency on spring constant is [20]:

$$f_{res} = \frac{1}{2\pi}\sqrt{\frac{k_{spring}}{n \cdot m_{cant}}} \qquad (2)$$

where m_{cant} is mass of the cantilever (calculated by length x width x thickness x density) and n is a correction factor taking into account that mass towards the free end of the cantilever contributes the most to resonance frequency. For rectangular beams, this factor is n = 0.24. The product of n and m_{cant} is also called effective mass. The density of silicon is about 2300 kg m^{-3}. The density of silicon nitride is above 3000 kg m^{-3} and depends on its stochiometric composition. Regarding the materials, silicon cantilevers are "softer" than silicon nitride cantilevers with identical dimensions. But since it is easier to fabricate thin silicon nitride films than thin silicon films, commercial silicon nitride AFM cantilevers normally have (due to their geometry) a smaller spring constant than silicon AFM cantilevers. A smaller spring constant results in a "softer" cantilever which is easier to bend and more sensitive. Making thinner and longer cantilevers improves their force sensitivity, but also renders them more fragile, more difficult to handle, and increases thermomechanical noise. A typical silicon cantilever used for stress measurements has dimensions of 500 x 100 x 1.2 μm resulting in a spring constant of 0.06 N m^{-1} and a resonance frequency of about 7 kHz.

To coat cantilevers with gold, an adhesion layer of about 1 nm of chromium or titanium has to be deposited first, otherwise the weakly bound gold layer will eventually come off. Gold deposition can be achieved with an electron beam evaporator in a high vacuum chamber at about 10^{-6} mbar with typical rates of deposition of 0.05 nm s^{-1}. The gold should be deposited on top of the adhesion layer, without breaking the vacuum, as slowly as possible, and at room temperature to avoid any intrinsic bending of the cantilevers. Especially soft and thin cantilevers tend to bend significantly during gold evaporation and as a consequence can sometimes no longer be used for sensing applications.

Cantilevers are connected to a die, a piece of original silicon wafer, typically several millimeters wide and long and about 0.5 mm thick. It serves as a handle with which to move and position cantilevers with tweezers. After functionalization the cantilever "chip" is mechanically mounted in a fluid cell. A fluid cell can for example consist of two flat pieces of plastic with a small rubber O-ring in between. One piece has a glass window so that an optical read-out scheme for cantilever bending can be used (see Section 9.2). The cell has an inlet and outlet to exchange fluids and its volume typically ranges from several hundreds to only a few microliters. Tubing and pumps are connected to deliver the samples.

9.2.2
Measuring Cantilever Bending

The standard method to measure cantilever bending in AFM is the beam bounce or beam deflection method. A laser beam is focused on the end of the cantilever and bounces off onto a position-sensitive photodetector which detects the position of the reflected laser spot (see Figure 9.1). When the cantilever bends the reflected spot moves accordingly and the degree of bending of the cantilever can readily be calculated from the motion of the spot on the photodetector. The photodetector gives either the absolute position (in the case of a linear position-sensitive photodetector) or the relative changes in position (e.g. for a split photodiode) of the spot. For relative changes the detector has to be calibrated first, by bending the cantilever by a known distance. It is important to take into account that a force applied at the end of a cantilever (as in AFM) will give a different cantilever bending profile compared to that resulting from a moment acting along the length of the cantilever [21]. The same deflection of the end of a cantilever therefore results in slightly different spot positions on the photodetector. This should be compensated for when calibrating surface stress measurements with AFM force measurements.

There are two other detection principles worth noting. Cantilever bending can be monitored interferometrically, which is about 10 times more sensitive than the beam deflection method [22]. For example, adjacent cantilevers can be designed with interdigitated fingers at their ends to form a diffraction grating. Focusing a laser beam on the grating will generate diffracted spots whose intensities indicate the relative distance between the two finger sets and hence the deflection of the cantilevers [23]. This method has the advantage that the photodetector needs only to be sensitive to intensity not position of a diffracted spot. Furthermore, this set-up directly delivers an intrinsic differential read-out signal and the whole sensor can be packaged in a small volume. Its drawback lies in the fact that the cantilever geometry, fabrication and functionalization are more complex.

Cantilever deflection can also be monitored with a piezoresistive read-out which does not require a laser source or a photodetector. The bending of a cantilever with an incorporated piezoresistor results in a change in its resistivity which can be detected electronically [24]. This set-up is still unreliable for long-term operation in a liquid environment and is so far less sensitive than a read-out obtained by beam deflection. But improvements to this method could lead to and elegant and direct electronic read-out for nanomechanical sensors.

The concepts for an optical read-out of linear cantilever arrays are straightforward. For example, one laser and one photodetector could be used for each cantilever. However, it is less complex to use one light source per cantilever and only a single large position-sensitive photodetector for the reflected light from all cantilevers (see Figure 9.1). The light sources are switched on and off within milliseconds and the signals from the detector are assigned to the cantilever whose light source was activated at that time. A quasi-parallel time-multiplexed read-out of several cantilevers can be accomplished. The light sources used are either pig-tailed laserdiodes with coupled fiber optics [25] or Vertical Cavity Surface Emitting Lasers (VCSELs) [26]. This type of set-up has been used for many experiments with cantilever arrays.

9.2.3 Differential Detection and Noise

In this section we focus on instrumental error sources and external noise sources. We are not discussing signal fluctuations caused by inhomogeneities and differences in sensor surface functionalization (see Section 9.3). In general using a cantilever array with two or more cantilevers has the advantage that some cantilevers can serve as reference. Detecting the difference in bending between a sensing and a reference cantilever helps to reduce unwanted background signals and increases the signal-to-noise ratio. It is becoming clear that for reliable biosensing experiments a differential set-up is absolutely necessary [9,23,27,28]. Since cantilevers, particularly those which are metal coated, are highly temperature sensitive, even small temperature fluctuations can lead to thermal bimetallic bending which could be mistaken for bending resulting from molecular adsorption [23]. Another potential source of background signals is the difference in refractive index between different buffers. Buffer changes can alter the light path of the read-out laser which would appear as bending [27,28]. Finally, it has been shown that cantilevers bend in response to preferential adsorption of ions onto the silicon oxide or silicon nitride surface [28]. A differential set-up will minimize these signals caused by external changes in the environment of the cantilever [9,27]. It reduces noise in the seconds timescale as well as long-term linear drifts. However, for a differential detection scheme to be precise, all cantilevers of an array should be mechanically identical. Cantilevers within one array of eight have been reported to deviate by only 0.5 to 2 % in their resonance frequencies which is sufficiently low for differential detection set-ups [5,23,29].

A source of systematic error between experiments carried out with different cantilevers is the determination of their spring constant. Cantilever geometry, especially thickness, can vary with fluctuations in local etch rates or other inhomogeneities in their fabrication process. Thereby, silicon nitride cantilevers or cantilevers made from silicon on insulator wafers have the most homogeneous thickness which varies only up to 2 % across a wafer. Nevertheless, proper calibration of cantilevers prior to the experiments is necessary and different calibration procedures have been reported with a typical accuracy of 10 % [19,30]. For example, the spring constant can be determined by measuring the thermomechanical noise spectrum of a cantilever in air. By analogy to Brownian motion, microfabricated devices are subjected to random thermal excitation resulting in thermal vibrations. The cantilever spring constant can be calculated by equating the thermal energy of a cantilever at room temperature ($^1/_2$ k_BT where k_B is Boltzmann constant, T is absolute temperature) with its mechanical energy determined from a thermal vibration spectrum ($^1/_2$ $k_{spring} < z^2 >$ where k_{spring} is cantilever spring constant, $<z^2>$ is mean square cantilever deflection) leads to the calculation of its spring constant [30]. Cantilever arrays have the advantage that all cantilevers are fabricated in close proximity and are more likely to be mechanically identical than single cantilevers from distant spots on a wafer or from different batches. A smaller source of systematic error in cantilever bending is the alignment of the laser spot on the cantilever using the beam deflection read-out. The closer the laser spot is focused to the base of the cantilever the smaller is the detected displacement for the same amount of overall bending.

Once proper measures are taken to avoid systematic errors, typical background noise for beam deflection (in aqueous solutions) is about 0.2 to 1 nm and signals detected to date are in the order of 10 to 100 nm giving a good signal-to-noise ratio. Major sources of noise are thermal motion of the cantilever, laser noise and detector noise [20]. Additional sources of noise while operating in liquid include, e.g. scattering of the laser beam by particles or aggregates in sample solutions.

Interestingly, the noise level of cantilever sensors increases for longer timescales in a 1/f manner [23]. The signal-to-noise ratio is also dependent on sampling rate and averaging of data points. The slowest timescale or maximum averaging to allow real-time measurements can be estimated from the binding parameters of the molecular system under investigation. The timescale of a molecular interaction is roughly half the time needed to reach equilibrium. The sampling rate should then clearly be faster than this timescale. Typical sampling rates are of the order of 1 to 10 Hz. In general, averaging and filtering of the raw data reduces high frequency noise and differential detection cancels out most of the correlated noise and drifts at longer timescales.

9.2.4 Mechanism of Cantilever Bending

To compare results from different stress experiments with different cantilever geometries and materials it is most convenient to present results in device geometry independent units of surface stress. The formula to calculate surface stress from the deflection z of a beam was introduced in 1909 by Stoney [31]:

$$\Delta\sigma = \frac{E \cdot t^2}{3(1-\nu) \cdot l^2} z \tag{3}$$

This equation takes into account the material propoerties of a cantilever such as Poisson ratio ν, Young's modulus *E*, and thickness *t*, and length. Typical values for silicon are $\nu \sim 0.3$, $E \sim 1.7 \times 10^{11}$N m^{-2}. Stoney's formula also shows that sensitivity to surface stress can be improved by making cantilevers longer and thinner. Changing the width of cantilever does not affect its sensitivity to surface stress (but can make them more sensitive to force measurements, see Equation 1).

Where does the surface stress originate from? The surface free energy or surface tension is the reversible work per unit area required to *create* a surface. It is the fundamental parameter that governs phenomena such as adsorption of substances onto solids, and adhesion or wetting of surfaces. Surface tension is normally measured in J m^{-2}. Surface stress is the reversible work *per* unit area required to *stretch* a solid surface elastically and is expressed in Nm^{-1} [32–34]. In contrast to liquid surfaces, the surface area of solid bodies cannot easily be altered to cope with a change in surface free energy. Solid surfaces can react to changes in surface free energy by elastic deformations described by a surface stress. This stress can either be a "compressive stress" resulting in expansion, or a "tensile stress" resulting in contraction of a surface. Surface stresses reported in biosensing experiments with cantilevers are almost always compressive.

It is known that adsorption of small molecules onto clean crystalline surfaces changes atomic configurations, bond angles, and electron distribution on the surface leading to surface stress [32]. It has been observed that for cantilevers covered with a polymeric layer, the polymer swells and thus creates a compressive stress on the surface when molecules diffuse in or bind to the polymer [26]. However, at a solid–liquid interface covered with biomolecules the situation is much more complicated and the mechanisms causing cantilever bending are still under discussion. For the experiments discussed here, the "surface" of a cantilever includes not only the surface atoms of the bulk cantilever material, but also a more or less ordered, porous, nanometer-thick layer of adsorbed complex biomolecules. In addition, an ionic double layer reaching several nanometers into the bulk liquid has to be taken into account. Therefore, it may be more appropriate to refer to an "interfacial stress" rather than a surface stress. So far, the following mechanisms have been proposed to explain the bending of a cantilever.

- The self-assembly of alkanethiols on gold produces a surface stress because of repulsive dipolar forces in the adsorbate–adsorbant system [17].
- Adsorption of molecules can completely change a surface from hydrophobic to hydrophilic, so that the change in surface free energy causes bending, e.g. during surface silanization or during protein adsorption [16,26].
- pH changes can alter surface charge by protonation or deprotonation of reactive groups causing surface stress due to charge repulsion [16,35–37].
- Hybridization of oligonucleotides on a surface can increase surface charge [5], or change molecular conformation [38] leading to bending.
- Changes in ion concentration will alter the double layer close to a surface; ions can bind to a surface or diffuse in molecular layers on the surface and cause bending [28,35].

A recent report describes in more detail the bending mechanism of cantilevers during DNA hybridization [39].

Of what order of magnitude are the stresses reported so far in biosensing applications? Surface stresses for adsorption of oxygen onto a silicon (111) surface are in the order of 7 N m^{-1} [40]. In the case of self-assembly of alkanethiols on a gold surface, stress changes of 0.2 N m^{-1} were reported [17]. The hybridization of oligonucleotides on a surface results in typical surface stresses of about 0.005 N m^{-1} [5]. The smallest detectable changes in surface stress reported so far are in the mN m^{-1} range. All reported surface stresses are relative stresses related to a change in surface stress and do not reflect an absolute total surface stress. Research is now directed towards a better and quantitative understanding of the underlying principles of surface stress in biorecognition experiments aiming especially to increase the sensitivity and the signal-to-noise ratio of the method. One promising avenue is to increase the surface area of a cantilever by creating a layer of porous gold as this was recently reported to significantly increase the response of a cantilever to a chemical stimulus [41].

9.3 Cantilever Functionalization

Functionalizing cantilever stress sensors is challenging because different molecules have to be immobilized on the top and bottom surface of a cantilever which is only 1 μm thick. For sensing applications one surface has to detect or bind specific molecules and the other should be less sensitive or be blocked against the adsorption of these molecules. Identical top and bottom surfaces would respond with the same surface stresses which would cancel each other out and the cantilever would therefore fail to bend. Only a difference in surface stress will lead to bending of a cantilever.

There are several strategies to functionalize cantilevers. Molecules can be physically "painted" onto cantilevers. With a microcapillary attached to a micromanipulator and filled with probe solution the cantilever can be gently touched with the liquid meniscus using a backward and forward motion of the micromanipulator to coat the entire surface. Cantilevers in an array have also been covered with polymers side by spray coating them individually through a shadow mask [26].

As mentioned above, cantilevers are often coated with a thin gold layer on one side to increase their reflectivity for optical beam deflection read-out. This coating can also be used to selectively functionalize or activate the top or bottom surface of the cantilever. A fresh clean gold surface can be functionalized via thiol chemistry [42]. Molecules containing a thiol group (e.g. 5'-thiol-modified oligonucleotides) can be directly self-assembled on a gold surface [5], or alternatively the gold can first be derivatized with a reactive monolayer of self-assembled thiols to which other molecules can be crosslinked for further functionalization [6]. The silicon surface can then be blocked against non-specific molecular adsorption simply by placing the whole cantilever into a solution of bovine serum albumin (BSA). On the other hand, the gold surface can first be blocked against molecular adsorption with a polyethylen glycol (PEG) thiol monolayer and then specific molecules can be immobilized on the silicon surface [5]. This can be done by direct adsorption or via silane chemistry and further cross-linking. Figure 9.3 summarizes the different approaches for cantilever functionalization reported so far.

Because of their geometry, cantilevers can be easily inserted into small separate reservoirs and thus be functionalized, e.g. in tips of micropipettes filled with the appropriate liquid or molecules (see Figure 9.4) [23, 43]. Cantilevers of an array can be introduced in parallel into a microcapillary array [27] or dipped into a microfluidic channel array [7] where they can be incubated for as long as necessary. Thereby, it is important to functionalize the whole length of cantilever since its bending results from the integral surface stress along its entire length. The functionalization of the hinge region of a cantilever close to the die is most critical since this region acts as a lever arm where a small bend is amplified into a large deflection of the tip of cantilever. An example of the functionalization of a cantilever array is shown in Figure 9.4.

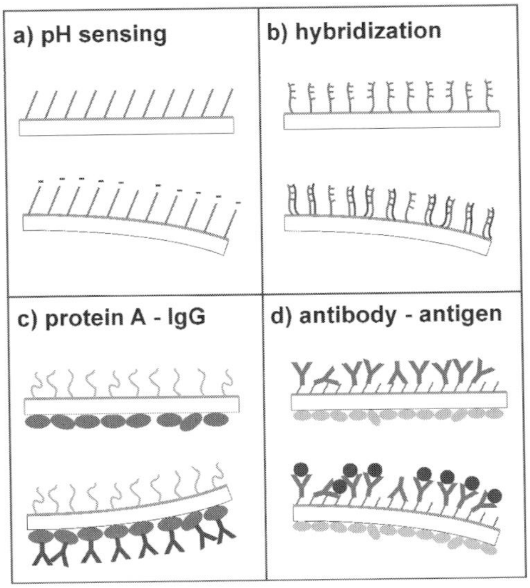

Figure 9.3 Surface functionalization strategies for cantilevers. The top surface of the cantilever consists of a thin layer of gold, the bottom side either of silicon oxide or silicon nitride. In all cases the side on which the analyte molecules bind is subjected to compressive surface stress. (a) Self-assembly of carboxythiols on the gold side of the cantilevers. By changing pH, carboxy groups become deprotonated exhibiting a negative charge [35]. (b) Self-assembly of 5'-thiol-modified oligonucleotides on the gold side of the cantilevers. Hybridization leads to an increase in surface charge and molecular density [5]. (c) The gold side of a cantilever is functionalized with a PEG thiol to prevent protein adsorption. The silicon can then be functionalized with protein A. Rabbit IgGs bind to protein A [5]. (d) Scheme for functionalizing cantilevers with antibodies. The gold can be functionalized with dithiobis(sulfosuccinimidyl propionate) [6] or dithiobis(succinimidyl undecanoate) [29] which provides an amine-reactive surface to bind proteins (for example, antibodies). The bottom side of the cantilever is blocked with BSA.

9.4 Experiments

9.4.1 Early Surface Stress Experiments

In 1995, early experiments investigating surface stress phenomena with microcantilevers addressed electrochemical problems and the interplay between surface stress and resonance frequency [44–46]. This was soon followed by a report outlining surface stress measurements for pH sensing, detecting changes in salt concentration, and for detecting non-specific binding of proteins on a cantilever surface [16]. Systematic surface stress measurements of the self-assembly of alkanethiols from the gas phase on gold were reported in 1997 [17] showing for the first time that the self-assembly of organic molecules on a solid surface can bend a microfabricated device. The bending of cantilevers was shown to increase with increasing chain length of the alkanethiols. The first use of cantilever arrays to detect gaseous

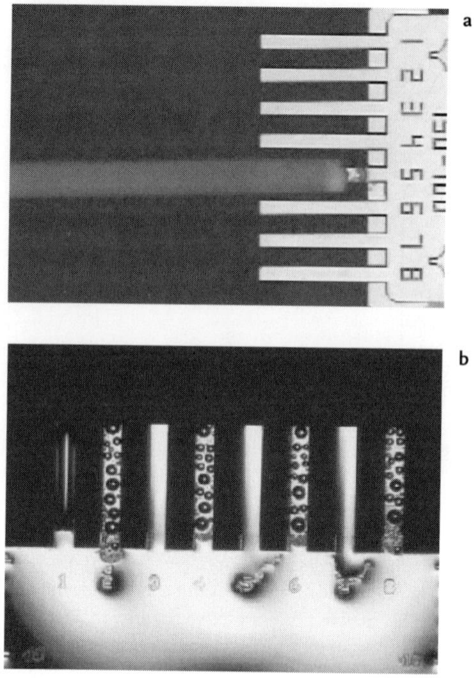

Figure 9.4 Cantilever functionalization. (a) Cantilevers can be functionalized by inserting them into microcapillaries. Here, the capillaries have an outer diameter of 240 μm and an inner diameter of 180 μm. The cantilevers are 750 μm long, 100 μm wide and 0.9 μm thick with a pitch of 250 μm. (b) Control of functionalization. The gold side of the cantilevers was functionalized with alkanethiols and carboxythiols. By cooling them below the dewpoint of air, water condenses on the surface and drops appear on the hydrophobic alkanethiol-covered cantilevers, while the hydrophilic carboxythiol-covered cantilevers show a closed water film [43].

analytes was then reported [47], leading later to a cantilever array-based artificial nose [26]. Cantilevers were covered with different polymers to detect various gaseous analytes such as solvents and baking aromas or clinically relevant traces of acetone in breath [9]. In 1999, detection of conformational changes of proteins on cantilevers [48] and antigen–antibody binding [49] via surface stress were reparted, but these early bioexperiments were undertaken without reference cantilevers. Differential detection of biomolecules was first used to observe protein adsorption [26]. Then experiments with a cantilever array, reported in 2000, demonstrated the detection of molecular recognition via surface stress including the detection of a single base mismatch in DNA hybridization, and the detection of a protein receptor–ligand interaction [5]. From this point on it was clear that biomolecular binding can indeed bend a cantilever. During the last 2 years, a growing number of surface stress measurements on different biological systems have been reported and these are discussed in detail below.

9.4.2
Nucleic Acids

Nucleic acids are easy to immobilize on gold surfaces when modified with a thiol group at their 5'-end. DNA oligonucleotides were the first biosystem investigated systematically with nanomechanical cantilever sensors. Detection of specific hybridization of 12mer and 16mer oligonucleotides to single-stranded probe oligonucleotides functionalized on cantilever surfaces was reported [5]. This report showed that the bending signal depends on the concentration of target oligonucleotides and a detection limit below 80 nM could be achieved. It was also shown that nanomechanical sensors can detect a single base mismatch within 12mer oligonucleotides. These experiments with silicon cantilevers resulted in deflections of 10 nm and compressive surface stresses of 0.005 N m^{-1} for hybridization at 80 nM of target oligonucleotide. The sensor was reusable once the oligonucleotide duplexes were denatured with urea, releasing the target molecules, and restoring the probes. Later the same group reported a hybridization experiments with arrays of eight cantilevers and they were also able to obtain kinetic data from nanomechanical sensors [27]. In these experiments a specific oligonucleotide sequence was detected within an 80-fold higher background of unrelated sequences. Figure 9.5 shows a typical differential hybridization experiment with an array of eight cantilevers. Unpublished results indicate that the sensitivity of nanomechanical sensors to oligonucleotides could be increased to below nanomolar concentrations. Experiments with single silicon nitride cantilevers undertaken by another group showed that stress sensors can distinguish different mismatch positions within oligonucleotide duplexes and that short target oligonucleotides can be detected with longer probe oligonucleotides [38,50]. They used 10mer to 25mer oligonucleotides and detected deflections up to 20 nm or surface stresses around 0.005 N m^{-1} with single silicon nitride cantilevers. Another interesting finding was that the direction of surface stress during hybridization, compressive or tensile, seems to depend on the buffer conditions in which the probe oligonucleotides were immobilized. This suggests that adjusting the experimental conditions could help tuning the sensitivity of nanomechanical sensors in the future. The adsorption of oligonucleotides onto gold has also been monitored with piezoresistive cantilevers although the detection of hybridization is still beyond the capabilities of this technology [51].

Finally, nanomechanical sensors have been applied to monitor enzymatic activities such as ligation and digestion of DNA on cantilever surfaces [52]. Because the use of these sensors allows real-time measurements, they are likely to contribute to improved characterization of the kinetics of various reactions involving nucleic acids.

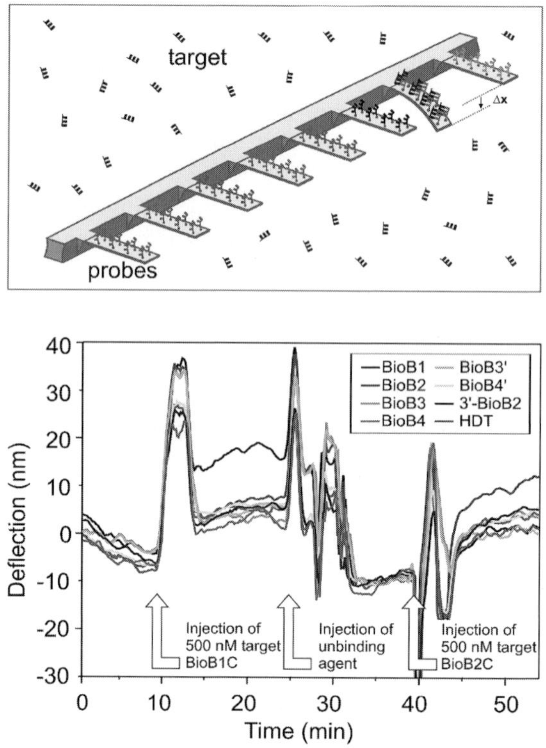

Figure 9.5 Hybridization experiment with an array of eight cantilevers. (a) Seven cantilevers were functionalized with seven different probe 12mer oligonucleotides and the eighth cantilever was functionalized with hexadecanethiol.
(b) Experimental data. By injecting one of the target sequences, only the cantilever covered with the complementary probe sequence responds. The array can be reset by injecting 30 % urea; the experiment is then repeated with another target sequence. (Figures kindly provided by H.P. Lang, University of Basel, reproduced with permission from the National Academy of Sciences, USA [27]).

9.4.3 Proteins

Proteins are much more sensitive than DNA to changes in their environment, chemical modification, or to immobilization on a surface. The development of protein detection using surface stress sensors is still in its infancy. A first receptor–ligand experiment with a differential set-up and direct negative controls within the same experiment was reported on a protein A–IgG system [5]. The gold side of silicon cantilevers was coated with a polyethylen glycol (PEG) thiol to block protein adsorption. Then the silicon side of one cantilever was covered with protein A and that of a second cantilever acting as a reference, with BSA (bee also Figure 9.3). Injection of goat IgGs showed no signal in the differential read-out but injection of rabbit IgGs showed a deflection of 10 nm or 0.005 N m^{-1} in surface stress. This reflects the known binding properties of IgGs from different species to protein A. Relativcly high concentrations of 0.1 mg mL^{-1} of IgGs were used in these experiments.

A first systematic study with proteins reported on the detection of prostate-specific antigen (PSA) at clinically relevant levels [6]. Single silicon nitride cantilevers were derivatized on their gold side with a thiol succinimide compound which self-assembles on gold and can also covalently bind to amino groups in proteins. The surface was then functionalized with anti-human PSA antibodies. The cantilever was washed with BSA solution to remove antibodies from the silicon nitride side of the cantilever and to prevent at the same time non-specific protein binding. Detection of 60 μg mL^{-1} to 0.2 ng mL^{-1} PSA in a 1 mg mL^{-1} background of other proteins was reported. Figure 9.6 shows the concentration dependence and the detection limit of this experiment.

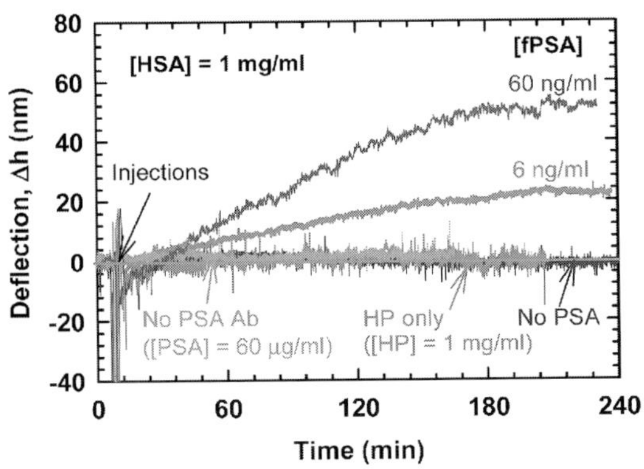

Figure 9.6 Detecting prostate-specific antigens with an antibody-covered cantilever. Detection of free prostate-specific antigen at ng mL^{-1} concentrations within a background of 1 mg mL^{-1} human serum albumin (HSA) and human plasminogen (HP) is concentration dependent. (Figure kindly provided by A. Majumdar, University of California, Berkeley, reproduced with permission from Nature Publishing Group [6]).

Using the same antibody immobilization strategy, an antibody–antigen experiment with a cantilever array using two different antibodies was recently reported [29]. Two cantilevers of an array were coated with antibodies against myoglobin and creatine kinase, respectively, two markers for myocardial infarction. A reference cantilever was coated with BSA. In a background of BSA each of the two antigens could be detected by their respective antibody-coated cantilever. In this case, the detection limit was observed to be 20 μg mL^{-1}. The cantilevers used were only 500 nm thick and the reported compressive stresses were about 0.004 N m^{-1}. Another finding was that the passivation or blocking of the non-functionalized side of the cantilever was necessary to observe the specific signals during the experiment. Otherwise a large bending from non-specific binding obscures the smaller bending resulting from specific interactions. Finally, another group has described the use of a nanomechanical cantilever sensor as a glucose sensor by functionalizing it with glucose oxidase, thus providing the first example of a mechanical detection of a small metabolite by an enzymatic reaction [53].

9.5
Conclusion and Outlook

We have reviewed a label-free biosensing technique, which has the potential to be developed into a label-free microarray for proteins and other biomolecules. As we have seen, experiments have already demonstrated the use of nanomechanical cantilever sensors for the detection of oligonucleotides and clinically relevant proteins. Experiments with eight cantilevers comparable to an eight-spot microarray have been reported. The particular appeal of this technique is that it can transduce virtually any biochemical interaction into a mechanical motion of a microfabricated sensor and it can do so without prior modification of the sample with labels or the use of reporter molecules.

Some of the ongoing research on nanomechanical cantilever sensors aims to combine these sensors with microfluidic devices to facilitate handling and automation. Other work aims to improve the sensitivity of the sensors. For application of this sensor technology to the microarray format the biggest technical challenge will be to extend one-dimensional cantilever arrays into two dimensions and to develop the necessary functionalization and read-out methods for a liquid environment. For data storage applications it has already been demonstrated that two-dimensional cantilever arrays can be produced and that cantilevers in such arrays can be addressed individually [54]. Several companies are already working on the commercialization of nanomechanical cantilever sensors for biosensing applications.

Acknowledgements

We thank Suzanne Gaudet and Hans Peter Lang for critically reading the manuscript and Cagri Savran for stimulating discussions. During preparation of this chapter J. Fritz was at the MIT Media Laboratory and be thanks Scott Manalis for his generous support.

References

1. Kricka, L.J. Nucleic acid detection technologies – labels, strategies, and formats. *Clin. Chem.* 1999, **45,** 453–458.
2. Kodadek, T. Protein microarrays: Prospects and problems. *Chem. Biol.* 2001, **8,** 105–115.
3. Schuck, P. Use of surface plasmon resonance to probe the equilibrium and dynamic aspects of interactions between biological macromolecules. *Annu. Rev. Biophys. Biomol. Struct.* 1997, **26,** 541–566.
4. Janshoff, A., Galla, H.-J., Steinem, C. Piezoelectric mass-sensing devices as biosensors - An alternative to optical biosensors? *Angew. Chem. Int. Ed.* 2000, **39,** 4004–4032.
5. Fritz, J., Baller, M.K., Lang, H.P., Rothuizen, H., Vettiger, P., Meyer, E., Güntherodt, H.-J., Gerber, Ch., Gimzewski, J.K. Translating biomolecular recognition into nanomechanics. *Science* 2000, **288,** 316–318.
6. Wu, G., Datar, R.H., Hansen, K.M., Thundat, T., Cote R.J., Majumdar, A. Bioassay of prostate-specific antigen (PSA) using microcantilevers. *Nature Biotech.* 2001, **19,** 856–860.
7. Raiteri, R., Grattarola, M., Berger, R. Micromechanics senses biomolecules. *Mat. Today* January 2002, 22–29.
8. Sepaniak, M., Datskos, P., Lavrik, N., Tipple, C. Microcantilever transducers: A new approach in sensor technology. *Anal. Chem.* November 1, 2002, 568A–575A.
9. Lang, H.P., Hegner, M., Meyer, E., Gerber, Ch. Nanomechanics from atomic resolution to molecular recognition based on atomic force microscopy technology. *Nanotech.* 2002, **13,** R29–R36.
10. Binnig, G., Quate, C.F., Gerber, Ch. Atomic force microscopy. *Phys. Rev. Lett.* 1986, **56,** 930–933.
11. Prescesky, S., Parameswaran, M., Rawicz, A., Turner, R.F.B., Reichl, U. Silicon micromachining technology for sub-nanogram discrete mass resonant biosensors. *Can. J. Phys.* 1992, **70,** 1178–1183.
12. Wachter, E.A., Thundat, T. Micromechanical sensors for chemical and physical measurements. *Rev. Sci. Instrum.* 1995, **66,** 3662–3667.
13. Tamayo, J., Humphris, A.D.L., Malloy, A.M., Miles, M.J. Chemical sensors and biosensors in liquid environment based on microcantilevers with amplified quality factor. *Ultramicroscopy* 2000, **86,** 167–173.
14. Yang, J., Ono, T., Esashi, M. Mechanical behavior of ultrathin microcantilever. *Sens. Actuators A* 2000, **82,** 102–107.
15. Barnes, J.R., Stephenson, R.J., Woodburn, C.N., O'Shea, S.J., Welland, M.E., Rayment, T., Gimzewski, J.K., Gerber, Ch. A femtojoule calorimeter using micromechanical sensors. *Rev. Sci. Instrum.* 1994, **65,** 3793–3798.
16. Butt, H.-J. A sensitive method to measure changes in the surface stress of solids. *J. Coll. Interface Sci.* 1996, **180,** 251–260.
17. Berger, R., Delamarche, E., Lang, H.P., Gerber, Ch., Gimzewski, J.K., Meyer, E., Güntherodt, H.-J. Surface stress in the self-assembly of alkanethiols on gold. *Science* 1997, **276,** 2021–2024.
18. Albrecht, T.R., Akamine, S., Carver, T.E., Quate, C.F. Microfabrication of cantilever styli for the atomic force microscope. *J. Vac. Sci., Technol. A* 1990, **8,** 3386–3396.
19. Sader, J.E. Calibration of atomic force microscope cantilevers. In: Hubbard, A. (Ed.) *Encyclopedia of Surface and Colloide Science.* Marcel Dekker, New York, USA, 2002, 846–856.
20. Sarid. D. *Scanning Force Microscopy.* Oxford University Press, New York, USA, 1994.
21. Miyatani, T., Fujihira, M. Calibration of surface stress measurements with atomic force microscopy. *J. Appl. Phys.* 1997, **81,** 7099–7115.
22. Manalis, S.R., Minne, S.C., Atalar, A., Quate, C.F. Interdigital cantilevers for atomic force microscopy. *Appl. Phys. Lett.* 1996, **69,** 3944–3946.
23. Savran, C.A., Sparks, A.W., Sihler, J., Li, J., Wu, W.-C., Berlin, D.E., Burg, T.P., Fritz, J., Schmidt, M.A., Manalis, S.R. Fabrication and characterization of a micromechanical sensor for differential detection of nanoscale motions. *J. Microelectromech. Sys.* 2002, **11,** 703–708.
24. Tortonese, M., Barrett, R.C., Quate, C.F. Atomic resolution with an atomic force microscope using piezoresistive detection. *Appl. Phys. Lett.* 1993, **62,** 834–836.

25. Lang, H.P., Berger, R., Andreoli, C., Brugger, J., Despont, M., Vettiger, P., Gerber, Ch., Gimzewski, J.K., Ramseyer, J.P., Meyer, E., Güntherodt, H.-J. Sequential position readout from arrays of micromechanical cantilever sensors. *Appl. Phys. Lett.* 1998, **72,** 383–385.

26. Baller, M.K., Lang, H.P., Fritz, J., Gerber, Ch., Gimzewski, J.K., Drechsler, U., Rothuizen, H., Despont, M., Vettiger, P., Battiston, F.M., Ramseyer, J.P., Fornaro, P., Meyer, E., Güntherodt, H.-J. A cantilever array-based artificial nose. *Ultramicroscopy* 2000, **82,** 1–9.

27. McKendry, R., Zhang, J., Arntz, Y., Strunz, T., Hegner, M., Lang, H.P., Baller, M.K., Certa, U., Meyer, E., Güntherodt, H.-J., Gerber, Ch. Multiple label-free biodetection and quantitative DNA-binding assays on a nanomechanical cantilever array. *Proc. Natl. Acad. Sci. USA* 2002, **99,** 9783–9788.

28. Cherian, S., Mehta, A., Thundat, T. Investigating the mechanical effects of adsorption of Ca2+ ions on a silicon nitride microcantilever surface. *Langmuir* 2002, **18,** 6935–6939.

29. Arntz, Y., Seelig, J.D., Lang, H.P., Zhang, J., Hunziker, P., Ramseyer, J.P., Meyer, E., Hegner, M., Gerber, Ch. Label-free protein assay based on a nanomechanical cantilever array. *Nanotech.* 2003, **14,** 86–90.

30. Gibson, C.T., Watson, G.S., Myhra, S. Scanning force microscopy - Calibrative procedure for "best practice". *Scanning* 1997, **19,** 564–581.

31. Stoney, G.G. The tension of metallic films deposited by electrolysis. *Proc. Royal Soc. Lond. A* 1909, **82,** 172–175.

32. Ibach, H. The role of surface stress in reconstruction, epitaxial growth and stabilization of mesoscopic structures. *Surf. Sci. Rep.* 1997, **29,** 193–263.

33. Butt, H.-J., Raiteri, R. Measurement of the surface tension and surface stress of solids. In: Milling, A.J. (Ed.) *Surface Characterization Methods.* Marcel Dekker, New York, USA, 1999, 1–36.

34. Haiss, W. Surface stress of clean and adsorbate-covered solids. *Rep. Prog. Phys.* 2001, **64,** 591–648.

35. Fritz, J., Baller, M.K., Lang, H.P., Strunz, T., Meyer, E., Güntherodt, H.-J., Delamarche, E., Gerber, Ch., Gimzewski, J.K. Stress at the solid–liquid interface of self-assembled monolayers on gold investigated with a nanomechanical sensor. *Langmuir* 2000, **16,** 9694–9696.

36. Raiteri, R., Butt, H.-J., Grattarola, M. Changes in surface stress at the liquid/solid interface measured with a microcantilever. *Electrochimica Acta* 2000, **46,** 157–163.

37. Ji, H.-F., Hansen, K.M., Hu, Z., Thundat, T. Detection of pH variation using modified microcantilever sensors. *Sens. Actuators B* 2001, **72,** 233–238.

38. Wu, G., Ji, H., Hansen, K., Thundat, T., Datar, R., Cote, R., Hagan, M.F., Chakraborty, A.K., Majumdar, A. Origin of nanomechanical cantilever motion generated from biomolecular interactions. *Proc. Natl. Acad. Sci.* USA 2001, **98,** 1560–1564.

39. Hagan, M.F., Majumdar, A., Chakraborty, A.K. Nanomechanical forces generated by surface grafted DNA. *J. Phys. Chem. B* 2002, **106,** 10163–10173.

40. Sander, D., Ibach, H. Experimental determination of adsorbate-induced surface stress: oxygen on Si(111) and Si(100). *Phys. Rev. B* 1991, **43,** 4263–4267.

41. Lavrik, N.V., Tipple, C.A., Sepaniak, M.J., Datskos, P.G. Enhanced chemi-mechanical transduction at nanostructured interfaces. *Chem. Phys. Lett.* 2001, **336,** 371–376.

42. Ulman, A. (Ed.). *Self-assembled Monolayers of Thiols.* Academic Press, San Diego, USA 1998.

43. Baller, M.K. Microcantilever sensor arrays: Nanomechanics, electronic noses and DNA chips. Thesis, University of Basel, Basel, Switzerland, 2001.

44. Chen, G.Y., Thundat, T., Wachter, E.A., Warmack, R.J. Adsorption-induced surface stress and its effects on resonance frequency of microcantilevers. *J. Appl. Phys.* 1995, **77,** 3618–3622.

45. Raiteri, R., Butt, H.-J. Measuring electrochemically induced surface stress with an atomic force microscope. *J. Phys. Chem.* 1995, **99,** 15728–15732.

46. O'Shea, S.J., Welland, M.E., Brunt, T.A., Ramadan, A.R., Rayment, T. Atomic force microscopy stress sensors for studies in liquids. *J. Vac. Sci. Technol. B* 1996, **14,** 1383–1385.

47. Lang, H.P., Berger, R., Battiston, F., Ramseyer, J.-P., Meyer, E., Andreoli, C., Brugger, J., Vettiger, P., Despont, M., Mezzacasa, T., Scandella, L., Güntherodt, H.-J., Gerber, Ch., Gimzewski, J.K. A chemical sensor based on a micromechanical cantilever array for the identification of gases and vapors. *Appl. Phys. A* 1998, **66,** S61–S64.

48. Moulin, A.M., O'Shea, S.J., Bradley, R.A., Doyle, P., Welland, M.E. Measuring surface-induced conformational changes in proteins. *Langmuir* 1999, **15,** 8776-8779.

49. Raiteri, R., Nelles, G., Butt, H.-J., Knoll, W., Skladal, P. Sensing of biological substances based on the bending of microfabricated cantilevers. *Sens. Actuators B* 1999, **61,** 213–217.

50. Hansen, K.M., Ji, H.-F., Wu, G., Datar, R., Cote, R., Majumdar, A., Thundat, T. Cantilever-based optical deflection assay for discrimination of DNA single-nucleotide mismatches. *Anal. Chem.* 2001, **73,** 1567–1571.

51. Marie, R., Jansenius, H., Thaysen, J., Christensen, C.B., Boisen, A. Adsorption kinetics and mechanical properties of thiol-modified DNA–oligos on gold investigated by microcantilever sensors. *Ultramicroscopy* 2002, **91,** 29–36.

52. Stevenson, K.A., Mehta, A., Sachenko, P., Hansen, K.M., Thundat, T. Nanomechanical effect of enzymatic manipulation of DNA on microcantilever surfaces. *Langmuir* 2002, **18,** 8732–8736.

53. Subramanian, A., Oden, P.I., Kennel, S.J., Jacobson, K.B., Warmack, R.J., Thundat, T., Doktycz, M.J. Glucose biosensing using an enzyme-coated microcantilever. *Appl. Phys. Lett.* 2002, **81,** 385–387.

54. Despond, M., Brugger, J., Drechsler, U., Dürig, U., Häberle, W., Lutwyche, M., Rothuizen, H., Stutz, R., Widmer, R., Binnig, G., Rohrer, H., Vettiger, P. VLSI-NEMS chip for parallel AFM data storage. *Sens. Actuators A* 2000, **80,** 100–107.

10
Image Analysis Issues and Solutions for High-Density Arrays

Anton Petrov and Soheil Shams

10.1
Introduction

DNA and protein array experiments, which employ printing of biological material onto solid or membrane surfaces and consequent hybridization or binding, require the ability to manage large quantities of data both prior to and following the actual experiment. There are three major issues involved in data management [1]. The first is to keep track of the information generated at the stages of chip production and binding during the experiment [2]. The second is to process microarray images to obtain the quantified signal intensity (expression) values from the arrays and the third is to mine the information from the signal data [3]. In this chapter we will concentrate on the issues related to array image analysis.

The goal of array image processing is to measure the intensity of the spots of labeled and printed biological material, quantify expression values based on these intensities, assess the reliability of the data, and generate warnings to the possible problems that may occur during the array production and/or binding phases. Microarray images consist of arrays of spots arranged in sub-grids. All the sub-grids usually, but not always, have the same numbers of rows and columns of spots. These "sub-grids", also called "quadrants", are arranged in relatively equal spacing with each other, forming a "meta-array". Ideally, a simple computer program could easily accomplish the image-processing task by superimposing a uniformly spaced array of circles with defined dimensions and spacing on the images. In theory, the pixels falling inside and outside these circles would be considered signal and background, respectively. In reality, however, the exact location of each sub-grid, and each individual spot within each sub-grid, may vary from slide to slide. There are a number of sources contributing to the problems, mainly due to mechanical constraints in the spotting process, immobilization and binding inconsistencies, and possible scanning problems. The problem of inconsistent spot location and size necessitates a reliable approach for automatic spot finding.

Contamination is also a major source of problems in the microarray image processing and especially for Proteomics in its current state. From the procedure of spotting to binding, dust particles floating in the air may land on the slides and these often appear as very bright spots on the scanned images. Certain abnormali-

Protein Microarray Technology. Edited by Dev Kambhampati

ISBN: 3-527-30597-1

ties in surface chemistry can also cause some damage to quantified expressions. An example of contamination in microarray images is shown in Figure 10.2. Small contamination spots distributed in the background and small contamination particles with an extremely high intensity of fluorescence inside the spots are commonly seen on such images. In addition to the comparatively unstable behavior of proteins, such defects can make measurement data completely useless. Image processing methods can be used to provide the best estimate of the true signal level by identifying the contamination and removing it before applying the measurements. Various segmentation algorithms can be used to identify the correct signal, background and contamination regions for every spot. Some of them will be discussed in this chapter.

Choice of a specific measurement in order to characterize the intensity level of a spot is also considered a very important step in microarray analysis. One can choose between mean, median, mode, or total intensity to represent the expression level. Moreover, when a microarray experiment involves two channels, a control (reference) channel as well as the experimental channel, the ratios between the channels for the above measures can also be calculated.

Additionally every researcher tends to be concerned with one or another kind of quality assessment of the measurements that result from image processing. A minimum requirement for an ideal image processing system would be to be able to exclude "low-quality" spots from further analysis. Even at this stage there is no agreement on how to define a "low-quality" spot. Imagine how difficult it becomes to assess a spot's quality quantitatively and to use this information further in the expression analysis step. Important results regarding our approach in producing a reliable quality control procedure for microarrays will be discussed later in the chapter.

Finally, a microarray experiment is a typical example of high-throughput research demanding reliable automated image processing to accelerate the discovery workflow and to reduce human error. Although several attempts in providing such systems have been made in commercial and academic software tools, the majority have had limited success in handling the wide range of microarray images. We will discuss why this is a particularly challenging problem and describe a successful attempt at meeting this challenge.

Microarray image processing involves a number of well-established steps. It begins with a captured microarray image as the input and finishes with an output of quantified protein expression values as well as potential quality measurements and quality flags. Figure 10.1 demonstrates this process. In the following sections of this chapter each step of this process will be discussed in detail.

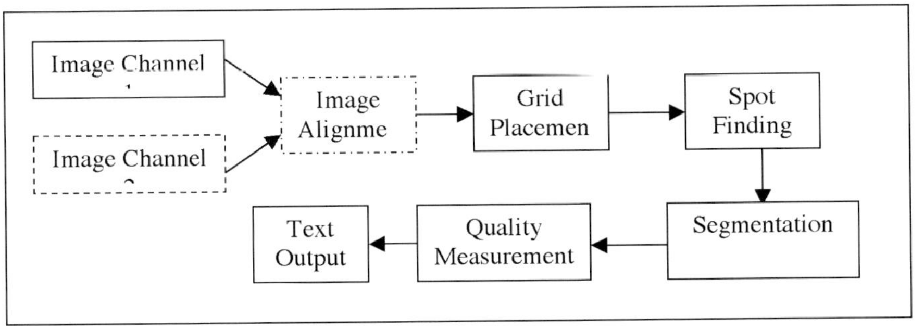

Figure 10.1 Microarray image analysis.

10.2 Image Alignment, Grid Placement, and Spot Location

10.2.1 Image Alignment

A common approach in microarray experiments is the use of a two-channel system inherited from DNA microarrays [4]. This type of system is usually used to analyze protein expression regulation in response to a specific treatment. In such experiments, a microarray is developed using two different types of label (for example Cy3/Cy5). One label is used for the control and the other is used for labeling the treated material. The scanner will then generate two images from the same array, one for each labeled channel. It is not usually necessary to process these images simultaneously. However, one might need to overlay these images for visual inspection or for pixel-by-pixel ratio quantification. For this reason, an initial image alignment stage might be necessary when dealing with two-channel images. Both manual and automatic image alignment can be carried out. In the manual mode the user will need access to the tools for two basic operations: image rotation and translation. In the automated mode both operations will be performed without user intervention. Various algorithms based on matched filtering and feature extraction can be utilized for automated operation. The results of such algorithms are usually very stable and reliable. This is explained by the fact that aligned images represent screenshots of the same array and are therefore relatively easy to match.

The goal of a spot finding operation is to locate the signal spots in images and estimate the size of each spot. There are two different levels of sophistication in the algorithms for spot finding, corresponding to the degree of human intervention in the process. These are described below in order of the degree of manual intervention.

10.2.2
Manual Spot Finding

This method is essentially a computer-aided image processing approach. The computer provides the tools which allow the user to identify the signal spots in the image. Typically, a grid frame is provided which the user can place manually on the image and manipulate to fit the spatial extent of the spots in the image. Because the spots in the image may not be evenly spaced, the user may need to adjust the grid lines individually to align them with the arrayed spots. The user may also have to adjust some or all of the grid points to superimpose the spots in the image. The size of each circle may also need to be manually adjusted so that it exactly matches the size of each individual spot. In the achievement of an accurate measurement, this method is prohibitively time consuming and labor intensive. Thus considerable inaccuracy of the data may be introduced due to human error. Historically, this was the first method of choice for microarray image analysis.

10.2.3
Automatic Spot Finding

The ultimate goal of array image processing is to build an automatic system, which utilizes advanced computer vision algorithms to find the spots without the need for any human intervention. This method would greatly reduce the human effort, minimize the potential for human error, and offer a great deal of consistency in the quality of data. Such a processing system would require the user to specify only the expected configuration of the array (e.g. number of rows and columns of spots and inter-spot spacing). The system would automatically search the image for the grid position. Most of the methods used for this purpose are based on some type of matched filtering. Matched filtering can be applied row- or column-wise (1-D) or for the whole grid (2-D).

Having found the approximate grid position, which specifies the center of each spot, the neighborhood can be examined to detect signal and background. Knowledge about the image characteristics should be incorporated to account for variability in microarray images. The spot location, size, and shape should be adjusted to accommodate noise, contamination and uneven distribution. The quality of data is high because this approach is unbiased. Human intervention and possibilities for errors are minimized.

10.3
Spatial Segmentation of Signal and Background Pixels

After the location of the spot has been determined in the image, a small patch around that location (target region) can be used to quantify the level of spot expression. The next step is to determine which pixels in the target region are due to the signal from the actual spot and which are background. This operation is called signal or image segmentation in computer vision terminology. A number of methods of image segmentation have been developed with different levels of sophistication.

10.3.1 Pure Space-based Signal Segmentation

Methods of this type use purely spatial information obtained from the location of the spot, to segment out signal pixels. After the spot location operation has been completed, the position and size of the spot is determined. A circular mask of the computed size is placed in the image at the determined position to separate the signal from the background. It is assumed that the pixels inside of the circle are due to the true signal and those outside are background. The measurements are then applied to these classified pixels. These types of methods are optimal when the spot location operation is effective, i.e. spots have been correctly located and sized, the spot shapes are close to perfect circles, and no contamination is present [5]. Knowing the configuration of the array and the spacing between spots, the user can specify the number of pixels around the spot that can be used to compute the background value. However, irregularity of the spot shape and size is more likely to be the rule rather than the exception in many microarray images. Whenever these conditions occur, the accuracy of the measurements is largely compromised. In addition, spot contamination is still a large issue in many microarray images. In reality, such a procedure, being computationally inexpensive on one hand, becomes critically inaccurate on the other hand (Figure 10.2).

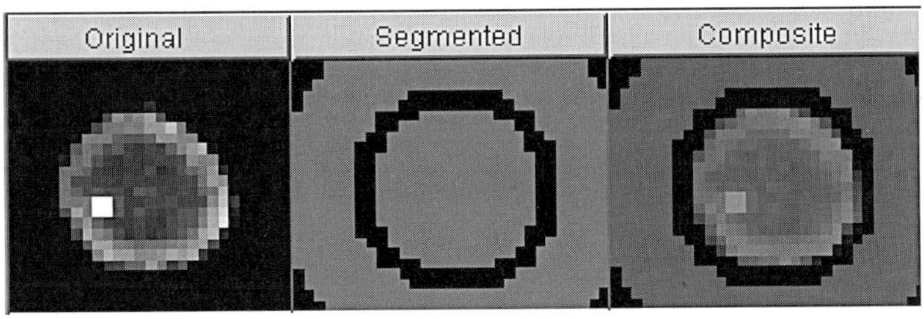

Figure 10.2 Results of the circular segmentation method. The red pixels denote the signal area and the green pixels denote background. Black colored pixels in the panel labeled "segmented" indicate ignored pixels. Note that the spot is not perfectly circular and falls partly outside the signal area.

10.3.2 Pure Intensity-based Signal Segmentation

Methods of this type use intensity information exclusively to segment out signal pixels from the background. They assume that the signal pixels are statistically brighter than the background pixels. A set percentage of pixels with the highest intensity values are considered to be the signal. Although this method was employed with early image analysis systems, it is currently not used in most systems due to the high error rates. An improvement over the simple intensity-only segmentation is to use various clustering procedures to group pixels into background and signal groups [6]. Such a method seems very natural at the first sight. One can use k-means, k-medians or hierarchical clustering to extract signal and background regions. This operation will find an optimal arrangement of pixel intensities to maximize the difference between cluster levels and minimize the within-cluster variance at the same time. Obviously, this approach greatly depends on the choice of distance metric in intensity space.

Such extraction will be robust with respect to the location of the spot within considered region and with respect to its shape. However, since this method completely ignores spatial information regarding the pixel arrangement, the outcome can be completely unpredictable in the case of a low SNR. In this case the decision on where to separate signal and background will rely mainly on metric properties. In addition, the number of clusters should be taken into account and those due to contamination should be noted; basically, the image should be analyzed for the presence of contamination. The decision as to which clusters are due to contamination can be very difficult as the contamination may be contained within several clusters.

10.3.3
Mann–Whitney Segmentation

Based on the result of the spot location operation, a circle is placed in the target region to include the spatial region of the spot. Because the pixels outside of the circle are assumed to be the background, the distribution of these background pixels can be used to determine which pixels inside the circle are signal pixels. Mann-Whitney test is used to obtain a threshold intensity level [7,8].

Briefly Mann–Whitney segmentation can be carried out as follows. A number of pixels from the background (say, 10 pixels) are randomly selected and their intensity values will form the first sample. The same number of pixels (10 pixels) is selected from the signal, but only those with the lowest intensity values within the signal mask. The two samples are then compared using the Mann–Whitney test to establish whether their mean values are the same. This test is selected because of its robustness with respect to assumptions regarding distribution. If the samples pass the test, a pre-defined number of pixels from the signal sample are replaced with those which are not yet in the sample and which have the lowest intensity values. Testing will continue until two samples fail the test. Pixels inside the circle having a higher intensity than the lowest of those in the signal sample are identified as being the signal.

This method works very well when the spot location is found correctly and there is no contamination on the image. However, when there are pixels due to contamination inside the circle, these will be identified as signal pixels. This is due to the fact that typical contamination will be seen as specular and this reflection has intensity a higher than that of the background. If there are contamination pixels outside the circle, or the spot location has been inaccurate so that some of the signal pixels are outside the circle, these high intensity pixels will raise the intensity threshold level. Consequently, those signal pixels which have an intensity that is lower than the threshold will be misclassified as background. This method also has its limitations when dealing with weak signals and noisy images. This list of flaws is similar to that associated with the pure intensity-based segmentation method.

10.3.4 The Method of Trimmed Measurements

This approach is another method that combines both spatial and intensity information to distinguish the signal pixels from the background. The logic of this method proceeds as follow. After the spot is localized and a target circle has been placed in the region, most of the pixels inside the circle are signal pixels and most of the background pixels are outside the circle. Due to the shape irregularity, some signal pixels may leak out of the circle and some background pixels may enter the circle. These pixels may be considered as outliers in the samples of the signal and background pixels. Similarly, contamination pixels may also be considered as outliers in the intensity domain. These outliers will severely change the measurement of the mean and total signal intensity. To remove the effect of outliers on these measurements, a fixed percentage of pixels from the intensity distribution of the pixels for both signal and background regions may be simply "trimmed-off" (see Figure 10.3). However, it cannot be determined *a priori* what percentage of outliers would be "right" for a spot and again, such an approach can produce inaccurate results.

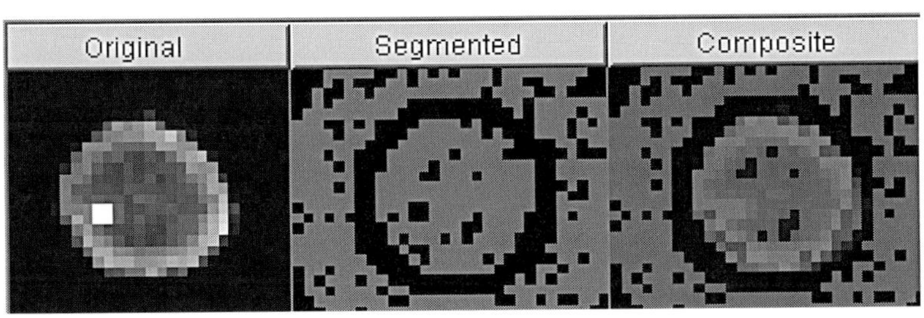

Figure 10.3 Results of "trimmed measurements" segmentation pprocedure. Note that the small amount of contamination in the signal area has been successfully removed along with other true signal pixels.

10.3.5 Integrating Spatial and Intensity Information for Signal Segmentation

The two methods discussed above use a minimal amount of spatial information, i.e. the target circle obtained from spot localization is not used to improve the detection of signal pixels. The priority in their design is to measure the intensity of the spots with minimal computation. These methods are useful in semi-automatic image processing because they are quick and the user can visually inspect the quality of data.

In a fully automated image processing system, the accuracy of the classification of the signal pixel becomes a central concern. Not only does the correct segmentation of signal pixels offer accurate measurement of the signal intensity, but it also permits multiple quality measurements to be carried out based on the geometric properties of the spots. These quality measures will alert the operator to spots having questionable quality values after the completion of an automated analysis. The correct classification of the signal pixels can be realized by algorithms which use information from both spatial and intensity domains (i.e. hybrid methods) [2–4,9]. Spatial information should be used not only to initialize the signal/background mask, but also to optimize the relative spatial arrangement of pixels from the same class. This is why one of the most popular approaches utilized in hybrid procedures is "region-growing" (see Figure 10.4) [9,10].

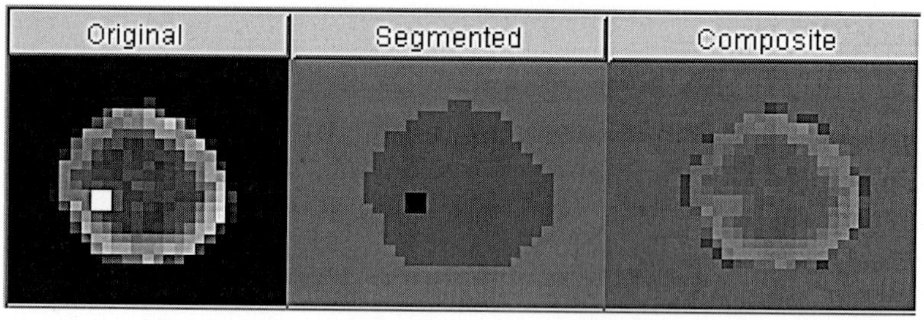

Figure 10.4 Results of the segmentation process using a "hybrid" method. Note that the pixels associated with the small amount of contaminant have been successfully removed from the signal area which has been matched to the non-circular spot shape to include all signal pixels. (see Colour Plate p. XXXII).

An experimental comparison between the hybrid and pure spatial segmentation methods is presented in Section 10.7.

10.4 Data Quantification

The key information which should be recorded from microarrays is the expression strength of each target. Under idealized conditions, the total florescent intensity from a spot is proportional to the expression strength. However, those idealized conditions almost never exist. Usually spots differ in their shapes and sizes even throughout the same array.

When the spot is contaminated, the contribution of the contaminated region to the signal intensity is not measurable. The image processing may not correctly identify all the signal pixels, thus, quantification methods should be designed to address these problems. The common values computed after segmentation are total, mean, median, mode, volume, intensity ratio, and the correlation ratio across two channels.

- The *total* signal intensity is the sum of the intensity values of all the pixels in the signal region. As has been indicated above, total intensity is sensitive to the variation in the amount of biological material deposited on the spot, the existence of contamination, and the anomalies in the image processing operation. Because these problems occur frequently, the *total intensity* may not be an accurate measurement.

- The *mean* signal intensity is the average intensity of the signal pixels. Using this parameter has certain advantages over the calculation of the total intensity. Very often the spot size correlates to the concentration of the sample in the wells during the spotting process. Measuring the mean will reduce the error caused by the variation in the amount of immobilized bait deposited on the spot. Since advanced image processing permits the accurate distinction of contamination pixels from the signal pixels, the mean should be one of the parameters of choice.

- The *median* of the signal intensity is the intensity value that splits the distribution of the signal pixels in half. The number of pixels above the median intensity is the same as those below. Thus, this value is a landmark in the intensity distribution profile. The advantage of choosing this landmark as the measurement is due to its robustness to outliers. As has been discussed in the last section, contamination and problems in the image processing operation introduce outliers in the sample of identified signal pixels. The mean measurement is very vulnerable to these outliers. Thus, if the image processing operation is not sophisticated enough to ensure the correct identification of signal, background, and contamination pixels, the median intensity seems to be a better choice of parameter than the *mean* intensity.

- The *mode* of the signal intensity is the "most likely" intensity value and can be measured as the intensity level corresponding to the peak of the intensity histogram. It is also a landmark in the intensity distribution. Thus, it enjoys the same robustness against outliers offered by the median. The trade-off is that the mode becomes a biased estimate when the distribution is multi-modal. This is because the mode value will be equal to one of the modes in the distribution depending on which one is the highest.

- The *volume* of signal intensity is the sum of the signal intensity above the background intensity. It may be computed as (mean of signal – mean of background) – area of the signal).

- If the experiments are carried out with two channels, then the *intensity ratio* between the channels might be the only quantified value of interest. This value will be insensitive to variations in the absolute amount of bait material applied since it is the ratio between the two channels which is being measured. This ratio can be obtained from the mean, median, or mode of the intensity measurement for each channel.

- Another way of computing the intensity ratio is to apply *correlation* analysis across the corresponding pixels in the two channels of the same slide. This ratio can be computed by fitting a straight line through a scatter plot of intensities of individual pixels. This line must pass through the origin (after the background has been corrected for) and the slope gives the intensity ratio between the two channels. This is also known as a regression ratio. The slope of the line can be estimated using various techniques (mean ratio, median ratio, weighed ratio etc). This method may be effective when the signal intensity is much higher than the background intensity. The motivation behind using this method is to bypass the signal pixel identification process (segmentation). However, for spots of moderate to low intensities, the background pixels may severely bias the ratio estimation of the signal towards the ratio of the background intensity. Then the advantage of applying this method becomes unclear and the procedure suffers the same complications as encountered in the signal pixel identification methods discussed above. Thus it may have no advantage over the intensity ratio method. One remedy to this problem is to identify the signal pixels first before undertaking correlation analysis. In this case it becomes unclear why the correlation ratio is used instead of the intensity ratio. However, some specialists prefer this approach to the previously described intensity ratio arguing that the correlation ratio has better distribution properties (log-normal) [11].

In estimating the true signal value, it is necessary to reduce the effects of nonspecific fluorescence, such as the auto-fluorescence of the glass slides. For most analysis calculations, the background intensity should be subtracted from the signal intensity before any ratio calculations are undertaken. The method for determining the background intensity can vary depending on the quality of the arrays and the spacing between individual spots. The same measurements as those discussed above, such as mean, median, mode, and total, can be used to compute the background intensity values.

The results of a comparative study of these various quantification approaches is presented in Section 10.7.

10.5 Quality Control

In a fully automated image processing system, the accuracy of the signal pixel classification becomes a central concern. Not only does the correct segmentation of signal pixels offer accurate measurement of the signal intensity, but it also permits multiple quality measurements based on the morphological properties of the spots. These quality measures can be used to alert the operator to the presence of spots having questionable quality values after the completion of an automated analysis. Sources of measurement inaccuracies may lie in both image imperfections and the spot image extraction procedure itself.

Different quality measures can be used for image-based quality assessment. The choice depends mainly on the particular microarray design, the sophistication of the equipment and the procedures used to calculate the measurements. The most widely used methods are the ratio of the signal standard deviation within the spot and its mean expression (CV), the deviation of the spot from its expected position in the grid and spot circularity measures (for example the ratio of the perimeter squared and the spot area). These measures used separately, or combined into some kind of a decision-making technique can be used to flag a spot as being of low quality.

10.5.1 Background Contamination

Background defects may appear in an arbitrary part of an image due to various reasons. Such an effect can influence the expression level of a large number of spots located in the contaminated area. As we mentioned earlier, we assume that the output of the segmentation procedure for each spot region delivers the signal (spot) pixels, background pixels around the spot and ignored pixels. The latter are usually isolated from the rest of the image to avoid local contamination (such as dust particles) influencing the expression measurement. Our current task is to identify possible abnormality in the background. Let us take the mean of the background intensity around the spot as the local background estimate. In an ideal situation, when no contamination occurs across the image, the background means will have a normal distribution (application of Central Limit Theorem).

A number of simple statistical tests can be applied to identify spots with significantly high or low background means. To illustrate our speculations we used a standard t-test technique. Quantitative comparison of the background mean for a current spot to the normal distribution with two moments estimated from overall distribution of image background means will produce a p-value. This p-value multiplied by 2 will give the confidence value for this spot being "clean" from any background anomaly. Low (close to 0) confidence values will alert the researcher. Setting the threshold for such a confidence value at some pre-selected level will provide automated flagging of suspicious spots. Of course, such an approach will produce a characteristic rate of false results depending on the threshold selected.

10.5.2 Signal Contamination

Another possible source of signal disturbance can be a non-homogeneous distribution of the sample within the spot. To assess this possibility we suggest analysis of signal volatility within the spot. High signal variation will lower confidence in the expression value. We suggest taking spot intensity variance as an estimate of signal volatility. The distribution of spot variance can also be approximated by a normal distribution, however experiments show that the parameters of such a probability will strongly depend on the expression level of the spot. Thus, to retrieve more accurate information regarding the distribution of signal variance, all the spots on the image should be grouped according to their expression levels. The spots can be sorted by their mean expression values and split into bins with equal numbers of spots. Finally the analysis of significance can be undertaken in a manner similar to that described in the previous paragraph, but for each bin separately. A flagging procedure, similar to that described previously can also be applied.

10.5.3 Position Offset

A combination of spot location and image segmentation procedures will provide estimates for the center location of each spot (for example, the center of mass of a spot region). In order to test the results produced by such a scheme, the so-called "expected position" of every spot in the grid should be obtained. Evidently it is possible to use the locations of the nodes in a pre-defined rigid grid corresponding to the ideal microarray structure for this purpose. However, in reality many different factors can introduce systematic irregularities into the grid structure. Irregularities such as variation in inter-row or inter-column distances or slight curvature in the rows or columns should not influence the decision made regarding spot quality. In such a situation the best solution would be a least square fit of straight lines to the spot centers row- and column-wise. As a result, the intersection points of these lines can be utilized as the expected spot locations. Once these coordinates have been obtained, the testing procedure becomes relatively simple. Spots with offset higher than a certain fraction of the average expected inter-spot distance would be flagged.

10.5.4 Percentage of Ignored Pixels

The signal contamination assessment we described previously does not take into the account any information about how many pixels were ignored during the segmentation procedure. Such information may give us an additional understanding of the quality of a spot and it may be extremely important if a region of ignored contamination is added to the area of the signal.

First compute the total number of pixels in the ignored regions directly neighboring the signal region for every spot. Then following ratio should be computed for each

$$\text{spot: } R = \frac{\text{No. of ignored pixels neighboring the signal}}{\text{No. of signal pixels + No. of ignored pixels neighboring the signal}} \times 100\% \qquad (1)$$

Different types of microarray will probably give different values for this ratio. Therefore, the threshold for flagging should be set at a appropriate level R_0 and all the spots with a ratio higher than this threshold should be flagged. An R value below 10–15% is usually acceptable.

10.5.5 Percentage of an Open Perimeter

Sometimes the signal area produced by segmentation appears to be close to the boundary of the patch dedicated to current spot. This may happen as a result of the spot shifting dramatically from its expected location or because of its unusual shape.

In this case the perimeter of the spot signal P_{total}, and the length of that part of signal boundary which coincides with the box indicating the boundary of the spot (the box around the spot, in which the segmentation procedure was performed) P_{open}, can be computed. The ratio $\widetilde{R} = \frac{P_{open}}{P_{total}}$ x 100 % will be a measure of the percentage of the "open" perimeter of the spot signal. The threshold for flagging of the spot should be set at a level $\widetilde{R}_0$. All the spots with a ratio higher than this threshold should be flagged. A value of $\widetilde{R}$ below 10% is usually acceptable, as it may be due to high spot density on the slide. But again, this threshold will be subject to adjustment depending on the current experimental conditions.

10.5.6 Shape regularity

As mentioned previously, several measures can be used for estimating the regularity of the shape of a spot. One of these will be the scaled ratio of the area of a spot and its perimeter squared. Such a measure seems natural since it can be scaled to 1 for an ideal circle and any non-circular shape will have lower value. Various measures can be used to establish "shape regularity".

Experiments with ImaGene 5.5 (BioDiscovery Inc.) have proved that the criteria outlined here when applied in a logical sequence, will provide reliable detection of different types of defects (see Figure 10.5). Furthermore, the performance of our experiments was generally superior to manual flagging. This is achieved by the use of the aforementioned statistics which detect slight abnormalities in spot measurements. Taking into account further sophisticated statistical analyses which are usually carried out on the data obtained, the elimination of such abnormalities may become significant.

10.6 Batch Automation

As mentioned in the Introduction, microarray experiments have becomes classic examples of a high-throughput discovery process. Full automation of at least the image processing stage would be of a great value to this industry. Many attempts have been made in this direction in both the freeware and commercial software sectors. Most of these however, have obvious drawbacks and restrictions making their full-scale utilization prohibitive.

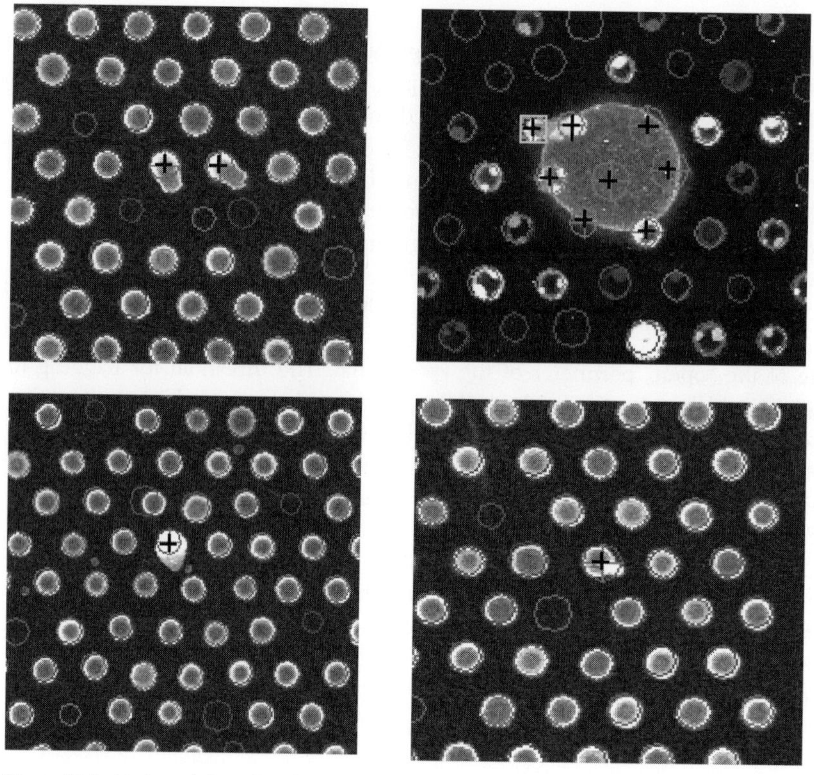

Figure 10.5 Various defects flagged by ImaGene 5.5. (see Colour Plate p. XXXII).

There are several obstacles to fully automating the microarray image operation.

- Grid location and spot adjustment procedures should be robust and flexible enough to accommodate all possible variations in spot locations and shapes.
- Image segmentation procedures should take into the account both the spatial and distributional information within the neighborhood of each spot in order to extract the signal, background and contamination areas as accurately as possible.
- A reliable automated quality control procedure should be incorporated to immediately detect any problem which may be encountered at either the spot, sub-grid or image level. Statistical errors of both types (Type I and Type II) should be minimized to insure optimal utilization of the microarray data.

Most of the existing software tools lack at least one of aforementioned characteristics. The last requirement appears to be especially difficult to implement. With the rich variety of microarray types available on the market it has become increasingly difficult to build a unified system of criteria that would give acceptable results for at least a significant majority of arrays. Additionally, many research laboratories have their own concepts of satisfactory spot quality. Thus the only way to construct an acceptable QC algorithm would be with the involvement of widely accepted generic statistical tests capable of identifying abnormalities in the data which may not be visible to the human eye.

10.7 Experimental Results

The results of several microarray experiments will be used here to illustrate the concepts and questions outlined throughout this chapter. The greatest difficulty in determining the optimum method for segmentation and quantification is the lack of knowledge about the "ground truth" in these experiments. Since it is almost impossible to know exactly the expression value for a particular spot in addition to the difficulty in eliminating non-image processing noise due to fabrication, binding and scanning conditions, we cannot easily compare the results of the various approaches.
In our first study here, we utilized an array image produced using Versalinx Microarray Technology by Prolinx, Inc. This array contains three rows of spots, each containing a specific concentration of the PDBA-HA.11 peptide modified via the N-terminus with PDBA-X-NHS (50, 75, 100 μg mL^{-1}). Each row contains 20 spots. The results described below were obtained using BioDiscovery's ImaGene 5.5 software.

10.7.1 Signal Estimation: Segmentation Method

The first set of measurements was designed to analyze the differences in performance of different spot segmentation algorithms. Therefore, we measured the standard deviation of the signal mean across spots representing the same concentration. In doing so, we arrived at three different numbers representing three different concentrations.

Figure 10.6 shows replicate standard deviation (SD) plots for the mean signal obtained with two segmentation methods: purely spatial (circular) segmentation with background buffer width equal to approximately 20 % of average spot diameter and hybrid intensity/spatial segmentation.

As can be seen from the graph, the hybrid method produced uniformly lower variation of replicate measurements. This indicates the potentially higher reproducibility of the hybrid method. This effect can be explained by the fact that the hybrid segmentation method is able to isolate all signal pixels even in cases where poor results were obtained from the spot location procedure and where there were deformed spot shapes. In this case the signal mean represents the actual "binding concentration" throughout the spot.

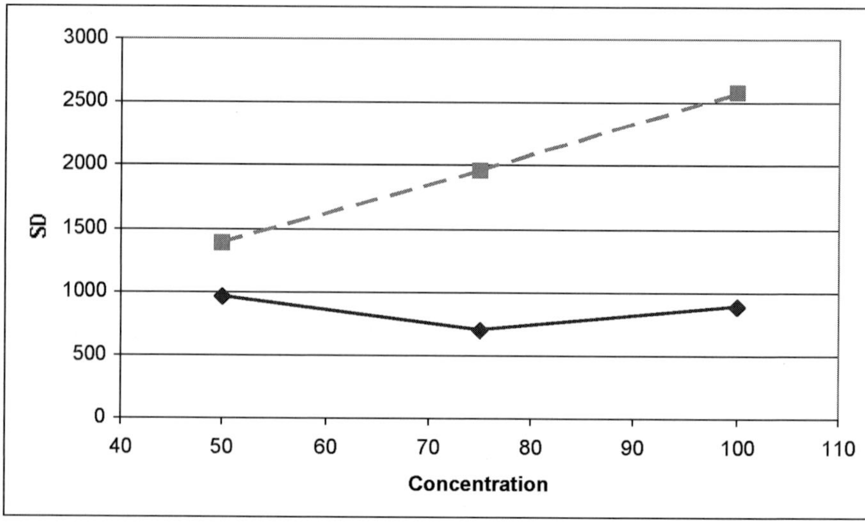

Figure 10.6 Plots of the standard deviations of the signal mean for three concentrations of sample using (a) circular segmentation (broken line), and (b) ImaGene 5.5 hybrid segmentation (solid line).

However the superior reproducibility of the hybrid method is significantly compromised if median and mode signals are used. Therefore the choice of methods for signal measurement will be discussed below.

10.7.2
Signal Estimation: Quantification Method

As mentioned earlier, the choice of method for measuring expression is also an important detail in any reliable microarray technique. On this occasion we used the same image to plot the SD of replicate signals using only the hybrid method, but using three different measures: mean, median and mode; the resulting graphs are shown in Figure 10.7. As can be seen, plotting the mean signal gives a result that is superior to the plots of the other two measures. Circular segmentation will give a similar relationship between the quantification measures, but the advantage of using the mean in the quantification method is not as significant or uniform.

Also if all possible combinations of segmentation types and quantification types are assessed, it will become apparent that the cross-concentration standard deviation plots for the mean signal using the hybrid segmentation method are uniformly the lowest. This justifies using the mean signal in combination with the hybrid segmentation approach.

It can be argued that the lower variance of the resulting measurement between the replicates is not necessarily an indicator of the superiority of the method used. Keeping this in mind we put together another experiment based on the same image. Let us assess how change of concentration reflects on the change in expression level given by every combination of quantification and segmentation method.

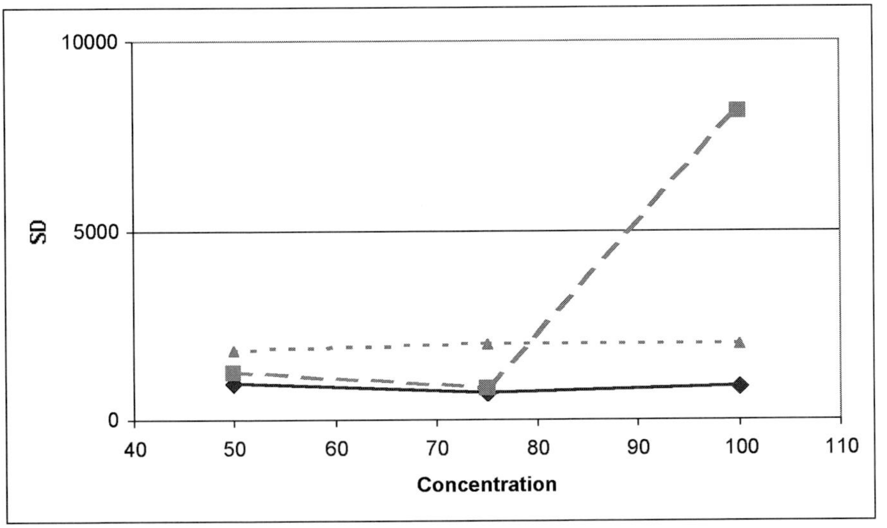

Figure 10.7 Plots of the standard deviations of replicates of the signal for three concentrations of sample using (a) signal mean (solid line), (b) signal median (broken line) and (c) signal mode (dotted line) with the ImaGene 5.5 hybrid segmentation method.

Table 10.1 shows the percentage gain in the replicate-wise average expression level when concentration is increased by 50 % (50 to 75 µg mL^{-1}). It can be seen that again the mean signal measurement combined with the hybrid segmentation method gives the most adequate response, bringing the expression level up by 51 % – the percentage closest to 50 % among all the values in the table. The results shown in Table 10.2 lead to similar conclusion based on 100 % increase in concentration (50 to 100 µg mL^{-1}). Signal mean based on the hybrid segmentation method provides a 103 % increase in expression, again very close to the expected 100 %.

Table 10.1 Percentage change in expression level when the concentration level is increased by 50 %.

	Mean	*Median*	*Mode*
Fixed circle	63%	67%	65%
Hybrid	51%	64%	67%

Table 10.2 Percentage change in expression level when the concentration level is increased by 100 %.

	Mean	*Median*	*Mode*
Fixed circle	138%	143%	140%
Hybrid	103%	114%	143%

10.7.3
Background Estimation: Segmentation Method

It is also important to be able to reliably measure the background level since it is a common practice to correct the signal for background. We applied the technique outlined in the above section regarding signal quantification to compare different types of measurements and segmentation techniques used for measuring the background value. Using plots of replicate-wise standard deviations for the mean background measurement under different concentrations we compared the performance of the hybrid method versus the fixed circle method. And again, it can be observed that the hybrid method produced more consistent estimates (Figure 10.8). In addition to the plots, we compared our measurements to the actual background level away from the spots. The hybrid method provided estimates closer to the actual level.

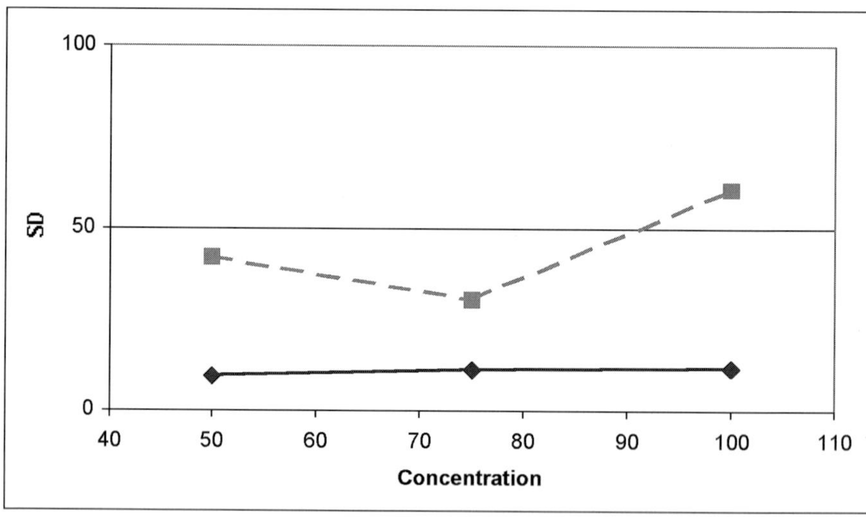

Figure 10.8 Plots of the standard deviations of the mean of the background for three concentrations of sample using (a) circular segmentation (broken line), and (b) ImaGene 5.5 hybrid segmentation (solid line) methods.

10.7.4
Background Estimation: Quantification Method

We also tested the performance of different quantification metrics during background estimation. Our metrics of choice were again mean, median, and mode. Figure 10.9 demonstrates that the background median is the optimum parameter to use for the correction of noise due to the background, as it gives the lowest variability. The background median also gave values closest to the actual background level in areas away from the spots.

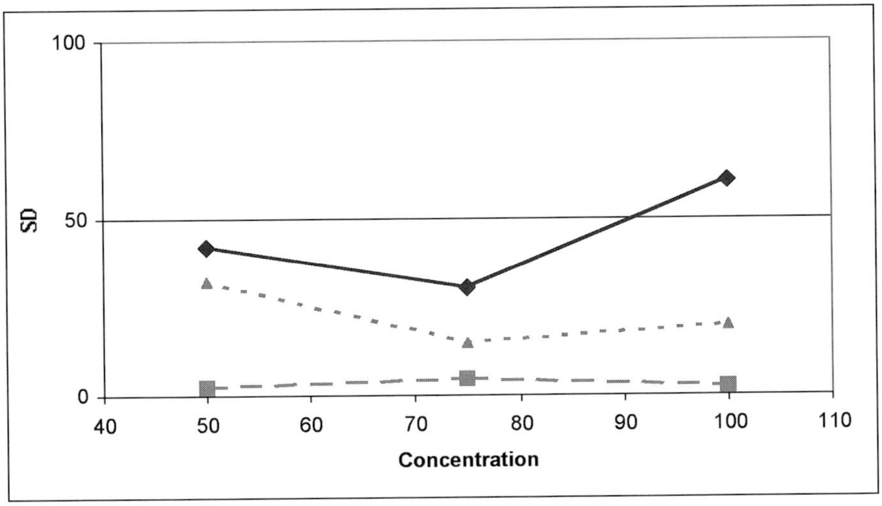

Figure 10.9 Plots of the standard deviations of replicates of the background values for three concentrations of sample using (a) mean (solid line), (b) median (broken line) and (c) mode (dotted line) with the ImaGene 5.5 hybrid segmentation method.

It should be noted that although we used hybrid segmentation for these plots, the relationship that was found also applies to the fixed circle technique. However, if all the possible combinations of segmentation/quantification techniques are tested, it will become apparent that the hybrid segmentation technique combined with the use of the median background metric gives the best results with regard to both reproducibility and standardization. Of course, the single experiment which we have mentioned throughout the above sections does not guarantee that our results will apply to all possible types of microarrays, but in our professional experience we have found that it is valid for the vast majority of experiments.

10.7.5
Ratio Estimation: Quantification Method

In order to analyze and compare the performance of the intensity ratio and the correlation ratio, a different set of images will be required. These images must have two channels. We used four sets of DNA microarray images, each containing two replicates of two-channel experiments (a total of four images). For the intensity ratio we chose the ratio of signal means (with local background medians subtracted) based on the results of hybrid image segmentation. For the correlation ratio we selected the mean of the pixel intensity ratios for the corresponding signal areas. It should be noted that we took into account only those pixels which where classified as signal in both channels, as the hybrid segmentation is performed in each channel separately and may produce two signal areas which differ slightly from each other. The local background median was subtracted from every intensity value before calculating the ratio. The results of our analysis indicated very similar average reproducibility (measured by replicate variance) between the two measures. However, the main question highlighted in the literature [11] concerns the comparative distributional properties of the two measures. It is widely accepted that log-transformed ratios of the expressions between two channels are approximately distributed according to Gaussian law in the absence of up- or downregulation.

Initially we conducted eight Kolmogorov–Smirnov tests for goodness of fit (at p-value level of 0.05), comparing residual distributions to a Gaussian PDF. Both types of log-ratios failed the tests for all four array types. This proves that strictly speaking Gaussian properties should never be applied to log-ratio microarray data in any statistical framework. However, we can still analyze which type of log-ratio measure can best be approximated by a Gaussian distribution. In order to do this, we plotted the two-tailed probabilities for four cut-off points (0.5 x SD, 1 x SD, 1.5 x SD and 2 x SD). A typical plot for the intensity ratio distribution, correlation ratio distribution and Gaussian distribution is shown in Figure 10.10. As can be seen, the two types of measurement demonstrate close distributional behavior. However, both appear to be far from an ideal Gaussian distribution. Again, although a graph for the distribution of log-ratio data may approximate a Gaussian curve, statistically speaking non-parametric methods should not used when making any inference about or based on such a distribution. It is a similar situation for intensity ratio versus correlation ratio when the median signal is used instead of the mean signal in the calculation of the ratios.

The outcome of this research is that it is evident from the point of view of their distributional properties that there would be no advantage in selecting the intensity ratio over the correlation ratio or vice versa.in The only obvious advantage of using

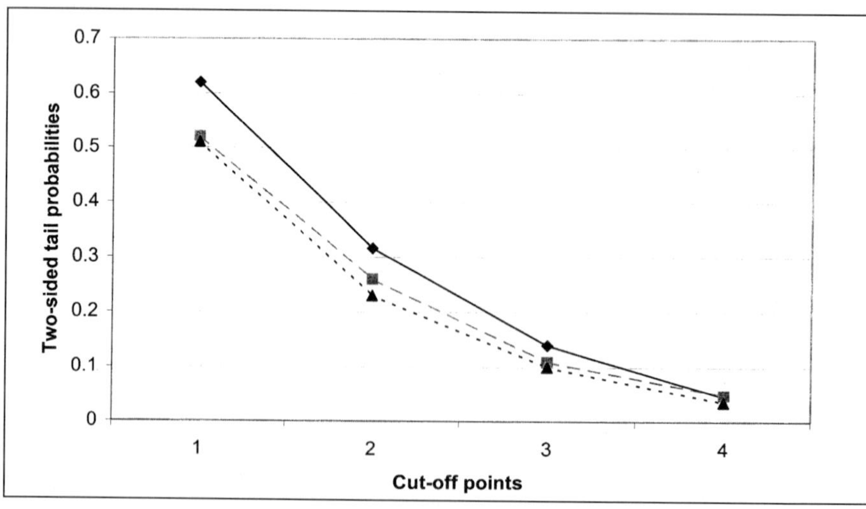

Figure 10.10 Two-tailed probabilities plot for four cut-off points (0.5 x SD, 1 x SD, 1.5 x SD and 2 x SD) of (a) a Gaussian distribution (solid line), (b) the residual distribution of the log intensity ratio (dotted line), and (c) the residual distribution of the log correlation ratio (broken line).

the correlation ratio is its ability to clean up the output when a fairly unsophisticated segmentation method is used (circular etc).

10.7.6
Quality Control

This is the last, and the most challenging stage of automated microarray image processing. To evaluate the performance of an automated quality flagging system, we compared manual flagging with our automated system output . This experiment was carried out using Amersham CodeLink system images of DNA microarrays where 250 images were analyzed using the quality flagging system implemented in ImaGene in its fully automated mode. The quality control criteria outlined in Section 10.5 were set up based on several training images prior to analysis. The same images were then viewed by human operators so that they could flag the spots displaying visible defects. The results of that comparison appeared to be very encouraging. A total of 90 % of the flags issued by ImaGene coincided with the manual flags. In addition, the human operators agreed that 99 % of the spots which were not flagged by ImaGene were "healthy". Moreover in cases of "conflict" careful review of the spots flagged by ImaGene revealed that in more than half the cases ImaGene had made a correct decision. The sensitivity of the ImaGene system to deviations in the background level and spot abnormalities led to the identification of problems which were sometimes not visible to the human eye.

Of course the quality control system described in this chapter needs fine tuning for each new type of array, making a set-up of automated image processing procedures somewhat complicated. However, taking into the account the numbers of single-type microarray images processed per day by some of the high-throughput laboratories, the significant reduction in time and resources may well be worth the effort.

10.8
Conclusion

Protein and DNA microarrays have proved to have similar image processing issues that can be addressed through the utilization of advanced spot location, segmentation and quality control techniques. Even though these technologies are known to be different and to have their own specifics (problem of the dynamic range for proteomics chips as an example), the basic workflow of expression extraction remains the same.

Based on our experimental data from processing proteomics and DNA microarray images we can conclude that it is possible to provide a reliable unsupervised system for measurement extraction. In this chapter, we have compared a number of different image segmentation and quantification schemes and have provided suggestions to solve the problems inherent in processing such images. We have also shown that application of advanced techniques when automating feature extraction and quality control can bring expression data to a new level of reliability and reproducibility.

References

1. Debouck, C., and Goodfellow, P. DNA microarrays in drug discovery and development. *Nature Genet.* 1999, **21** (Suppl.), 48.
2. Hallborn, J., and Carlsson, R. Automated screening procedure for high-throughput generation of antibody fragments. *BioTechniques,* 2002, **33,** S30–S37.
3. Zhou, Y-X., Kalocsai, P., Chen, J-Y., and Shams, S. Information processing issues and solutions associated with microarray technology. In: *Microarray Biochip Technology.* Schena, M. (Ed.) Eaton Publishing, Natick, Massachusetts, 2000, 167–200.
4. Sreekumar, A., Nyati, M. K., Varambally, S., Barrette, T. B., Ghosh, D., Lawrence, T. S., and Chinnaiyan, A. M. Profiling of cancer cells using protein microarrays: discovery of novel radiation-regulated proteins. *Cancer Res.* 2001, **61,** 7585–7593.
5. Yang, Y.H., Buckley, M. J., Dudoit, S., and Speed, T. P. Comparison of methods for image analysis on cDNA microarray data. Technical report No. 584, Department of Statistics, University of California, Berkeley, 2000.
6. Bozinov, D., and Rahnenführer, J. Unsupervised technique for robust target separation and analysis of microarray spots through adaptive pixel clustering. *Bioinformatics* 2002, **18,** 747–756.
7. Duda, R.O., and Hart, P.E. Pattern *Classification and Scene Analysis.* J. Wiley, New York 1973.
8. Chen, Y., Dougherty, E.R., and Bittner, M.L. Ratio-based decisions and the quantitative analysis of cDNA microarray images. *J. Biomed. Optic* 1997, **2.** 641–647.
9. Adams, R., and Bischof, L. Seeded region growing. *IEEE Trans. Pattern Anal. Machine Intelligence* 1994, **16.**
10. Castleman, K.R. *Digital Image Processing.* Prentice Hall, 1996.
11. Brody, J., Williams, B., Wold, B., and Quake, S. Significance and statistical errors in the analysis of DNA microarray data. *Proc. Natl Acad. Sci. USA* 2002, **99,** 12975–12978.

Index

Protein Microarray Technology. Edited by Dev Kambhampati

ISBN: 3-527-30597-1